T0302081

WATER QUALITY IMPACTS OF THE ENERGY-WATER NEXUS

Energy and water have been fundamental to powering the global economy and building modern society. This cross-disciplinary book provides an integrated assessment of the different scientific and policy tools around the energy–water nexus. It focuses on the ways in which water use, as well as wastewater and waste solids generated from fossil fuel production and their utilization for power affect water quality and quantity. Summarizing cutting-edge research in the context of current policy and regulations, it describes the geochemical techniques for detecting contamination sources. The authors highlight the growing evidence that fossil fuel production, from both conventional and unconventional sources, leads to water quality degradation, while regulations for the water and energy sector remain fractured and highly variable across and within countries. This volume will be a key reference for scholars, industry professionals, environmental consultants, and policymakers seeking information on the risks associated with the energy cycle and its impact on the environment, particularly water resources.

AVNER VENGOSH is Distinguished Professor of Environmental Quality at Duke University's Nicholas School of the Environment. His research focuses on evaluating the prevalence and sources of contaminants in global groundwater resources, and the impact of energy development such as hydraulic fracturing and coal combustion on the quantity and quality of water resources. In 2019, 2020 and 2021 he was named on the list of the world's Most Highly Cited Researchers.

ERIKA WEINTHAL is Professor of Environmental Policy and Public Policy at Duke University's Nicholas School of the Environment. She specializes in global environmental politics with an emphasis on water and energy. She has coauthored *Oil Is Not a Curse: Ownership Structure and Institutions in Soviet Successor States* (Cambridge University Press, 2010) and coedited *The Oxford Handbook on Water Politics and Policy* (2017).

WATER QUALITY IMPACTS OF THE ENERGY-WATER NEXUS

AVNER VENGOSH

Duke University

ERIKA WEINTHAL

Duke University

CAMBRIDGE
UNIVERSITY PRESS

CAMBRIDGE
UNIVERSITY PRESS

University Printing House, Cambridge CB2 8BS, United Kingdom

One Liberty Plaza, 20th Floor, New York, NY 10006, USA

477 Williamstown Road, Port Melbourne, VIC 3207, Australia

314–321, 3rd Floor, Plot 3, Splendor Forum, Jasola District Centre, New Delhi – 110025, India

103 Penang Road, #05–06/07, Visioncrest Commercial, Singapore 238467

Cambridge University Press is part of the University of Cambridge.

It furthers the University's mission by disseminating knowledge in the pursuit of education, learning, and research at the highest international levels of excellence.

www.cambridge.org
Information on this title: www.cambridge.org/9781107061637
DOI: 10.1017/9781107448063

First published 2022

A catalogue record for this publication is available from the British Library.

Library of Congress Cataloging-in-Publication Data
Names: Vengosh, Avner, author. | Weinthal, Erika, author.
Title: Water quality impacts of the energy-water nexus / Avner Vengosh, Erika Weinthal.
Description: New York : Cambridge University Press, [2021] | Includes bibliographical references and index.
Identifiers: LCCN 2021037873 (print) | LCCN 2021037874 (ebook) | ISBN 9781107061637 (hardback) | ISBN 9781107672055 (paperback) | ISBN 9781107448063 (epub)
Subjects: LCSH: Water quality. | Water quality management. | Energy policy–Effect of water quality on. | BISAC: SCIENCE / Earth Sciences / Hydrology
Classification: LCC TD365 .V46 2021 (print) | LCC TD365 (ebook) | DDC 363.6/1–dc23
LC record available at https://lccn.loc.gov/2021037873
LC ebook record available at https://lccn.loc.gov/2021037874

ISBN 978-1-107-06163-7 Hardback

Contents

Acknowledgments   *page* vii

1   Introduction   1
     1.1   Introduction   1
     1.2   The Energy–Water Nexus   4
     1.3   The Energy–Water Quality Nexus   7
     1.4   Governing the Water–Energy Nexus   9
     1.5   Structure of the Book   10

2   Quantification of Energy and Water Flows   14
     2.1   Introduction   14
     2.2   Energy and Water Metric   16
     2.3   Global Fossil Fuel Production   18
     2.4   The Energy–Water Nexus Concept   22
     2.5   The Example of the US Energy–Water Quantity Nexus   25

3   The Coal–Water Nexus   29
     3.1   Introduction   29
     3.2   What's in Coal   34
     3.3   Coal Mining and Processing   38
     3.4   Electricity Generation   52
     3.5   Coal Combustion Residuals   60
     3.6   Integration: The Combined Water Footprint
           of Coal Utilization   72
     3.7   Environmental Regulations and Coal   74
     3.8   Coal Transitions   77

4   Conventional Crude Oil and Tight Oil–Water Nexus   81
     4.1   Introduction   81
     4.2   The Distinction between Conventional and Unconventional
           (Tight) Oil Exploration   86
     4.3   Water Use for Enhanced Oil Recovery and Hydraulic Fracturing 90
     4.4   Flowback and Produced Waters   95
     4.5   Oil Sands   116
     4.6   Refining Crude Oil   119
     4.7   Oil Spills   121
     4.8   Oil Combustion and Fugitive Atmospheric Emission   125
     4.9   Integration: The Combined Water Footprint of Crude
           Oil Utilization   127
     4.10 Policy and Regulations   130

5   Conventional and Unconventional Natural Gas–Water Nexus   136
     5.1   Introduction   136
     5.2   The Origin of Natural Gas   141
     5.3   Water Use for Natural Gas Production   153
     5.4   The Quality of Frac Water   161
     5.5   Flowback and Produced Water   163
     5.6   Stray Gas Contamination of Groundwater Resources Induced
           from Hydraulic Fracturing   189
     5.7   Coalbed Methane   194
     5.8   Integration of Water Use and Produced Water Generated
           from Natural Gas   207
     5.9   Policy and Regulations   209

6   Integration: The Role of Energy in the Anthropogenic Global
     Water Cycles   215
     6.1   Introduction to the Anthropogenic Global Water Cycle   215
     6.2   Global Water Use for Fossil Fuels   220
     6.3   The Global Impaired Water Intensity   229
     6.4   The Emission of Greenhouse Gases from Fossil Fuels   233
     6.5   The Energy Use for Water   249
     6.6   Clean Future and the Legacy of Fossil Fuels   256

References   262
Index   301
Colour plates can be found between pages 184 and 185.

Acknowledgments

This book reflects nearly two decades of our individual and collaborative research on both water and energy and increasingly on the water–energy nexus. It would not have been possible without the dedicated work and contributions of many individuals, including our graduate students, postdocs, colleagues from within and outside Duke University, environmental activists, state officials, and above all, community members who live near energy producing sites and helped us to conduct the research. Funding from different federal and private agencies has supported earlier research that has informed this book.

We are particularly grateful to our former and current graduate students who conducted research on the energy–water quality nexus, including Laura Ruhl, Nathaniel Warner, David Vinson, Hadas Raanan, Brittany Merola, Jennie Harkness, Nancy Lauer, Andrew Kondash, Rachel Coyte, Zheng Wang, Efrat Farber, Mengjun Yu, and Daniella Hirschfeld. Many of them are now colleagues. We are also thankful to our former postdocs and now colleagues, Tom Darrah, Stephen Osborn, Tewodros Godebo, and Kate Neville; our colleagues at Duke University, Gary Dwyer, Robert Jackson, Helen Hsu-Kim, Dalia Patino-Echeverri, Emily Bernhardt, Rich Di Giulio, Jon Karr, Emily Klein, Bill Chameides, and Bill Schlesinger; colleagues outside of Duke, Bill Mitch, Tom Johnson, Jim Hower, Ellen Cowan, Jessica Brandt, Pete McMahon, Justin Kulongoski, Autumn Romanski, Danny Smith, Amrica Deonarine, Grace Schwartz, Gideon Bartov, Robert Poreda, Yunyan Ni, Dan Liu, Shifeng Dai, Kim Parker, Wolfram Kloppmann, Romain Millot, Collin Whyte, Myles Moore, Jane Baka, Martine Savard, Jason Ahad, Yousif Kharaka, Kirin Furst, Jennifer Hoponick Redmon, Chris Kassotis, Karl Muehlenbachs, Ori Lahav, Jeannie Sowers, Karen Bakker, and Jen Baka.

Special thanks to Andrew Kondash who made many of the figures in the book and helped to better quantify the energy–water nexus parameters.

Finally, none of this would have happened without the support and the love of our children, Tom, Noga, Emma, and Adam.

1

Introduction

1.1 Introduction

Energy and water are vital for powering the global economy and sustaining modern society. Throughout the world, energy access and water availability are limiting factors for fostering economic growth, promoting sustainable development, and securing public health. Yet, after a century of heavy exploration and consumption of fossil fuels, the unique balance between energy and water has changed; water resources and availability are rapidly declining, and water quality, particularly in areas associated with substantial water exploitation and energy development, is deteriorating. At the same time, many parts of the developing world lack access to 24/7 electricity, and the ability of countries to transfer or treat water, such as wastewater, or to develop desalination is largely dependent on energy availability. That energy and water are intricately interconnected is increasingly apparent to individuals, communities, businesses, and governments. This interdependence of energy and water is called the "energy–water nexus" and is the focus of this book.

Research on the energy–water nexus concept has centered on attempts to evaluate the amount of water needed for exploration (e.g., coal mining, hydraulic fracturing), electricity production or conversion (e.g., using water for cooling thermoelectric plants), and waste production (e.g., oil and gas wastewater). Other analyses have examined the amount of electricity needed for water management (e.g., pumping, transport, treatment). This book aims to explore an additional component – the impact of fossil fuel exploration and electricity production on the *quality* of water resources. Through systematic evaluation of major fossil fuel consumption (coal, oil, natural gas), this book demonstrates that the impact on water resources is far greater once the water quality component is considered. The *energy–water quality nexus* thus undergirds our discussion of the broader energy–water nexus.

Understanding the energy–water quality nexus is critical for devising far-reaching socioeconomic and environmental policies due to society's consumptive behavior and its continued reliance on fossil fuels for supporting economic activity. This reliance upon fossil fuels has taken place over an atypically short period of time when viewed through the lens of geologic time. Yet, because the impacts of fossil fuel use have been so dramatic, causing shifts in the global climate and mass extinctions of plant and animal species, scientists have coined the term the Anthropocene to account for this new geological epoch that some trace back to the Industrial Revolution[1, 2, 3]. Scientists have tracked these global changes through the increase in atmospheric carbon dioxide from a preindustrial value of 270–275 ppm to about 310 ppm by 1950, accelerating over the latter half of the twentieth century[4] and crossing the 400 ppm mark during the early 2010s (in May 2021, the level was 419.13 ppm; www.co2.earth/). The role of fossil fuels in the environment is thus associated with carbon emissions and accumulation of carbon dioxide, methane, and other hydrocarbons in the atmosphere that have made the world warmer. The average global temperature in 2017 was 0.84°C above the twentieth-century average of 13.9°C , making it the third-warmest year on record behind 2016 (warmest) and 2015 (second warmest) since 1880[5].

Most notably, the utilization of fossil fuels – primarily coal and petroleum – has dramatically altered the carbon dioxide budget of the atmosphere. Reliance upon coal as a major global energy source, in particular, is often signaled out as a driver of anthropogenic-induced climate change. Since the first use of coal to generate electricity for homes and factories in the 1880s[6], coal has been the major energy source for electricity production in the USA and continues to constitute a major energy source. However, due to the rise of unconventional shale gas during the last decade, coal's contribution to US electricity production has been declining; in 2010, coal contributed 45% to US electricity production and was reduced to 33% in 2015 and 30% in 2017. During the same period, natural gas, mostly from unconventional shale gas, rose from 24% to 33% (data from the US Energy Information Agency [EIA][7]). Yet in spite of the reduction in coal combustion, coal mining within the USA has not dropped since a significant portion of mined coal is exported from the USA. Countries importing US coal include India, Japan, Brazil, South Korea, and the Netherlands; the latter also serves as a transit hub for other countries in Europe[8]. Since 2010, the export of US coal has increased from 72 to 108 million (metric) tons[9], which constitutes ~13% of annual coal production in the USA (a total of ~640 million tons in 2019). The EIA projects that US coal production and export will remain relatively stable for at least the next 50 years[7]. Coal exploration and combustion are also expected to continue to be predominant energy sources across the world. China extracts coal at a level that is equivalent to about 50% of global coal production (about 3,846 million tons in 2019[10]); about 60% of the total energy mix is derived from coal (in 2016 it was 62%), whereas coal constitutes 80% of the electricity sector[11]. In India, coal production has rapidly increased (756 million tons in 2019[10]) and constitutes 80% of national energy extraction. In the Middle East, several countries already have large coal-fired electricity generation, such as Turkey and Israel, but other countries such as Egypt, Oman, Iran, Jordan, and the United Arab Emirates (UAE), with no current coal-fired electricity generation, plan to build coal capacity in the near future. Likewise, Pakistan is gearing up to install several large-scale coal plants[12], many of which are through Chinese investments to expand the China–Pakistan Economic Corridor[13]. Overall, fossil fuels (oil, natural gas, coal, bitumen) comprise about 80% of the current total annual global energy production. The EIA predicts that in spite of the expected rise in renewable fuels, fossil fuels will still account for 77% of energy use in 2040.

While coal production will remain constant, natural gas mostly from tight gas, shale gas, and coalbed methane sources is projected to be the fastest-growing fossil fuel[14]. Increasing reliance upon natural gas at the beginning of the twenty-first century highlights the importance of examining global energy and water resources together. The development of new drilling technologies and production strategies such as horizontal drilling and hydraulic fracturing has significantly improved the production of shale gas and tight oil by stimulating fluid flow from impermeable source rocks that long ago were not considered as viable energy sources. Since the mid-2000s, these developments have spurred the exponential growth of shale gas and tight oil-well drilling across the USA and Canada; shale gas was increasingly seen as an alternative fuel source in parts of the world, such as in South America and China. In the USA, shale gas increased from zero in the early 2000s to 624 billion cubic meter (BCM) in 2018, which was the equivalent of 63% of the US annual natural gas production in 2018 (a total of 1,051 BCM[15]). Likewise, in Sichuan Basin in China, shale gas production has increased to 7.8 BCM in 2017, on a projection to a national target of 100 BCM by 2030[16].

While shale gas exploration was initially touted as a means to reduce greenhouse gas emissions and hence decelerate climate change, the wide-ranging socioeconomic and environmental impacts of using hydraulic fracturing and horizontal drilling technologies have elicited local protests across the USA and Europe, as well as in Latin America and South Africa. Emerging evidence for atmospheric methane emission[17–22], large volumes of water utilization[23, 24], and water contamination[19, 25–27] has altered public perceptions about hydraulic fracturing and unconventional energy development. The "shale revolution", as it has been referred to in the popular press, has underscored that such magic bullets come with unforeseen consequences – in this case, a wide array of impacts upon water resources. As this book will illuminate, although water and energy are intertwined and essential for modern life, compartmentalizing energy from water obscures the ways in which the energy–water nexus cannot simply focus only on water quantity and availability, but also must directly address the impacts upon *water quality*.

1.2 The Energy–Water Nexus

The notion that water and energy are intimately linked is central to the concept of the water-energy nexus (Figure 1.1)[24, 28–38]. The mere exploration and production of a wide array of energy resources – including hydropower, oil and gas, coal, nuclear, and biofuels – requires significant amounts of water. Conversely, energy is necessary for the production and purification of various water sources, including desalination and wastewater treatment, and for its pumping from groundwater sources, transportation from one region to another, and use in irrigation. Industrial and agricultural activity both depend on the consumption of water and energy, and they become increasingly coupled when water is scarce and energy is nonrenewable[37]. In the arid southwestern USA, cities such as Phoenix, Arizona and Las Vegas, Nevada grew rapidly because of the use of large amounts of water and electricity; the expansion of intensive agriculture in the dry southwest has also depended upon energy resources to move water[39–42].

Research on the water–energy nexus has sought to elucidate the amount of energy needed to produce water resources, especially where water is scarce. The Middle East – one of the world's most water-poor regions – has been a focus of

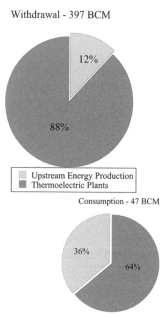

Figure 1.1 The relative proportions of global water withdrawals (water used that returns to the hydrosphere) and consumption (water used that is lost and not returned to the hydrosphere) for primary energy production (drilling, mining, hydraulic fracturing) and electricity generation (cooling thermoelectric plants). Data were taken from the IEA[45].

study to understand how much energy is needed to produce water. Siddiqi and Anadon (2011)[43], for example, found that in Saudi Arabia up to 9% of total annual electrical consumption may be attributed to groundwater pumping and desalination. The energy-producing states of the Gulf have been able to overcome their water deficit through a reliance upon cheap energy. Owing to advances in reverse osmosis seawater desalination, the price of desalination has continued to decline, as has been the case in Israel, which integrated its reverse osmosis seawater desalination with the national electricity system. Yet, as this book will further articulate, desalination is, nevertheless, a prohibitively costly strategy for countries that are energy poor.

Because water is unevenly distributed both spatially and temporally, energy becomes all the more important for augmenting water's accessibility and use for populations that may not be near the water source. In countries where water is abundant in some areas and scarce in others, governments must provide water to areas in greater need for both human consumption and to support agricultural and industrial activities, or to construct large storage systems to cope with temporal availability. Projects such as the Disi Water Conveyance Project in southern Jordan, which extracts fossil groundwater and then transports it to the capital, Amman, require large energy inputs. China, in particular, has invested heavily in water transfer projects to bring water from south-central China to the water-stressed heavily populated and industrialized northern central plain[44].

Water, in turn, is a major ingredient necessary for generating energy. Specifically, water is used for primary energy extraction (mining, drilling, hydraulic fracturing), processing (coal cleaning, oil refining, transport), and conversion for electricity generation, using water for cooling thermoelectric plants. The International Energy Agency (IEA) estimated that about 400 BCM was globally withdrawn for energy production in 2014[45]. About 50 BCM was estimated to be consumed, which means that this amount of water was not returned to the hydrological system[45]. Much of the global water withdrawal was for cooling power plants for electricity generation (350 BCM; 88%), while water withdrawals for primary energy production (such as drilling or mining) were only 47 BCM (12%). In contrast, the majority of water consumption was for primary energy production (30 BCM; 64%), while water consumption for cooling constituted only 34% (17 BCM) of the total water consumed (Figure 1.1)[45]. We therefore see different ways in which global water is utilized for energy generation; the largest water volume is used for electricity production but mostly returned to the hydrological system, while a smaller volume is used for primary energy production such as coal mining or hydraulic fracturing, but most of it is lost from the water cycle. Overall, global water withdrawal and consumption for fossil fuels were estimated as 251 BCM (63% of total global water withdrawal for energy production) and 31 BCM (65% of total water consumption), respectively[45]. Through using more updated data on global energy production and water intensity data, this book shows (Chapter 6) much higher estimates for global water withdrawal and consumption associated with fossil fuel consumption.

In the USA, the total water withdrawal and consumption in 2016 were 224 BCM and 16.3 BCM, respectively, from which fresh water (180 BCM for withdrawal and 13 BCM for consumption) was the major water source. Water withdrawal and consumption for fossil fuels in the United States were 137 BCM (62%) and 7.9 BCM (48%)[29]. Overall, the annual water withdrawal for the energy sector in the United States is about 40% of the national water withdrawal, mostly for cooling thermoelectric plants, similar to the amount of water used for the agricultural sector[46]. About 10 BCM of fresh water that is used for cooling thermoelectric plants is annually lost,[29] mostly through evaporation. To put this is in a broader perspective, this volume is significantly higher than the amount of annual water consumption of many counties in arid zones (e.g., Jordan with less than 2 BCM per year).

As shown above, cooling thermoelectric plants requires the largest volume of water withdrawal, while primary energy production consumes more water[45, 46]. One aim of this book is, then, to explicate how the source of energy (i.e., coal, natural gas, nuclear) can generate differential water use and consumption during the different stages of extraction (e.g., coal mining versus hydraulic fracturing of shale gas) and electricity production (i.e., differential water use for different types of thermoelectric plants)[28, 31, 34, 40, 45, 47, 48]. The volume of water that is used for cooling thermoelectric plants varies significantly based on the source of energy and the cooling technology. Numerous studies have examined the energy–water quantity nexus[28, 30, 31, 34-36, 38, 40, 41, 45-52], with attempts to quantify the volume of water that is withdrawn and that is consumed during the different energy production stages. Many of these studies have aimed to quantify the amount of water used per unit of energy, defined as "water intensity," typically described as volume per energy unit (e.g., liter per giga joule [L/GJ] or cubic meter to giga joule [m^3/GJ]). The amount of water required for the different energy sources varies across the different stages of production (extraction, processing, and generation or combustion).

This book moves beyond these earlier studies that have evaluated the volume of water used in different stages of energy production. This book also aims to examine the volume of water that is lost due to degradation of the water quality directly linked to energy production. Many studies have looked at water quality degradation associated with primary energy production (e.g., contamination from coal mining, hydraulic fracturing) or electricity generation (thermal pollution of cooling water discharged to the environment[53, 54]), yet the overall water quantity lost due to water quality degradation is not as well developed. As such, we further expand the water–energy nexus to address the water quality dimension. A focus on water quality has broader implications for many countries that seek to ensure access to clean water for their populations, especially as they work to meet the Sustainable Development Goals[55].

1.3 The Energy–Water Quality Nexus

Living in the Anthropocene has also meant that unprecedented changes to our global climate from industrial and agricultural activity, coupled with population growth, have resulted in the diminishment and contamination of global fresh water resources on a massive scale. As we lay out in this book, the Anthropocene presents new challenges for governments, communities, and industries to address global climate change caused by our voracious consumption of water and energy, but to do so requires an integrated and intersectoral approach to understanding water and energy use and their impacts. Climate change models have shown that global warming is expected to increase overall renewable water availability – reflected in increasing precipitation in water-rich countries[56–64]. Yet, in arid and semi-arid zones, studies have also found that global warming significantly reduces precipitation and hence increases water scarcity[65–69], as demonstrated in the Middle East[70–74] as well as in the United States, such as in California[75–79]. The decoupling of global warming and growing water utilization in water-scarce areas envisages severe water scarcity and regional crises directly linked to exacerbating the water crisis.

Likewise, growing populations and rising water demands in countries facing increasing water scarcity are fodder for further aggravating the global water crisis. The 2014 United Nations World Water Development Report (WWDR) estimated that more than a billion people live in areas of water stress, which is expected to triple by 2025[80]. Another study by Mekonnen and Hoekstra (2016) showed that two-thirds of the global population (4.0 billion people) live under conditions of severe water scarcity at least one month of the year, while half a billion people in the world face severe water scarcity all year round[81]. For example, in the summer of 2018, India, which extracts the largest volume of worldwide groundwater (about 50% of global groundwater extraction[82]) experienced a severe water shortage in which 600 million people were exposed to extreme water stress; more pertinently, 200,000 people die every year in India due to inadequate access to safe water[83]. At the same time, global warming was linked to a reduction of precipitation in northern India[84]. The overexploitation of groundwater resources has led to declining groundwater levels in northwestern India[85,86], which, in turn, also induces water quality degradation, including large-scale uranium contamination[87]. In short, India demonstrates that the water–energy nexus is not limited to only direct impacts but is also more complex, in which global warming is known to reduce precipitation and water availability, and consequently result in the

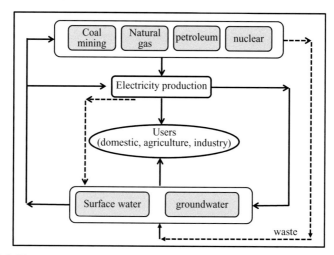

Figure 1.2 The water–energy nexus schematic illustration. The solid and dashed lines reflect energy, water, and wastewater flows and the interrelations between water and energy.

degradation of water quality. Likewise, in the Middle East and North Africa, the increased aridity and concurrent droughts have reduced water availability[70] and degraded water quality in the form of rising salinity[88].

That fossil fuel use has been a driver of climatic change is undisputed[89]. Yet, in many countries throughout the world, government policies continue to incentivize the use of fossil fuels for energy consumption. Few have comprehensively addressed the pernicious effects of fossil fuels exploration on water quality[90]. *This book aims to fill this gap and illuminate how conventional and unconventional explorations of fossil fuels affect the quality of water resources, with a particular focus on the effects of energy exploration and production on water quality* (Figure 1.2). The book provides an overview of the impacts of energy development through the different stages of exploration, production, distribution, and waste management on the quantity and quality of water resources that are associated with these different stages. Thus, while the traditional energy–water approach has focused on quantifying the amount of water directly used or consumed for the different stages of energy production[28], this book takes a much more holistic approach and evaluates the water quality impact that is associated with the different stages of energy exploration and production, as well as the legacy of contamination from fossil fuels. In addition, this book explores the indirect effect of fossil fuel exploration and climate change on the availability and quality of water resources.

Where this book differs from other water–energy nexus scholarship is that it pivots toward water quality impacts. For example, the book examines the water quality impacts during exploration (e.g., selenium contamination of streams in areas of mountaintop mining, stray gas contamination of groundwater near shale gas wells, and surface water contamination from spills associated with hydraulic fracturing) and post-exploration or production (e.g., acid-mine drainage from abandoned coal mines, coal ash disposal, legacy of oil produced water spills). Furthermore, the book explores all the aspects of water quality impacts during the lifetime of energy production and provides information for the long-term effects, beyond the direct impact of the energy–water nexus. As part of our inquiry, we offer a new water–energy paradigm – *water impaired* – that aims to quantify the volume of water that is lost due to water quality degradation, in addition to the direct water withdrawal and consumption.

1.4 Governing the Water–Energy Nexus

While much of the book will lay out the quantitative and qualitative measurements for understanding the water–energy nexus, this book is also designed to put the nexus within a governance and policy framework. There is a great deal of literature that has sought to study both water and energy in the social sciences. On the one hand, there is significant research on the "resource curse" that argues oil (vis-à-vis its presence or reliance) leads to a number of negative economic and political effects, ranging from poor/unbalanced economic growth to low levels of human development, authoritarian regimes, and a greater propensity for conflict[91]. On the other hand, the field of international water has sought to understand how states manage and prevent conflict over their international waters, especially where states may have different interests for water development – hydroelectric vs. agriculture. Yet, in both of these fields, water and energy are intimately linked. For example, the social sciences have taken a more critical look at the role of oil and gas exploration owing to the broader socioeconomic and environmental impacts the exploration and production of oil and gas have had on communities in which these processes take place. Case studies from countries such as Nigeria and Ecuador have brought to the forefront the devastating environmental and health impacts on communities that live close to production operations, ensuing from water contamination and the flaring of natural gas[92]. Likewise, many industrialized countries have not been immune from environmental disasters, as witnessed by such disastrous oil spills as that in Santa Barbara in 1969, the Exxon Valdez oil spill in Alaska in 1989, and the Deep Horizon oil spill in 2010 in the Gulf of Mexico, all of which have had deleterious effects on marine ecosystems and communities, which depend upon fishing for livelihoods.

Many economic decisions have been driven by this connection between water and energy, and yet policymakers have been slower to realize these connections. In the field of international water, international organizations failed to realize that water and energy needed to be managed simultaneously, such as in the Aral Sea Basin[93]. In the USA, it has become increasingly apparent to analysts that "we can no longer consider the formulation of rational energy policy and water policy to be independent" (Gleick 1994, 295[28]). Countries increasingly are taking steps to recognize the connection between water and energy. Take, for example, the US Energy Policy Act of 1992 (Public Law-486, 102nd Congress, Washington, DC, October 24) which established water-efficiency standards for household water fixtures owing to the implications for energy of water use[94].

Addressing the energy–water quality nexus as a policy objective has not been straightforward owing to the long-standing environmental contestation over the links between energy production and water contamination issues. For example, the development of new technologies such as hydraulic fracturing and horizontal drilling that led to the exploration of shale gas, tight oil, coalbed methane, and oil sand, combined with environmental issues associated with coal mining and coal ash disposal, are all associated with different aspects of water contamination, and are the focus of current societal debates about the sustainability and the actual impacts of these processes. In many cases, the perception in the media about the environmental impacts from unconventional oil and gas production, for example, has been detached from the scientific literature. The disconnect at times between perceptions of water contamination and what is *known* is a major challenge for policy design and has led to numerous debates over what approach is best for regulating the energy–water quality nexus between the many vested stakeholders, which include environmental groups, homeowners, industry, and government institutions. This book aims to provide a scientific basis to inform these public debates. This is, no doubt, an ambitious task. As such, this book relies largely upon peer-reviewed literature to provide the factual evidence (or lack thereof) for water contamination associated with the extraction and energy generation of fossil fuels.

1.5 Structure of the Book

In order to evaluate the energy–water quality nexus from different angles and perspectives, this book integrates energy data, water flow, water quality data through integrated physical science, including geochemistry, hydrology, and hydrogeology, combined with social science, including socioeconomic analysis and policy implications. The chapters of this book are organized as follows. In Chapter 2 we present an outlook of the evolution of the different fossil fuel sources, including coal, conventional and unconventional (tight) oil, oil sand, conventional natural gas, unconventional shale gas, and coalbed methane. The chapter provides the data on energy flows and evaluates global energy consumption for understanding the magnitude of these developments. The chapter also presents the traditional water–energy concept; the amount of water withdrawal and consumption for fossil fuel exploration in the recovery stage (coal mining, oil and gas drilling, mining, hydraulic fracturing, oil enhancement), processing (coal washing, oil refinery), conversion (electricity production), and post-conversion (waste disposal). The chapter examines the water-intensity metrics by normalizing water volume per energy unit or electricity, with an emphasis on the distinction between water withdrawals and water consumption, and considering the complete lifetime cycles of water extraction for energy exploration, processing, and generation.

Chapter 3 explores the coal–water nexus with evaluation of the chemical hazards in coals and the magnitude of their emission with coal combustion, the quantity of water use for coal mining and processing, and implications for water quality, with an emphasis on the formation of coal mine effluents such as acid mine drainage and large water-impaired intensity. The chapter describes the different technologies for cooling thermoelectric plants and the associated water intensity and thermal pollution upon the discharge of the cooling water to the hydrosphere. The chapter explores the formation of coal combustion residuals after coal combustion, the enrichment of contaminants in coal combustion residuals, the volume and quality of effluents associated with coal ash impoundments, and the environmental risks associated with the management of coal ash and leaking coal ash impoundments. Finally, the chapter looks at the different regulations in different countries with respect to coal mining, combustion, and disposal, as well as the transition from coal to different energy sources.

Chapter 4 explores the water–energy nexus associated with conventional and unconventional (tight) oil exploration. The chapter first explains the distinction between conventional and unconventional (tight) oil exploration, provides updated values of the water intensity associated with conventional oil drilling and enhanced oil recovery, as well as hydraulic fracturing for extraction of unconventional tight oil. The chapter explores the source, volume, geochemistry, and water quality issues of oil-produced water generated from conventional oil exploration as compared to flowback and produced water generated from unconventional oil wells. The chapter explains the nature and origin of oil field brines, and the inorganic (salts, metals, and naturally occurring radioactive elements) and organic constituents that are associated with oil- and gas-produced waters. The chapter provides examples of possible mechanisms of water contamination (spills, disposal) and the long-term environmental legacy of oil drilling. The chapter evaluates the impact of oil-produced water on the environment and quantifies the associated impaired water intensity, as well as the ability to reuse oil-produced water for beneficial use. The chapter explores the water use and quality associated with oil sand extraction and processing and possible environmental effects. The chapter describes the water use and impact on water quality from oil refining. The chapter evaluates the impact of oil spills on water resources and the magnitude of water loss due to oil spills. Finally, the chapter examines policy mechanisms that influence the extent of environmental externalities that have emerged from oil production and exploration, including government regulations, forms of corporate society responsibility, and types of contracts.

Chapter 5 evaluates the origin of natural gas with the classification of its different types and chemistry. The chapter explains the principles of hydraulic fracturing and formation of shale gas. The chapter systematically evaluates the magnitude of the water use for hydraulic fracturing as compared to conventional gas drilling, and the water intensity associated with natural gas processing and utilization for electricity generation. The chapter presents the changes of the water footprint of hydraulic fracturing and water intensity over time in the USA and China, and shows how the intensification of hydraulic fracturing has increased both water use and wastewater production from unconventional shale gas exploration. The chapter draws upon data from emerging scientific reports in the USA to explore the global implications of unconventional energy for other countries with shale gas potential. The chapter presents the major organic constituents associated with frack water used for hydraulic fracturing. The origin, geochemistry, and volume of flowback and produced water from shale gas and conventional natural gas operations are evaluated, with special analysis of the potential impact on surface water and groundwater resources and associated impaired water intensity. The chapter discusses the chemical composition of produced water and flowback waters originated from shale gas wells, the different types of liquid and solid wastes, the contaminants in these wastes (salts, trace elements, radioactivity, man-made organic constituents), and the possible effects of the different disposal practices and treatment technologies, as well as the risks of injection wastewater into deep aquifers to induce seismicity and earthquakes. The chapter explores the debate on the effect of stray gas contamination induced from leaking shale gas as opposed to naturally occurring methane in groundwater. The chapter discusses the mechanisms of potential groundwater and surface water contamination directly and indirectly associated with shale gas development and hydraulic fracturing. The chapter explains the distinction between groundwater pollution induced directly from shale gas drilling relative to other sources of water contamination and the methods that have been employed for delineating this distinction. The chapter evaluates the mechanism of coalbed methane extraction, the quality of produced water generated with coalbed methane, and environmental implications. Finally, the chapter discusses the different regulations and possible safeguards to protect fresh water resources from hydraulic fracturing. Particular attention is given to the devolution of policy to the states in the USA, whereby different states have introduced different regulations (including moratoria), contracts and financing mechanisms, and policy spaces for the private sector to introduce voluntary regulations (e.g., voluntary disclosures of chemicals used in the hydraulic fracturing process).

Finally, Chapter 6 presents the Anthropogenic Global Water Cycle in which overuse of natural water resources has led to water deficit in many areas across the globe. The water deficit has been exacerbated by global warming and increased drought intensity induced by fossil fuel use, combined with large water use for fossil fuels, increased energy generation for water transfer and treatment, and consequently further water use and impact on depleting water resources. The chapter presents a new assessment of the global water withdrawal and consumption for the different stages of fossil fuel production and consumption, using the most up-to-date information on water intensity and energy production (up to 2019). The chapter also presents a new assessment of the global wastewater volume that is generated from coal, natural gas and oil production, and consumption. The chapter highlights the high magnitude of the impaired water intensity derived from water contamination on a global scale and the importance of the magnitude of the water loss due to water contamination, in addition to direct water use for fossil fuels. The chapter explores the global emission of carbon dioxide, water vapor, and methane from fossil fuel operations, including natural gas flaring and the impact of the accumulation of greenhouse gasses in the atmosphere. The chapter evaluates the energy use for water transport, wastewater treatments, and desalination with an emphasis on future water quantity and quality deterioration that would require additional treated water and thus additional energy sources that would further exacerbate the Anthropogenic Water Cycle. Finally, the chapter discusses the options for a clean future (and transition) to renewable energy sources with much lower water intensity and environmental impacts.

All told, due to the depletion and degradation of global water resources induced by overuse and climate change, the energy–water nexus will become more pronounced as policymakers and societies must consider what types of energy sources to turn to in the twenty-first century, especially when water scarcity is becoming a more limiting factor. The book thus provides an accumulated water footprint of fossil fuels that includes estimation of the *volume of the impaired water*, in addition to the direct withdrawal and consumed waters linked to energy development and electricity production. The book provides a summary of the scientific and policy issues facing countries in the twenty-first century as countries navigate decisions to exploit conventional and unconventional energy resources and protect their water resources. It concludes with a forward-looking discussion of emerging challenges, controversies, and scientific and policy dilemmas, taking climate change into account.

2

Quantification of Energy and Water Flows

2.1 Introduction

Over the twentieth century, two sources of fossil fuels have dominated world economies – petroleum and coal. While coal is often associated with fueling the Industrial Revolution in England in the eighteenth century with the advent of the steam engine, it began to play a role in local livelihoods in England toward the end of the thirteenth century when blacksmiths and artisans began to substitute coal for wood as a source of fuel[95]. The birth of the oil industry came much later with the discovery of oil in the Absheron Peninsula in Azerbaijan in 1848 and then near Titusville, Pennsylvania, in 1859 by "Colonel" Edwin Drake. For the next 30 years, oil production was concentrated in Pennsylvania in the United States[96] such that by the end of the nineteenth century, Pennsylvania comprised the entire US oil industry[97].

Coal is the less glamorous of these two fossil fuels. From its beginnings, it has been seen as a dirty, putrid resource, causing unpleasant smells and poor air quality in urban centers. Dating back to the seventeen century, the smoke from the burning of coal was known to damage the architecture in cities such as London[98]. The intensification of poor air quality from the smoke and thick fog that accompanied coal burning not only was a public nuisance in cities by sullying clothing[99] but also led to a rise in respiratory diseases such as bronchitis in Britain by the beginning of the twentieth century[98]. Mining coal is also a dangerous enterprise for workers who are exposed to treacherous working conditions and long-term health effects, including black lung disease. Yet, because coal is so plentiful, it continues to be used for energy production across the world in the twenty-first century and is expected to continue so for the next decades[100–102].

Oil, in contrast, is much more of a twentieth-century fuel and differs from coal in a number of ways, including how concentrated the oil sector is. More so, the oil sector is defined by its volatility whereby its prices have fluctuated widely since the industry began, resulting in numerous boom-and-bust cycles in major energy-producing states. From its earliest days, the USA experienced booms and busts in the energy sectors. Owing to instability in global energy supplies (particularly oil), rising oil prices and increasing demand have elicited fear that the world's oil production would reach maximum capacity, or what has become known as "peak oil"[97]. The most famous illustration of this concept is referred to as Hubbert's Peak, named after a geologist who sought to calculate in the mid-twentieth century how long it would be before the USA hit peak oil production.

For years, "peak oil" – the notion that oil and gas are finite resources and that soon the rate of exploration would exceed the rate of production and would reach the point where the available oil and gas resources could no longer meet the increasing demands – has been considered the trigger and incentive for the search for new oil fields, technological advancements in drilling technologies, improvements in efficiency, and over time also alternative energy sources.

The development of unconventional energy resources in the twenty-first century – often colloquially referred to as the "shale revolution" – has drastically altered economic dependence upon conventional coal and oil and gas; fossil fuels increasingly are being extracted from shale gas and tight sand oil, including the extraction of heavy oil from tar sand and natural gas from coalbed formations (i.e., coalbed methane). While coal mining still remains a significant form of fossil fuels, unconventional oil and gas exploration has boosted the extraction of fossil fuels, at least to the first half of the next millennium. In 2016, natural gas primarily composed of shale gas became the larger energy source (32%) of the electricity sector in the United States. Coal, which had been the predominant source for decades, became only secondary. In 2019, shale gas production increased to 75% of the total natural gas production, which contributed 39% to the electric power sector generation, while coal's contribution was reduced to 24% of US electricity generation. While the shale revolution has taken place mainly in the USA and Canada, it began to spread globally with rising exploration in China, South America, and Australia[16, 103–106]. Hydraulic fracturing has triggered exploration of tight oil that made the USA a large worldwide oil producer. At the same time, global coal production has continued, and in spite of a temporary decline between 2014 and 2017, it rebounded to reach peak global production of 8.1 billion tons in 2019 (data from BP global dataset[10]).

For most of the history of fossil fuel extraction, the focus has been on its impact on air quality and human health, increasingly linked to global climate change. Less attention has, however, examined the role of water, excluding its use in the production processes. This chapter provides a basic exposition of the metric used to quantify energy and water flows, a survey of global energy trends, and an introduction to the water–energy nexus parameters that will be used in this book. We use the interconnection of energy and water flows in the USA to demonstrate the magnitude and the dependence of these fluxes. Because fossil fuel extraction is not expected to diminish significantly in the near future unless countries commit to an energy future premised upon renewables, this chapter sets the stage for understanding how countries' reliance on fossil fuel production affects water quality.

2.2 Energy and Water Metric

When evaluating energy production, the metric can be the volume or mass of the energy source; for example, mass (tons, or short tons in the USA) for coal, or volume for natural gas (cubic meter, or cubic feet in the USA) and petroleum (barrels or cubic meter) production. The mass and volume of fossil fuel sources are often converted to an energy unit (joules [J] in the international metric system [SI units] and British Thermal Units [BTUs] in the USA) that conveys the amount of energy generated from the mass or volume of fossil fuels. Table 2.1 presents the common conversion of mass or volume to energy unit based on the heat content of the fossil fuels. We present both the SI and the US units; BTUs are commonly presented in US publications and by the US Energy Information Administration (EIA), while the metric system is reported by the BP global energy dataset and other international organizations. The conversion of mass or volume to energy depends on the heat capacity of the fossil fuels, which varies between sources (oil, natural gas) and within a source. For example, different coals would have different energy (heat) contents based on their carbon contents, with the highest coal grades (anthracite, bituminous) having about double the heat capacity relative to low-grade coals. For natural gas energy content, the volume is converted for the natural gas and possible presence of natural gas liquids and residual oil condensates. For crude oil energy content, the conversion is for oil and the possible presence of wet gas steam, following the method reported in Kondash and Vengosh (2015)[23]. For electricity generation from thermoelectric power plants, the basic unit (in both the US and SI systems) is kilowatt-hour (kWh), which can also be converted to energy unit, and 277.7 kWh equals 1 gigajoule (joules $\times 10^9$, or GJ). Since the efficiency of power generation in coal-fired plants and natural gas plants is relatively lower than the potential heat capacity of coal or natural gas, the actual electricity generation from a given weight of coal or volume of natural gas is lower. Table 2.1 presents empirical data from electricity generation in the USA, China, and India that demonstrate the lower energy production of coal and natural gas as compared to their potential heat capacity. These relationships are important, as they help to determine common energy production for different energy utilizations and also to normalize water use to this common energy generation (e.g., cubic meter per gigajoule). For large-scale national and global energy fluxes, the US Department of Energy (DOE) utilizes the unit *quad*, which is equivalent to 1.055×10^{18} joules

Table 2.1. *Conversion of the commonly used energy units in the SI and US systems and comparison between the energy (heat) capacity of the different fossil fuels to actual electricity generation values**

	SI system			US system			
Source	Unit	Amount	Gigajoules (GJ)	Unit	Amount	BTU	Megawatt per hour (MWh)*
Coal	Metric Ton	1	21.26**	Short ton	1.1023	2.02×10^7	5.91
Natural gas	Thousand cubic meter	1	38.25	Thousand cubic feet (MFC)	35.31	3.63×10^7	10.62
Crude oil	Barrels (BBL)	1	6.12	Barrels (BBL)	1	5.8×10^6	1.69
	Ton of coal equivalent (TCE)	1	29.39			27.8×10^6	8.14
	Ton of oil equivalent (TOE)	1	41.86			39.7×10^6	11.63

* The empirical data of the relationships between weight (coal) and volume (natural gas) to electricity production (MWh) were generated from annual coal and natural gas utilized for the electricity sector normalized to the annual electricity generation.

** 27–30 for bituminous/anthracite coals, 15–19 for lignite/sub-bituminous coals.

(1.055 exajoules [EJ]) or 10^{15} (1 quadrillion) BTUs. The total primary US energy production in 2015 was 88 quads, which is equivalent to 92.84 EJ[107].

Other commonly used energy units are *ton of coal equivalent* (TCE) and *ton of oil equivalent* (TOE), which are used to describe the relative energy flows from different fossil fuel sources. The conversion of TCE and TOE to SI and US energy units is provided in Table 2.1. The SI system for water volume is liter (L) or cubic meter (m^3), while in the USA the unit is gallon (3.8 L = 1 gallon). It is common to express the use of water as volume per time, and typical reports for large water fluxes or use are in cubic meter per year, while in the USA, it is described as gallon per day (e.g., billion gallon per day [BGD]). The common practice to quantify the energy–water nexus is to evaluate the volume of water consumed or withdrawn per energy unit; for example, liter per gigajoules (L/GJ) and cubic meter per megawatt per hour (m^3/MWh). The amount of water use per energy is defined as *water intensity*, which varies during the different stages of energy exploration and electricity. Assessing the overall water intensity of a fossil fuel source will require adding the different components of the life cycles of energy production. These stages include (1) water use for mining (for coal), drilling (for oil and gas), or hydraulic fracturing (for unconventional oil and gas); (2) water use for processing (e.g., coal cleaning, oil refinement); and (3) water use for electricity generation (typically for cooling thermoelectric plants commonly expressed as m^3/MWh). The other component that is part of the water–energy nexus is the volume of wastewater that is generated, normalized to energy unit (e.g., L/GJ), which is defined as *wastewater intensity*. As we will show later, we use the wastewater intensity to quantify the volume of impaired water affected by the discharge of low-quality wastewater and the volume of water needed to dilute the contaminated water. Finally, we define another component, the *electricity intensity*, which quantifies the amount of electricity for generating water (through pumping, water transport and distribution, wastewater treatment, and reverse osmosis desalination), which is kilowatt-hour per cubic meter (kWh/m^3).

2.3 Global Fossil Fuel Production

Global energy consumption has increased continuously over the last 45 years by 3.7-fold at a rate of 7.5 EJ per year, up to the level of 583.9 EJ in 2019 (Figure 2.1). Based on the BP global dataset (2020)[10], the major sources of global energy consumption in 2019 were crude oil, coal, and natural gas that, respectively, contributed 33%, 27%, and 24% to global energy consumption, reflecting the predominance (84%) of fossil fuels in global energy production (Figure 2.2). Petroleum has been the predominant global energy source. Figure 2.3

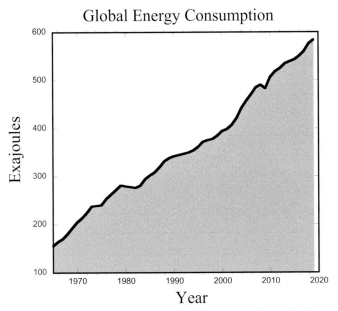

Figure 2.1 Variations of total global energy consumption (in exajoules) over time. Data were retrieved from the BP global energy dataset (2020)[10].

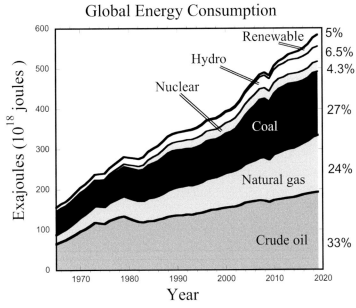

Figure 2.2 Variations of global energy consumption (in exajoules) of the major energy sources over time. Percentage values of the relative contribution of the different energy sources of total global energy consumption in 2019 are marked. Note the predominance of fossil fuels that constituted in 2019 about 84% of total global energy production.
Data were retrieved from the BP global energy dataset (2020)[10].

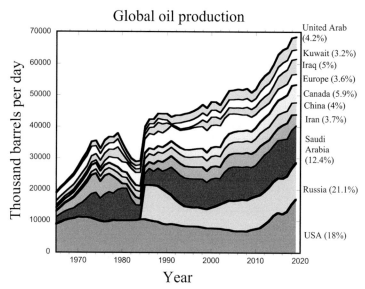

Figure 2.3 Variations of global crude oil production (in thousand barrels per day) of major regional producers over time. Note the predominance of oil production in the USA, Saudi Arabia, and Russia during the last three decades. The sharp rise of US petroleum production in the late 2000s marks the beginning of the extraction of unconventional tight oil.

Data were retrieved from the BP global energy dataset (2020)[10].

presents the distribution of major countries that produce petroleum: the USA in the early 1980s was a major producer, but production steadily declined until 2006 when the discovery of unconventional shale oil and tight oil increased production rates once again. Between 2012 and 2013, the USA became the world's largest petroleum producer (18% of global oil production in 2019), followed by Saudi Arabia (12.4%), and Russia (12.1%). The next tier of oil-producing regions include Canada (5.9%), Iraq (5%), United Arab Emirates (4.2%), China (4%), Iran (3.7%), Europe (mostly Norway; 3.6%), and Kuwait (3.2%) (Figure 2.3).

The coal data show a large increase in global coal production since 2000; from a global annual production of 4500 million metric tons during the 1990s when the USA and China were the major coal producers, global coal production has doubled and reached a peak of an annual production of 8255 million metric tons in 2013. Following a temporary reduction in annual production between 2013 and 2016, world coal production rose again in 2017–2019. China became the major global coal producer (47% of global production in 2019), followed by India (9.3%), the USA (7.9%), Indonesia (7.5%), Australia (6.2%), and Russia (5.4%) (Figure 2.4). Coal production in the USA reached a maximum in 2007, with a production of

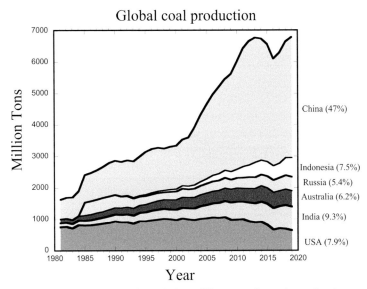

Figure 2.4 Variations of global coal (in million metric ton) production and key coal-producing countries. Note the predominance of coal production in China that extracts about half of global coal exploration.
Data from Global Energy Statistical Yearbook 2018[108].

over 1,000 million tons, but since 2008 this has declined. While US coal production has been declining, coal production in India (717 million metric tons in 2017) has increased to constitute almost 10% of global coal production, exceeding US production rates. It is interesting to note that a temporal reduction in coal production in China (2013–2016) was reversed in 2017, with a 3% increase (Figure 2.3). Global natural gas production has been increasing steadily, and since the early 1970s, global natural gas production has increased fourfold, up to almost 4,000 BCM per year in 2019 (Figure 2.5). The USA (23.1% of global natural gas production in 2019) and Russia (17%) have been the major producers of natural gas, followed by Iran (6.1%), China (4.5%), Qatar (4.5%), Australia (3.9%), and Saudi Arabia (2.9%) (Figure 2.5). The rise of natural gas production in the USA is directly associated with the advent of hydraulic fracturing and unconventional shale gas exploration in the early 2000s.

Overall, despite increasing awareness of the climate crisis and the expansion of renewables worldwide, the beginning of the twenty-first century is, nonetheless, marked by increasing rates of global fossil fuel production, including a growing role for unconventional shale gas and tight oil production. As of 2019, renewable energy sources of solar and wind only account for 5% of total energy production in contrast to fossil fuels, which constitute 84%.

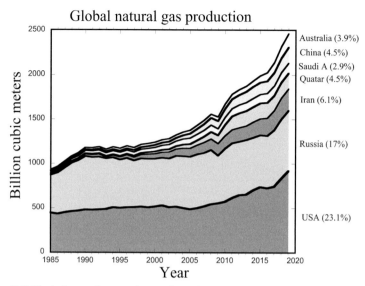

Figure 2.5 Variations of natural gas (in BCM) production in major natural gas-producing countries. Note the rise of natural gas production in the mid-2000s, which made the USA the largest natural gas world producer.
Data from Global Energy Statistical Yearbook 2018[108].

2.4 The Energy–Water Nexus Concept

The traditional energy–water nexus literature is focused on the quantification of the use of water for the different life-cycle stages of energy extraction, including primary (or *upstream*) energy production (mining, drilling, hydraulic fracturing), processing (coal cleaning, oil refinement), electricity generation (cooling systems in thermoelectric plants), and a *downstream* section that involves the disposal or treatment of solid and liquid wastes. The literature has distinguished between water *withdrawal* (water that is extracted for different energy cycles) and water *consumption* (water that is actually consumed during these processes and does not return to the hydrological cycle)[28, 30, 31, 33, 45, 47, 62]. In those terms, direct water use is calculated in addition to the indirect water footprint derived from each of the energy stages, such as transport. The International Energy Agency (IEA)[45] estimates that in 2014 the annual global water withdrawal for energy production was 350 BCM, from which water withdrawal for energy production from fossil fuels constituted 58% (230 BCM) (Figure 2.6). Global water consumption was estimated as 48 BCM, out of which the fossil fuel fraction was only 27% (13 BCM)[45]. For primary energy production, which, respectively, composed 11.8% and 62.5% for global water withdrawal and consumption, fossil fuels

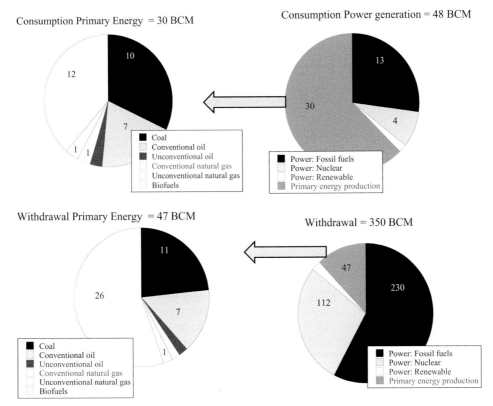

Figure 2.6 Distribution of the global water withdrawal and consumption volumes for energy production.
Data from International Energy Agency[45].

(coal, conventional and unconventional oil, conventional and unconventional natural gas) constituted 45% and 61% for 2014 water withdrawals and consumption, respectively (Figure 2.6). For comparison, the annual overall renewable water in all Middle East and North Africa (MENA) countries is estimated as 632 BCM, and countries such as Jordan and Israel consume less than 2 BCM per year[70, 109].

In this book we further expand the energy–water nexus concept and add a third component, defined as *water impaired*, which represents the impact of each of the energy-producing stages on water quality (Figure 2.7). Numerous studies have shown the impact of fossil fuel development on water quality (see review for example in Allen et al., 2011)[90], but the overall integrated impact has not been determined. In the following chapters we present detailed information on the impact and implications of energy production on water quality. We present the water footprints of fossil fuel extraction through analysis of the different life-cycle

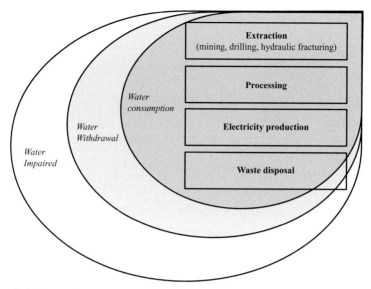

Figure 2.7 Illustration of the conceptual three components of the energy–water nexus for the different stages of energy exploration, processing, electricity production, and waste disposal, which include utilized water directly consumed for each cycle ("water consumption"), withdrawal water extracted from water resources ("water withdrawal"), and a third new concept of "impaired water" that also includes the water quality component of the energy–water nexus.

stages of energy extraction and production and will evaluate the three layers of impact: water withdrawal, water consumption, and water impaired by fossil fuel exploration and production (Figure 2.7). While establishing the water metric for withdrawal and consumption of water for energy is directly linked to measurable parameters such as volume of water used per extracted mass of coal, or volume of water used for kilowatt energy produced in a thermoelectric plant, the impaired water quantification is much more complicated and thus challenging. We show that water withdrawal involves not only the actual volume of water extracted for energy exploration or production but also the hydrological and water quality consequences resulting from these operations (mining, drilling, waste disposal). For example, mining for coal includes the water used directly for the mining process as well as the volume of water that is contaminated due to mining for both surface (e.g., mountaintop coal mining) and subsurface coals. The impaired water is the estimated volume of water that could be impacted from all stages of energy exploration and production, including the potential water contamination associated with the disposal of solid (e.g., coal ash) and liquid wastes (e.g., flowback and produced water from oil and gas exploration). *Thus, expanding the concept of the water–energy nexus with the addition of the water quality component offers a new paradigm, which is the focus of this book.*

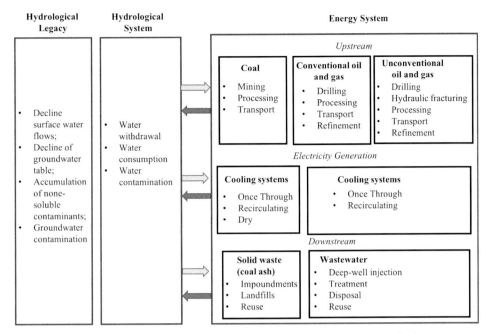

Figure 2.8 Illustration of the relationships between the active hydrological system, the energy system for fossil fuels composed of upstream, electricity generation, and downstream sections, and the hydrological legacy system that reflects the long-term effects of water use and water contamination.

Quantification and the establishment of a water volume metric for water quality degradation is not a simple task, but we integrate known and updated values of the water intensities of the different fossil fuels with new insights concerning the key elements for evaluating and quantifying the magnitude of the *impaired water* through the different stages of fossil fuel production. Figure 2.8 illustrates the different stages of the energy system with respect to direct impact (water use, contamination) on the current hydrological system as well as the long-term effects and legacy of the hydrological system affected by energy development. In the following chapters we provide metric values for the fluxes presented in Figure 2.8.

2.5 The Example of the US Energy–Water Quantity Nexus

Two documents of the US Department of Energy[110] and US Geological Survey[111] present the total water and energy flows in the USA. We use these analyses to introduce the dynamics and interrelationships between energy and water fluxes, using the US case as an example to illustrate the complexity of the energy–water nexus. Figure 2.9 presents the water (BCM per year) and energy (quad = 1015 BTU per year) sources, and Figure 2.10 presents their flows and the interconnection

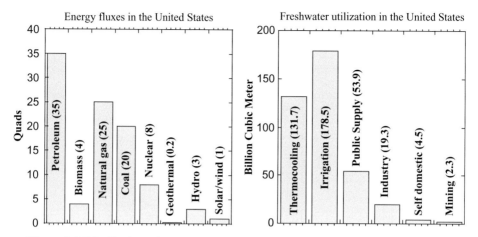

Figure 2.9 Distribution of the annual water (right) and energy (left) sources in the USA based on integration of DOE (for energy[110]) and USGS (for water[111]) reports. Values for energy and water fluxes are in quad (1,015 BTU) and BCM per year, respectively.

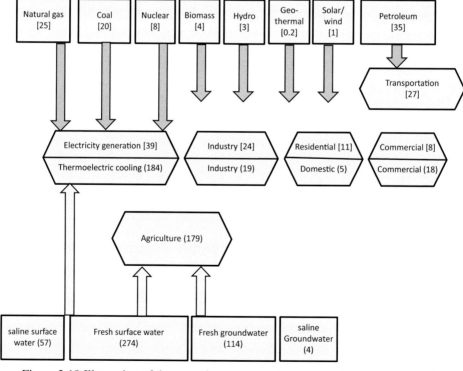

Figure 2.10 Illustration of the annual energy (grey) and water (white) fluxes in the USA based on integration of DOE (for energy[110]) and US Geological Survey (for water[111]) reports. Values for energy and water fluxes are in quad (1,015 BTU) and BCM per year, respectively. Note that electricity generation is the primary flux of both energy and water fluxes in the USA.

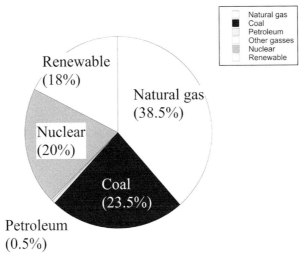

Figure 2.11 Distribution of the electricity sources in the USA in 2019. Data from US Energy Information Administration[112].

between water and energy flows. We use the water volume data for 2014 presented by Dieter et al. (2018)[111] and energy flow data presented by the DOE (2014)[110] report. Of a total of 444.8 BCM water extracted in the USA, a total of 388.2 BCM of freshwater was annually withdraw from surface water (273.6 BCM) and groundwater (114.4 BCM), and was utilized for thermoelectric cooling (388 BCM total, from which 131.7 BCM was freshwater, equivalent to 33.9% of total freshwater utilization), irrigation (178.5 BCM; 45%), public supply (53.9 BCM; 13.8%), residential homes (4.5 BCM; 1.1%), industry (19.3 BCM; 4.9%), and mining (2.3 BCM; 0.6%). The breakdown of the water flow shows that irrigation (45% of total freshwater withdrawal) and cooling thermoelectric plants (34%) are the major factors in water consumption in the USA (Figure 2.10). Most of the water used for thermoelectric cooling is returned to the surface after use, while the majority of water for agriculture is consumed. The energy sources in the USA in 2014 were petroleum (35 quads per year, 36.4% of total energy produced in the USA), biomass (4 quads, 4.2%), natural gas (25 quads, 26%), coal (20 quads, 20.8%), nuclear (8 quads, 8.3%), geothermal (0.2 quads, 0.2%), hydropower (3 quads, 3.1%), and wind/solar energy (1 quad, 1%). Electricity generation consumes 39 quads, which was 40% of total energy production in the USA, mostly from natural gas, coal, nuclear, geothermal, hydropower, and wind/solar energy (Figure 2.11). Consequently, electricity generation consumed large fractions of annual water and energy (38% and 40%, respectively) production in the USA. In 2019, 4.12 trillion kWh of electricity were generated at utility-scale electricity generation facilities in the USA, from which 63% was from fossil fuels – coal, natural gas, and petroleum – 20% was from nuclear energy, and about 18% was from renewable energy sources (Figure 2.11). Natural gas generated 1,582 billion kWh (38.4%), coal 966 kWh (23.5%), and petroleum 19 kWh (0.5%) of electricity, respectively[112].

Chapter 2 Take-Home Messages

- The common methods of quantification of the water used for the different life cycles of energy production (recovery, processing, power generation, waste) are *water withdrawal* (water that is used and returned to the hydrosphere) and *water consumption* (water that is not returned to the hydrosphere).
- In addition to the volume of water that is directly used for energy production, water quality degradation induced from energy production requires a certain volume of clean water to balance the contamination impact; that volume is defined here as *impaired water*.
- Quantification of the water use for energy production is through normalization of the water volume to energy use, defined as *water intensity*.
- Despite of the rise of renewable energy, by the end of the second decade of the 2000s, fossil fuels became the predominant energy sources.
- In the USA, the water volume mostly used for cooling thermoelectric plants exceeds the water used for the agriculture sector, reflecting the critical role of water availability on energy production.

3

The Coal–Water Nexus

3.1 Introduction

Coal is one of the dominant energy sources for electricity generation worldwide. Throughout the world, it has been the primary fossil fuel that has accelerated the process of industrialization, initially in the USA and Europe, but also later in China. Coal emerged as an important source of energy in the mid-1800s in the USA, as steam engines and steel mills began to switch from wood to coal[113]. By 1880, coal already made up about 40% of the total US energy supply and by the end of the nineteenth century had grown to 70% with the introduction of coal-fired power plants[113].

In 2019, coal production accounted for 27% of the world's primary energy supply (Figure 2.2) and was mined across the globe (Figure 3.1). British Petroleum (BP) estimated that by the end of 2019, proven reserves of global coal were 1,070 billion metric tons, from which the annual global coal production in 2019 was 8.13 billion metric tons[10] (i.e., 0.76% of global reserves; Figure 3.2). Owing to the vast supplies of coal worldwide, it is a readily available source of energy; 40% of global electricity is fueled by coal, and in many countries, coal is the predominant source for electricity, as in Poland (94%), South Africa (92%), and China (77%)[114]. About 23% of the global proven coal reserves are in the USA (Figure 3.2). During most of the twentieth century, coal was the major energy source for the electricity sector in the USA, providing 50% of electricity during the 1990s. Yet, with the rise of unconventional natural gas, coal as part of the overall fuel mix in the USA began to decline such that, by 2017, coal only contributed about 30% of the electricity sector relative to 32% by natural gas (Figure 2.4). In spite of these changes in the USA, the global coal supply has not changed[115], and since the early 2000s, global coal production has rapidly increased, reaching a peak in 2013 of over 8,000 million metric tons (Figure 3.3). After 2013, global coal production reduced and increased again in 2017–2019 (Figure 3.3). In 2019, the global annual coal production was 8,130 million metric tons, where China

Global Coal Production in 2019

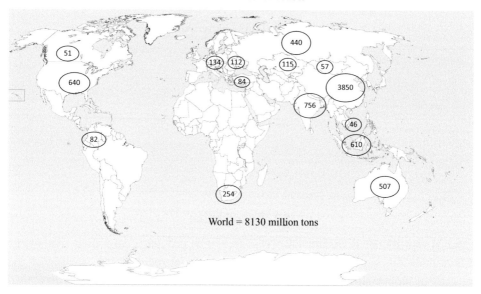

Figure 3.1 Map of global coal production. Values in millions metric tons per year (for 2019) and were retrieved from BP global dataset[10].

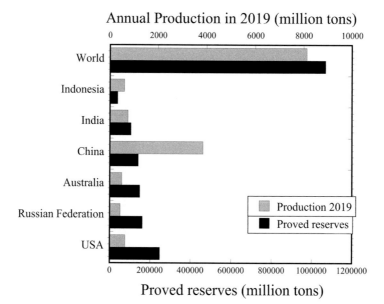

Figure 3.2 Global coal reserves and annual production (million metric tons) of the major coal-producing countries. Data from BP global dataset[10].

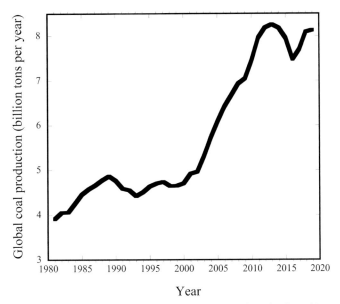

Figure 3.3 Global coal production (in metric tons) during the last 40 years. Data from BP global dataset[10].

(44%), India (9.6%), the USA (9.3%), and Australia (6.6%) were among the major coal production nations[116] (Figure 3.2). While Germany – once a major coal producer and consumer – has pledged to phase out coal by 2038, and despite closing down its hard coal mines in 2018, Germany continued to rely on hard coal and lignite for power production in 2018 (35.3% compared to 35.2% from renewables, 11.7% from nuclear, and 12.8% from natural gas[117]).

In the USA, 90% of coal is consumed to generate electricity, while in China about half of the coal is used for the electricity sector and the other half is used for the industrial sector (e.g., high temperature baking of coal generates coke used in the steel industry). In the USA, about 37% of coal mined is extracted via surface mining in high or extremely high water-stressed areas (i.e., >80% of watershed water is used) in the western arid USA (Figure 3.4)[118]. China has 15,119 coal mines with a total annual productivity of 3.69 billion metric tons, of which 70% are located in water-scarce regions and 40% have severe water shortage[119]. In the USA, coal production decreased by 69% since 2011, down to 686 million metric tons in 2018[120] (Figure 3.5). Yet since 2016 coal production has stabilized, with the proportion of exported coal increasing (from 10% in 2011 to 15% in 2018). Most of the US coal is mined in the Western basins (55%), followed by the eastern Appalachian Basin (mostly in West Virginia; 27%) and the Interior (Illinois) Basin (18%)[120, 121] (Figure 3.5). Where once coal mining was an important part of the local economies in the states of North Rhine-Westphalia and the Saarland in

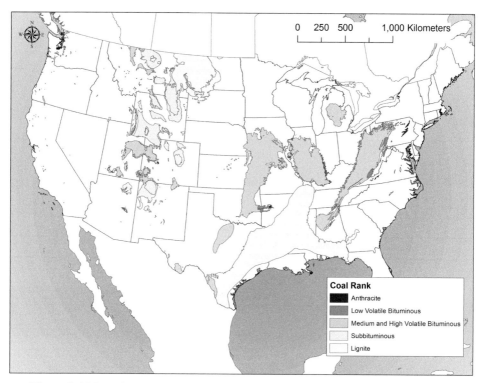

Figure 3.4 Map of major coal basins in the USA. Map generated by A.J. Kondash (Duke University) based on US Geological Survey and US Energy Information Administration (EIA)[130]. (A black-and-white version of this figure will appear in some formats. For the colour version, refer to the plate section.)

Germany, as in Appalachia in the USA, with the decline of the coal economy, as of December 2018, Germany has relied on hard coal imports primarily from Russia (35%), the United States (18%), Australia (13%), and Columbia (11%)[117].

Unlike the limited natural gas and oil resources, coal deposits are vast, with an estimated 984 billion metric tons of proven coal reserves worldwide (Figure 3.2), which suggests sufficient supply for the next 200 years[114]. Yet, emissions from coal combustion account for 40% of global carbon emissions[122], and thus remain a major source of CO_2 atmospheric accumulation and global warming (see Chapter 6). In addition, the rising coal combustion in countries like China and India has resulted in increasing atmospheric pollution and poor air quality in the major cities of these countries due to the emission of SO_2, NOx, and particulate matter (PM)[123, 124]. Coal combustion is also the major source for atmospheric mercury emission[125, 126]. Consequently, breaking the link between economic growth and coal production is key for reducing the dependency on cheap coal. Qi et al. (2016)[127] have demonstrated that "coal peaks" in the UK, USA, and China

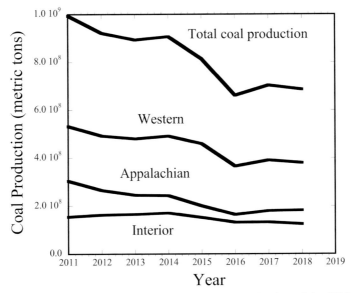

Figure 3.5 Changes of coal production in the major coal basins of the USA during the last decade. Data from Acharya and Kharel (2020)[120].

reflect the rise of coal production parallel to the per-capita GDP growth to the point where the economy is no longer dependent on manufacturing and construction and has shifted towards the service sector. For example, between 2000 and 2013, the annual coal consumption in China grew from 1.36 billion metric tons to over 4.24 billion, which was reduced to 3.97 billion metric tons in 2015[127]. According to Qi et al. (2016)[127], this reduction in coal production in China, in spite of the growth of per-capita GDP, reflects the "coal peak," which was shown also in the UK in 1956 and the USA in 2007. Yet the reduction trend shifted after 2016, and coal production in China grew again, reaching a peak of 3.8 billion metric tons in 2019[10].

Dramatically reducing coal production and thus carbon emissions is critical for current international agreements to address successfully the climate crisis, with some scholars pushing further for a treaty that would keep coal in the ground[128, 129]. To date, however, different technological approaches have been promoted to mitigate carbon emissions; for example, sequestration of carbon to the subsurface instead of releases to the atmosphere is the principle of the Carbon Capture and Storage (CCS) technology that has been developed to address this issue[122]. Yet, CCS is still at a developmental stage, and many have questioned the ability of this technology to remove the sufficient magnitude of carbon necessary for reducing global warming. Other countries have sought simply to limit coal as part of a broader approach to reducing greenhouse gas emissions, as has been the case in Germany with its energy transition strategy. Nonetheless, in spite of increasing recognition of the importance of the contribution of coal combustion to climate change and global warming, global coal production did not decline, and after a short period of reduction between 2013 to 2016, rose again, up to 8.13 billion metric tons in 2019 (Figure 3.3).

In this chapter we follow the different stages of coal utilization and evaluate the magnitudes of water withdrawal, consumption, and impairment through the different life cycles of coal extraction, including surface and subsurface mining, processing such as washing to reduce impurities, cooling for combustion, and the management and disposal of coal combustion residues (CCRs, or coal ash; the solids that are left after coal combustion). While the primary focus of this chapter is on impacts of coal on water – especially the volume of the impacted water – this chapter concludes with a brief discussion of countries' strategies to move away from coal not only because of both the short and long-term impacts on the natural environment, but also because of high carbon emissions linked to the use of coal.

3.2 What's in Coal

In order to understand the risks of coal mining and combustion, it is essential to know what is in coal deposits. Coal rocks originate from organic-rich sediments deposited in low-oxygen environments such as swamps, deltas, alluvial plains, and coastal areas that allow for the burial and preservation of organic matter in the sediments, known as peat. The geological evolution (defined as diagenesis or coalification) of peat depends on pressure and temperature as well as the oxygen content that results in differential carbon contents in the coal rocks. The economic properties of coal depend primarily on the carbon content of the coal, as well as the ash content (i.e., the relative volume of solids residuals after combustion), moisture content, and elemental composition (e.g., sulfur). The subdivision of coal deposits is based on these properties; the lowest coal rank is lignite, also known as "brown coal," and comprises 17% of the world coal reserves with low carbon and high moisture and residual ash contents. The higher-rank coals are "hard coal" that includes sub-bituminous (30% of the world coal reserves), bituminous (50%), and anthracite (1%), with higher carbon and low moisture and ash contents[114]. While lignite and sub-bituminous coals are used primarily for power generation, about half of the highly ranked bituminous coal is also used for metallurgical (coking coal) purposes – for steel and other industries[114].

Numerous studies have evaluated the contents of trace elements in coal[131–144]. Following the Goldschmidt (1935)[145] method for calculating the "enrichment coefficients" of coal by a comparison of element content in coal and the Earth's crust average values, Ketris and Yudovich (2009)[139] compared the average

Table 3.1. *Data on the average concentrations (in mg/kg) of trace elements in Chinese coals[138], US coals[146], and world coals[139] as compared to the average sedimentary rocks reported in Ketris and Yudovich (2009)[139] and the upper continental crust reported in Rudnick and Gao (2014)[147]*

Element	Upper crust	Sedimentary rocks	Chinese coal	US coal	World coal
Transition metals					
Sc	14	10	4	4	4
V	97	91	35	22	25
Cr	92	58	15	15	16
Co	17	14	7	6	5
Ni	47	37	14	14	13
Cu	28	31	18	16	16
Zn	67	43	41	53	23
Alkali and alkaline earth					
Li	21	33	32	16	12
Rb	84	94	9	21	14
Cs	49	8	1.1	1.1	1
Be	2	2	2.1	2.2	1.6
Sr	320	270	140	130	110
Ba	624	410	159	170	150
Actinides and heavy metals					
Tl	1	0.9	0.5	1.2	0.6
Pb	17	12	15.1	11	7.8
Bi	0.2	0.3	0.8	N/A	1
Th	11	8	5.8	3.2	3.3
U	2.7	3.4	2.4	2.1	2.4
High-field strength					
Zr	193	170	90	27	36
Hf	5	4	4	1	1.2
Nb	12	8	9	3	4
Ta	1	1	0.6	0.2	0.3
Mo	1	2	3	3	2.2
W	2	2	1	1	1
Siderophile and Chalcophile					
Cd	0.1	0.8	0.3	0.5	0.2
In	0.06	0.04	0.05	N/A	0.03
Sn	2	3	2.1	1.3	1.1
Sb	0.4	1	0.8	1.2	0.9
Au	1.5	6	N/A	N/A	3.7
Hg	0.05	0.07	0.2	0.2	0.1

Table 3.1. (*cont.*)

Element	Upper crust	Sedimentary rocks	Chinese coal	US coal	World coal
Halogens					
Cl	370	2700	255	614	180
Br	1.6	44	N/A	N/A	5.2
I	1.4	1100	N/A	N/A	1.9
Metalloids					
B	17	72	53	49	52
As	5	7.6	3.8	24	8.3
Se	0.1	0.3	2.5	2.8	1.3

composition of trace elements in world coals to the average of these elements in sedimentary rocks. Table 3.1 presents data of the average values of Chinese coals[138], US coals[146], and world coals[139], as compared to the average sedimentary rocks reported in Ketris and Yudovich (2009)[139]. The comparison of the trace elements in coals, sorted by their chemical properties, to average sedimentary rocks is presented in Figure 3.6. The comparison shows distinctive (ratio >1) enrichment in metalloids (boron, arsenic, selenium), beryllium, molybdenum, and mercury in coals relative to sedimentary rocks. As we show later in this chapter, during coal combustion, the residual coal solids are further enriched in trace elements, and thus their concentrations are much higher than common sedimentary rocks. Table 3.1 presents the actual concentrations of trace elements and their enrichment factors in coals (average values from China, the USA, the world) relative to those in sedimentary rocks and the composition of the upper crust reported in Rudnick and Gao (2014)[147].

In addition to toxic elements, coal contains radioactive elements, in particular nuclides that are part of the uranium- and thorium-decay series. The co-precipitation of the radioactive elements of uranium and thorium in the original peat sediments that evolved over million of years to coal formations resulted in decay and accumulation of different radioactive elements that are part of the uranium- and thorium-decay chains. Over time, the concentrations of radio-nuclides, defined as "activity" (i.e., concentrations in radioactive units that reflect the degree of radioactivity or decay over time) of the different elements become equal (i.e., the decay rate of a parent nuclide is equilibrated with the accumulation and decay rates of the daughter nuclide), a process that is defined as "secular equilibrium." Given the long age of coal rocks, the activity of radionuclides from the same decay chain become equal to the activity of the initial uranium or thorium

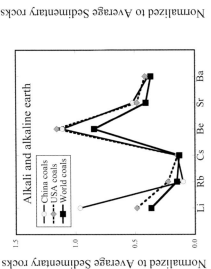

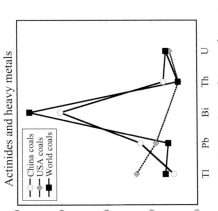

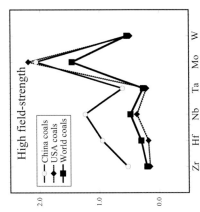

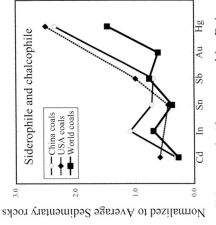

Figure 3.6 Comparison of trace elements (in mg/kg) sorted by their chemical properties in Chinese coals (data reported by Dai et al., 2012[138]), US coals (Finkelman, 1993[146]), and world coals (Ketris and Yudovich 2009[139]) as compared to average sedimentary rocks.[139] Elements like selenium, bismuth, molybdenum, and mercury in coals are enriched (i.e. ratio >1) relative to the average composition of sedimentary rocks.

that was deposited in the original peat sediments. While most US and Chinese coals are known to have relatively lower uranium concentrations, typically between 2 and 3 ppm[138], some coal deposits in China[148–155], Czech Republic, and South Africa[156, 157] are characterized by much higher uranium and thorium concentrations. The high concentrations of uranium and thorium and thus their decayed-radioactive elements in coals imply much higher concentrations in the residual solids after combustion, and therefore coal ash originating from particularly uranium-rich coals poses high risks for nuclide mobilization and contamination of the environment.

In 2013, global coal production peaked with an annual production of 8.2 billion metric tons, declining to 7.4 billion metric tons by 2016. Taking into account the average concentration of trace elements in world coals reported by Ketris and Yudovich (2009)[139], the global fluxes of these elements from coal combustion are presented in Table 3.2. As shown for mercury[158–165], boron,[166] and vanadium[167], the global combustion of coal generates large contaminant fluxes and therefore constitutes a large contribution to the anthropogenic-derived elements in the environment, in some cases exceeding the naturally occurring global fluxes of geological/hydrological processes[168].

3.3 Coal Mining and Processing

Coal extraction includes mining and pre-combustion treatment that includes preparation of the coal to minimize contaminant emissions during the combustion process. This section focuses on the water use (withdrawal and consumption) for surface and subsurface coal mining, the pre-combustion treatment processes, the quantity and quality of wastewater that is generated from these activities, and the quantity and quality coal mine drainage effluents generate from interactions of water with residual rocks in the mines. Each of these stages involves actual and virtual (indirect) water use, in addition to the release of wastewater to the environment, causing degradation of a larger volume of water than what is directly used in the coal extraction process.

3.3.1 Water Use for Coal Mining and Processing

Traditionally, coal mining was carried out below ground (i.e., subsurface mining; Figure 3.7), and yet modern exploration has increasingly extracted coal through surface mining, including strip mining and mountaintop removal mining (MTM), which extracts surface coal from ridges and mountaintops in steep terrains, such as in Central Appalachia in the eastern USA (including Kentucky, West Virginia,

Table 3.2. *Estimates of the global elemental fluxes from global coal combustion (×10^{10} gram per year) using the average concentrations of elements in world coals reported by Ketris and Yudovich (2009[139]) and the annual global coal production in 2013 (8.2 billion tons)*

Element	World coal (mg/kg)	Element global flux (× 10^{10} gram per year)
Transition metals		
Sc	3.9	3.2
V	25	20.5
Cr	16	13.1
Co	5	4.2
Ni	13	10.7
Cu	16	13.1
Zn	23	18.9
Alkali and alkaline earth		
Li	12	9.9
Rb	14	11.5
Cs	1	0.8
Be	1.6	1.3
Sr	110	90.3
Ba	150	123
Actinides and heavy metals		
Tl	0.6	0.5
Pb	7.8	6.4
Bi	1	0.8
Th	3.3	2.7
U	2.4	2
High-field strength		
Zr	36	29.6
Hf	1.2	0.9
Nb	3.7	3
Ta	0.3	0.2
Mo	2.2	1.8
W	1.1	0.9
Siderophile and Chalcophile		
Cd	0.2	0.2
In	0.03	0.02
Sn	1.1	0.9
Sb	0.9	0.8
Hg	0.1	0.08
Halogens		
Cl	180	148
Metalloids		
B	52	42.7
As	8.3	6.8
Se	1.3	1.1

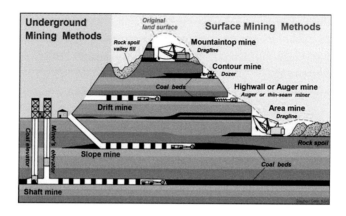

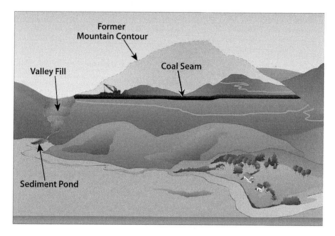

Figure 3.7 Illustration of surface and subsurface coal mining and mountaintop mining in the Appalachian Basin. Data are from US EPA website: www.epa.gov/sc-mining/basic-information-about-surface-coal-mining-appalachia. (A black-and-white version of this figure will appear in some formats. For the colour version, refer to the plate section.)

Virginia, and Tennessee) (Figure 3.7). MTM involves removal of a large volume of rocks and soil overlying the coal layers. The World Coal Association estimates that surface mining accounts for 40% of global coal mines relative to 60% of subsurface mines. In the USA, 74% of produced coals originates from surface mining and only 36% is derived from subsurface mining (the average value of coal production evaluated for 2013 to 2016[169]; Figure 3.7). In China, in contrast, 95% of coal extraction is through subsurface mining[119].

The water use for coal mining is mainly for (1) reduction of dust particles, particularly in subsurface mines; (2) reclamation of vegetation in areas of surface mining; and (3) processing coal to remove impurities and extraneous materials, such as for sulfur-rich coals. In some areas, water is also used to transport coal slurry to a coal power plant, which increases the water consumption. In China, approximately 500 L of water is used to produce one to two metric tons of coal. The annual coal production in 2018 was 3.5 billion metric tons[170], which suggests utilization of the range of 0.9–1.8 BCM for coal mining in China. In addition, about 2,500 L is used for washing one metric ton of coal[170] in China, and since 56% of coals are washed in China, this would account for 5.1 BCM per year. Zhang and Anadon (2013)[170] reported annual water withdrawals and consumption of 1.37 and 0.79 BCM, respectively. Zhang et al. (2016)[171] reported higher volume of annual water use for coal mining in China as 7.1 BCM in 2015, which is one of the highest yet most up-to-date estimates. Grubert and Sanders (2018)[29] estimated the total water withdrawal and consumption for US coal mining as 1.35 and 0.66 BCM per year, respectively.

The volume of water use for subsurface mining is much larger than that of surface coal mining. Data on water use for surface and subsurface coal mining in the USA shows that while subsurface mining consists of 36% relative to surface mining with 74% of the total coal mining, the relative proportion of water withdrawal of subsurface mining (1.39 BCM per year; 47%) is higher than that of surface coal mining in the USA (1.03 BCM per year)[118]. Most of the surface mining in the western USA is conducted in moderate- to extreme-water stress areas (Figure 3.8). The different proportions of utilization of subsurface and surface coal mining between 2013 and 2016 in the USA is reflected in changes in the water withdrawal and consumption for coal mining (Figure 3.9)[118].

Values for water intensity of coal mining and processing vary considerably (Table 3.3); early studies suggested low values of 2 and 17 liter per gigajoules (L/GJ) for surface and subsurface mining[28], while later estimates show much higher values of 300 L/GJ and 763 L/GJ, respectively[31]. Using the US EIA data, Spang et al. (2014)[34, 48] showed lower values of 43 L/GJ, which is consistent with the empirical values reported by Zheng et al. (2013)[170] that are equivalent to 54 L/GJ. Based on the value of 7.1 BCM reported by Zhang et al. (2016)[171] for coal mining in China, the water intensity is calculated to 92.9 L/GJ (taking into account the coal production in 2015 in China was 76.4 Exajoules[10]). In the USA, the water intensity for coal washing was estimated as 245 L/GJ[31], while the empirical data for washing coals in China corresponds to water intensity of 360 L/GJ. More recent data for coal water use in the USA was presented by Grubert and Sanders (2018)[29] whereby they estimated the water intensity of 86.5 L/GJ for coal production for both withdrawal and consumption, while coal processing had a water intensity of 194 L/GJ for withdrawal and 95 L/GJ for consumption. Taking into account both the direct and indirect water use through transportation, fuels, and electricity use for pumping, Kondash et al. (2019)[118] estimated a larger water intensity for water withdrawal of subsurface coal mining

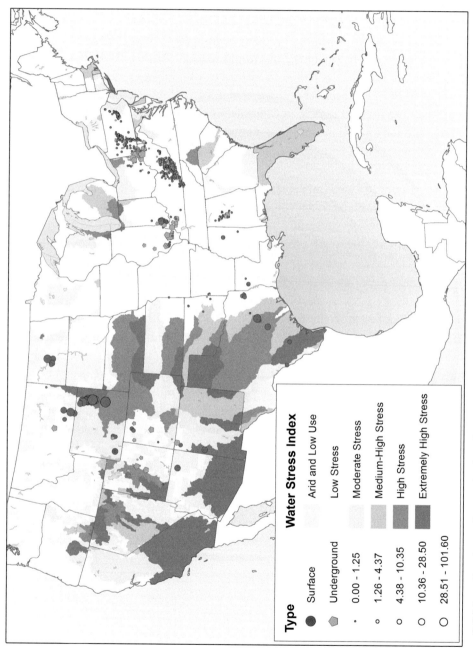

Figure 3.8 Map of major surface and subsurface coal mining on the background of the water-stress map across the USA. The map shows that the larger surface mining with higher water footprint occurs in highly stressed areas in the western USA. Map was created by A.J. Kondash (Duke University) based on EIA-7A, Coal Production and Preparation Report[172] and Aqueduct Water Stress Projections[173] . (A black-and-white version of this figure will appear in some formats. For the colour version, refer to the plate section.)

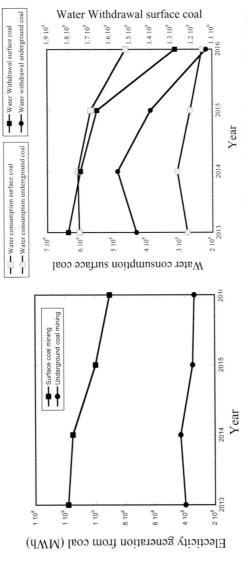

Figure 3.9 Changes in electricity production from surface and subsurface coal mining in the USA (expressed in MWh units: left) and the volume of water withdrawal and consumption (cubic meter, right) of subsurface and surface coal mining in the USA between 2013 and 2016. Data are from Kondash et al. (2019)[169] .

Table 3.3. *Water intensity (L/GJ) variations (median values) for different stages and types of coal mining and processing as reported in different studies*

Type of coal	Water consumption (L/GJ)	Water withdrawal (L/GJ)	Reference
Surface coal mining	300		Meldrum et al. (2013)
Subsurface coal mining	763		Meldrum et al. (2013)
Coal cleaning	245		Meldrum et al. (2013)
Coal mining	43		Spang et al. (2014)
Coal mining (China)	54		Zheng et al. (2013)
Wastewater from coal mining (China)	384		Zheng et al. (2013)
Coal cleaning (China)	360		Zheng et al. (2013)
Coke for industry	360		Zheng et al. (2013)
Coal mining (China)	50	125	Zheng and Anadon (2013)
Coke production (China)	122	254	Zheng and Anadon (2013)
Coal mining (China)	93		Zhang et al. (2016)
Coal mining (USA, dewatering)	87	87	Grubert and Sanders (2018)
Coal processing (USA)	9	107	Grubert and Sanders (2018)
Combined coal mining and processing (USA)	95	194	Grubert and Sanders (2018)
Subsurface coal mining (USA)	156	450	Kondash et al. (2019)
Surface coal mining (USA)	210	1080	Kondash et al. (2019)

(450 L/GJ) and surface mining (1,080 L/GJ), as well as water consumption of subsurface coal mining (156 L/GJ) and surface mining (210 L/GJ) (average values for 2013 to 2016)[118]. Zheng and Anadon[170] also distinguished between direct water withdrawal (18.4 L/GJ) to life-cycle water withdrawal (106.4 L/GJ) with differentiated direct water consumption (8.9 L/GJ) and life-cycle water consumption (41.5 L/GJ) for coal mining in China. Overall, estimates for water use associated with coal mining and processing have changed over time as more data have become available, mostly for the USA and China. The most recent studies have highlighted the indirect water use and additional water withdrawals that were not identified in earlier studies. Table 3.3 summarizes the water intensity values for coal mining reported in the literature.

While many studies have attempted to quantify the direct and indirect water use for coal mining, the extraction of groundwater, which is the principle water source for the majority of coal mines worldwide (e.g., over 95% in China[119]) has resulted in over-pumping and the drawdown of the groundwater table in the local aquifers, particularly in water-scarce areas. In northeast China, it was estimated that extraction of each metric ton of subsurface coal mining is equivalent to 1.07 cubic m of groundwater that is depleted from the aquifer[119]. Assuming that the northeast of China extracts ~40% of the total of China's coal (equivalent to 1.36 billion metric tons), we estimate that ~1.5 BCM of groundwater is depleted every year only in the northeastern part of China, which has suffered from severe water scarcity. Consequently, the groundwater depletion adds an additional water intensity component; using the groundwater depletion-to-coal ratios demonstrated for northeastern China[119], and the relationship between mass of coal used in China to generate electricity, we estimate an additional water intensity of 139 L/GJ due to groundwater depletion. Groundwater overexploitation commonly induces numerous water quality issues such as mineralization and seawater intrusion into coastal aquifers. In sum, the water quality deterioration induced by overuse of aquifers for coal mining is an important component of the impaired water, yet difficult to quantify on a global scale. This implies that all conventional estimates for water intensity of coal mining underestimate the actual impact as they do not include the additional impact induced from groundwater extraction and possible contamination processes that further increase the amount of water loss induced from coal mining.

3.3.2 The Quantity and Quality of Coal Mining Wastewater and the Long-Term Effects of Coal Mining

Coal mining and processing induce major changes in the water quality in associated watersheds and aquifers. The deterioration in water quality is a result of two factors tied to coal mining and processing: (1) man-made chemical additives that are used for coal cleaning/processing that are released in the wastewater; and (2) reactions of water with residual coals and non-coal rocks in subsurface and surface coal mines, generating typically mineralized and often acid effluents that discharge to the environment and can cause long-term contamination of the associated watershed and groundwater.

The coal cleaning process aims to reduce the contaminant levels in non-carbon materials associated with coal and thus reduce their emission during combustion. The process involves the physical separation of coexisting coal and non-coal materials and leaching out elements that are retained to the surface of the coal rocks[174]. For example, the separation of rocks containing sulfur minerals from coal can reduce the sulfur emission during coal combustion. A large fraction (60–80%) of ash and sulfur can be removed from coal by the physical preparation methods, thus avoiding the production of large volumes of coal combustion residuals and sulfur emission[175, 176]. The use of the term "clean coal" came about because of the development and implementation of such cleaning technologies that sought to reduce the occurrence of contaminants in the residual coal, and thus reduce the potential atmospheric emission of these contaminants during coal combustion[174]. Yet the cleaning process has resulted in the mobilization of these contaminants into the wastewater. The higher efficiency of the treatment technology to prevent atmospheric emission, the higher also the transport of contaminants to the wastewater that becomes a major outcome of the coal cleaning process (see Section 3.4).

Among the man-made chemicals used for the physical separation of coal and non-coal rock materials is 4-methylcyclohexane methanol (defined as MCHM)[177], with known high toxicity[178–180]. The release of wastewater containing MCHM poses major human health risks[180]. In 2014, approximately 38,000 liters of a liquid mixture containing crude MCHM were accidently released into the Elk River in West Virginia. Because of the proximity to a drinking water treatment facility, the contaminated water was distributed to approximately 300,000 residents that primarily depend upon the river water as their major source of drinking water[181]. In spite of the relatively small volume, this spill caused a major shut down of a drinking water source for a large population and highlights the potential high toxicity of the chemical additives used in the coal cleaning process[181].

Given the high concentrations of contaminants derived from the leaching out of coal and residual rocks as well as man-made additives like MCHM, the wastewater that is generated from washing and the treatment of coal poses a major risk to the environment and human health. In China, it is estimated that the volume of wastewater from coal mine wastewater is 2.2 BCM, with an average ratio of 4 cubic m of wastewater per metric ton of extracted coal[119]. This is equivalent to a water intensity of 576 L/GJ. The elevated levels of contaminants in coal wastewater imply that disposal of that water would further cause contamination of the receiving water. The magnitude of the impaired water depends on the contaminants' levels. Using the electrical conductivity of wastewater from Pennsylvania (median value ~1,300 μS/cm; Figure 3.10), which is fourfold higher than the minimum required level of 300 μS/cm (see explanation below), we propose that the water intensity of the impaired water induced from coal wastewater is $576 \times 4 = {\sim}2,500$ L/GJ, which is larger than the water intensity values estimated for both water withdrawals and consumption for coal mining (see Table 3.3).

In addition to the direct discharge of wastewater from coal cleaning and processing, there are additional processes that cause coal mining to affect the water quality and produce a large volume of impaired water. One of the major factors that controls the reactivity of coal and associated rocks with water is the presence of iron sulfide minerals (e.g., pyrite FeS_2) in the coals and associated rocks. The exposure of rocks with even a small amount of pyrite (1–5%) during and after the mining process causes these minerals to weather and oxidate, which leads to mineralization (i.e., enrichment of dissolved salts such as sulfate and bicarbonate) and an increase in the acidity (lower pH) and metals content of the water associated with coal mining[182–190]. The residual water, known as Acid Mine Drainage (AMD) is a major environmental problem worldwide that adversely affects both

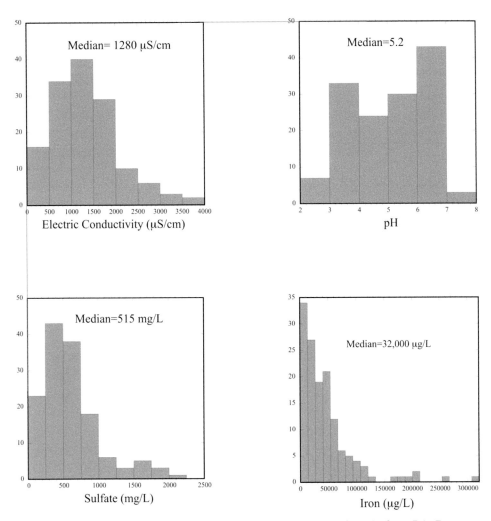

Figure 3.10 Histogram of major water quality parameters of AMD from PA. Data from Cravotta[185].

surface and ground waters[191, 192]. In cases where coals and associated rocks have low contents of sulfur or are composed of carbonate rocks that buffer the acidity, weathering of the residual rocks generates alkaline (high pH) water that is still associated with mobilization of salts and metalloids such as selenium[191, 193–202]. Consequently, it is common to define such wastewater as Coal Mine Drainage (CMD) that can have both high and low pH. The distribution of pH in AMD sites across the northern Appalachian Basin, USA is illustrated in Figure 3.11. Typically, AMD is generated in the watershed and subsurface where groundwater interacts with the residual mining materials. During active mining periods, groundwater is typically pumped to keep groundwater from flooding the mines; yet after the closure of mines or their abandonment, the rebounded groundwater

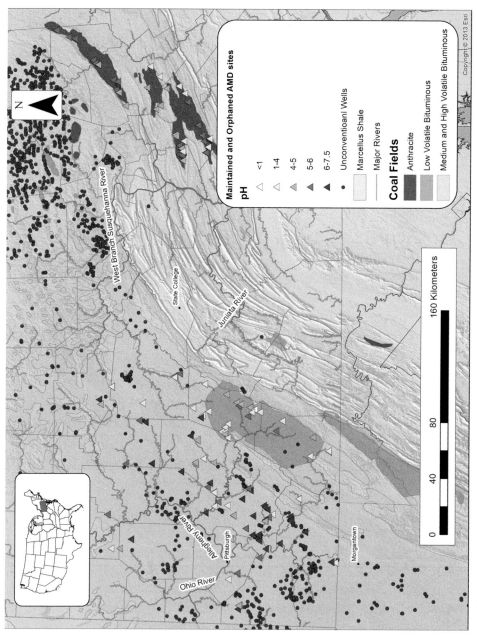

Figure 3.11 Maintained and orphaned AMD sites across the northern Appalachian Basin, USA. The pH of effluents generated from AMD (marked by triangles) are sorted by their values. The different coal types are presented. Map was prepared by A.J. Kondash (Duke University) based on ESRI Topographic Basemap[230] and EIA Layer Information for interactive state maps[130]. (A black-and-white version of this figure will appear in some formats. For the colour version, refer to the plate section.)

interacts with the exposed walls and ceilings of underground mine chambers, leading to the formation of AMD[203]. CMD is also generated from the interaction of surface water with residual rocks and tailing in the mine area. Such rocks have a greater surface area than regular rocks given the smaller grain size and thus are more prone to chemical reactions.

Overall, CMD drainage is a complex pollutant characterized by elevated concentrations of iron and sulfate concentrations, low and high pH, and elevated concentrations of a wide variety of metals depending on the host rock geology[189–191]. The weathering is an ongoing process, and thus CMD is continuously generated in mines and the surrounding environment long after the actual mine and reclamation activities are completed[204]. The degree of mineralization and acidity of CMD is primarily a function of the mineralogy of the rock material and the availability of water and oxygen. Yet the prediction of the magnitude of volume and contamination level is challenging[205]. Numerous studies have shown that CMD formation is long-term and endemic to coal mining, causing numerous cases of acute water quality degradation in hundreds of streams associated with coal mining[185–187, 195, 198, 199, 201, 204, 206]. In spite of numerous attempts to develop treatment technologies, the ability to stop or reduce the environmental impact of CMD is almost impossible[205]. Cravotta et al.[185–187] showed wide variations both in time and space in pH (between 3 and 7; median value of 5.2), salinity, sulfate, and some metals like iron, aluminum, and cobalt in CMD from the Appalachian Basin (Figure 3.10).

In the eastern USA, surface mountaintop mining (MTM) (Figure 3.7) has become the dominant technique for coal mining, particularly in the steep terrain of the Central Appalachian coalfields of Kentucky, Virginia, West Virginia, and Tennessee, which cover about 48,000 square km[206, 207]. During MTM, several overburden layers of sedimentary rocks are removed to access coal layers. The residual mined rock is typically placed in valleys adjacent to the surface mines, generating valley fills, which become the source for headwater feeding the local streams[208–216, 193, 209, 217]. Since the 1970s, surface mining for coal in the Central Appalachian region of the USA has generated a cumulative mining footprint of 5,900 km^2, and each metric ton of coal is associated with 12 m^2 of actively mined land[218] with estimates that surface coal mining resulted in a buried stream length of about 4,000 km[198]. The interaction of the acid AMD with residual rocks that are part of the valley fills causes dissolution of carbonate and other base minerals and the formation of alkaline (pH typically >7) saline effluents that are discharge from the valley fills. The alkaline effluents are characterized by high concentrations of dissolved elements with TDS values of up to 1,500 mg per liter (mg/L). The saline effluents are also associated with elevated levels of trace elements such as Se, As, Sr, and Mn to the levels that pose serious risks to the ecological system, such as

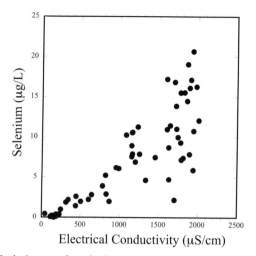

Figure 3.12 Variations of selenium concentrations versus the electrical Conductivity (the overall salinity) of the Mud River in West Virginia that flows through MTM areas and receives inflows from MTM effluents. Data from Vengosh et al. (2013)[197].

toxic effects in fish and birds in watersheds affected by MTM activities[207]. The US Environmental Protection Agency (EPA)[219] issued guidelines for permitting requirements for Appalachian mountaintop removal and other surface coal mining projects based on stream electrical conductivity levels (i.e., total dissolved salts expressed by electrical conductivity measurements). While the EPA considers that in-stream conductivity levels should be maintained at or below 300 μS/cm, in order to meet water quality standards, a value of 500 μS/cm was defined as a limited value[219]. It has been shown that selenium concentrations in streams that discharge from valley fills are associated with the high mineralization and sulfate content and overall salinity[198, 197] (Figure 3.12). Furthermore, it has been shown that the salinity, sulfate, and selenium contents in rivers increase with the percentage of mining in the watershed[198]. For example, a study in West Virginia has shown that the selenium occurrences in headwaters that originated from valley fills and streams in Mud River, West Virginia were highly correlated with elemental strontium concentrations and strontium isotope ($^{87}Sr/^{86}Sr$) ratios, which also coincided with the isotope fingerprint of specific geological formations and coals in the watershed, and thus provided a direct link between the selenium contamination of streams that flow into MTM areas and the residual rocks and coals that are the sources for this contamination[197].

The magnitude and prevalence of CMD are extremely large; over 7,000 km of rivers and streams in the eastern USA have been adversely contaminated by CMD[192, 185–187, 220]. Likewise, numerous groundwater and surface water resources near areas associated with coal mining in India[221–226] and China[227, 228] have been severely affected by AMD contamination. Evaluating the impaired water quantity induced from CMD impact is not a simple task; the volume and quality of CMD are dependent on (1) the climate, precipitation, and water replenishment (i.e., the volume of water that is recharged into active and abandoned mines and associated waste solids); (2) the structure of the mines and connection between shallow groundwater and surface water; and (3) the mineral composition in the rocks that controls the chemical reactions and, consequently, the mobilization of contaminants that generate CMD. Unlike other types of water contamination, CMD is an ongoing process in which new recharged water interacts with the residual coals and rocks and becomes contaminated with metals and metalloids. Some indications suggest that the magnitude of metals and metalloids mobilization during the weathering process decreases with time, and yet the formation of CMD is a long-term process that reflects both current and historic mining. As shown for both subsurface and surface mining, the size of the mine areas determines the magnitude and volume of CMD, and thus its effect on water resources[198, 201].

Most of the available data on the volume of CMD are related to surface water, as water discharge can be monitored at streams downstream from coal mines. Coal mines in Pennsylvania are particularly illustrative; the annual CMD discharge rates from anthracite coal mines in eastern Pennsylvania[229] and abandoned anthracite and bituminous coal mines in western Pennsylvania[185] were estimated to 847 and 380 million cubic m per year, respectively. Yet Cravotta from the US Geological Survey who has been working on CMD suggests that these CMD discharge rates do not fully include the whole volume of the annual CMD discharge in Pennsylvania. The Department of Environmental Quality (DEQ) in Pennsylvania estimates that underlying coal mining has occurred on 10,360 km^2 of Pennsylvania's surface, which equates to about 9% of the state's total surface area. Assuming 30% recharge and annual precipitation of 1,000 mm, the coal mine area across Pennsylvania is equivalent to CMD formation of 3.1 BCM per year (Cravotta, personal communication).

Given that CMD is a product of historical mining, an attempt to evaluate the water intensity of CMD should consider historical coal mining, rather than the current coal mining rates. The US EIA data show that during the 1960s and 1980s annual coal production in Pennsylvania was between 80 to 95 million tons, which decreased to 60–70 million tons in the 1990s and 2000s. In 2014, annual coal production was 60 million tons. Assuming an annual production of 3.1 BCM of CMD water, we used the relatively high historical rate of 90 million tons per year to calculate the expected energy (637 million GJ), which implies a water intensity value of 4,900 L/GJ for CMD in such a high precipitation area (annual rainfall >1,000 mm) as Pennsylvania. We predict that the magnitude of CMD volume and water intensity may be lower in areas with lower precipitation and replenishment rates. It is important to underscore that the additional water intensity induced from long-term and legacy coal mining has typically not been considered in estimates to quantify the water intensity of coal mining.

When considering the actual volume of the impaired water associated with CMD, the water volume intensity evaluated here is only a baseline for a much larger impact. The electrical conductivity (EC; a general measurement to describe water salinity) of effluents discharge from abounded subsurface coal mines in Pennsylvania and valley fills near MTM sites in West Virginia are ~1,300 μS/cm[182–190] and 1,500 μS/cm (median values)[197], respectively. The US EPA defines a "safe" EC level of water quality in streams as 300 μS/cm. Therefore, the EC values of CMD and AMD imply that in order to reduce the concentrations of impacted streams to safe values, the CMD volume must be diluted by at least ~5 times. Therefore, the volume of impaired water affected by discharge of CMD is the volume of CMD discharge multiplied by five. This is a conservative estimate since the differences between trace element concentrations in CMD and baseline water are much larger, and therefore the dilution factor to restore the quality of contaminated water would be much higher than five. Taking this factor into consideration, the water intensity of the impaired water derived from long-term formation and discharge of CMD is $4,900 \times 5 = 24,300$ L/GJ. As shown for the water intensity of impaired water derived from coal wastewater, the estimated water intensity of impaired water impacted by the legacy of coal mining is far larger than the direct (measured) water use for mining.

3.4 Electricity Generation

3.4.1 Water Intensity of Electricity Generation

By far, water use for cooling thermoelectric plants is the largest component of both water withdrawal and consumption along the life cycle of coal utilization[45]. The volume of water withdrawal for cooling thermoelectric plants in the USA (Figure 3.14) consists of almost 40% of the country's water use[111]. In China, thermal power generation consumes about 84% of the total water withdrawn for energy, 99% of which for in coal-fired power generation (data updated for 2010)[231, 232], although during the last few years other energy sources such as wind and nuclear have been added, reducing the percentage that coal makes up of power generation in China from 99% in 2000 to 91% in 2015[233]. In 2019 coal combustion contributed 65% of the total electricity generated in China (7,503.4 TWh), whereas renewable energy sources consisted of only 9.6% of the national electricity production (data from BP data set on global electricity[10]). On a global scale, in 2019 coal contributed 36% of global electricity production (27,004 TWh[10]).

Water is an essential component of coal plant operation; in many plants the heat generated by coal combustion converts water to steam, which operates the turbines to a generator that produces electricity. The steam is condensed and cooled by external water (Figure 3.13). The volume of water needed for cooling for coal plants depends on the cooling system; while older systems utilize larger volumes of water for the "once-through" or "open loop" cooling system, more recent coal plants are more efficient and use less water for cooling by recirculation or "closed loop" systems[29–31, 34, 40, 45, 48, 52]. In the more traditional system that is still commonly utilized in many coal plants in the USA and China, and predominantly in India, fresh water is used for cooling and then returned to the environment at typically higher temperatures, resulting in the *thermal pollution* of aquatic life in the receiving water (see Section 3.4.2). A small fraction of the water is lost due to evaporation, and thus an open loop system requires large water withdrawal but only a small consumption of water, as the majority of the water is returned to the hydrosphere. In contrast, in a coal plant that utilizes a closed loop cooling system, the water that flows through a steam condenser is cooled in a wet tower or a pond, and then reused for cooling in a closed loop. The volume for withdrawal of this system is much lower, but the water consumption is higher since the water is not returned to the environment. In the USA, the annual volumes of water withdrawal for cooling coal plants in 2014 and 2016 were estimated as 110^{29} and 91.5^{234} BCM and water consumption volumes were 2.0^{29} and 1.7^{234} BCM, respectively, with most of the coal plants located in the eastern USA (Figure 3.14). The majority (95%) of water use for cooling coal plants in the USA is fresh water[29]. In China, the total water consumption and withdrawal for power production increased from 1.25 and 40.75 BCM in 2000 to 4.86 and 124.06 BCM in 2015[233]. Zhang et al. (2016)[235] reported a higher water withdrawal volume of 145 BCM in 2011, in which 53% of the water used for cooling was seawater. Therefore, the fresh water withdrawal in China was estimated at only 68.2 BCM[235]. In India, in 2010 the water withdrawal and consumption for power generation was estimated as 34 and 4 BCM[236], of which 40% of the thermal plants are located in water-scarce areas[237]. Countries that use coal predominantly for power generation essentially are also relying upon water for cooling power generation. The amount of water needed for running a thermal plant is significant; the volumes from the USA, China, and India far exceed the annual whole water utilization of many countries in the Middle East (e.g., Israel with ~2 BCM per year).

The water intensity of cooling water for coal combustion depends on the type of the cooling system – open- vs. closed-loop systems – as well as the thermal capacity of the coal plant (Table 3.4). Throughout this book, we report water intensity values in L/GJ unit, which reflects the energy production of the fossil fuel source. Yet for the water intensity of electricity generation, we use the m^3/MWh unit to normalize the water quantity to the electricity generated from the power plants. In order to demonstrate the differential effects of water intensity of water

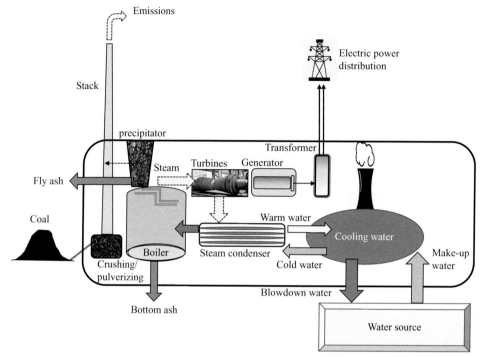

Figure 3.13 A schematic illustration of a closed-loop (recirculating) cooling tower system for thermoelectric coal-fire plants. Illustration follows the example of Plant Scherer in Georgia, USA, one of the largest coal-fire plants in the USA with a capacity of 520 MW[239]. (A black-and-white version of this figure will appear in some formats. For the colour version, refer to the plate section.)

use for cooling coal plants, we use the examples of China and the USA. While the annual water withdrawal for cooling coal plants in the USA (92–110 BCMs in 2014–2016[29, 234]) was comparable to that of China (145 BCM in 2015[235]), the electricity generated from coal in the USA (1,346 TWh) was threefold lower than that of China (4,046 TWh in 2015; data from BP dataset[10]), reflecting the much higher water intensity of the US cooling systems. Similarly, water consumption in the USA (~2 BCM[29, 234]) is only twofold lower than water consumption volume in China (~4.9 BCM[235]), in spite of threefold lower electricity production.

Several studies have attempted to quantify the water intensity for coal combustion, mostly in the USA and China (Table 3.4). Macknick et al. (2012)[47] have estimated that the water withdrawal intensity for open loop (once-through) cooling systems in the USA is 137.6 m^3/MWh (mean value, with some variations dependent on the coal sources; Table 3.4) relative to closed-loop recirculation cooling systems with water withdrawal of 3.8 m^3/MWh (recirculation of cooling towers) and 46.3 m^3/MWh (recirculation of ponds). Kondash et al. (2019)[118] used empirical data to estimate water withdrawal intensity of 152–582 m^3/MWh for

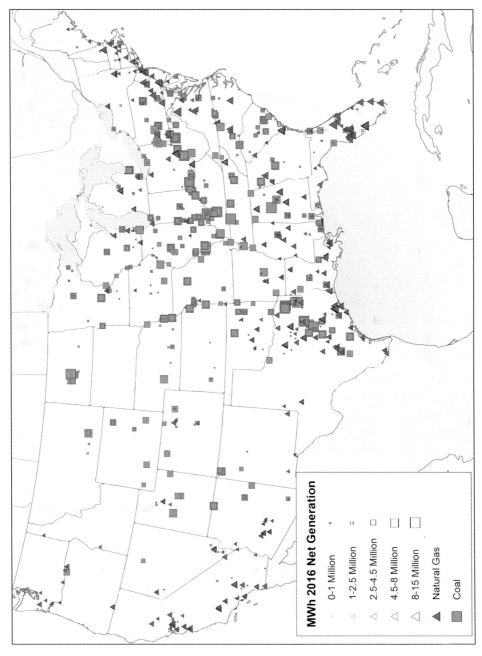

Figure 3.14 Distribution of major coal plants sorted by their capacity in the USA. Map was created by A.J. Kondash (Duke University) and is based on Kondash et al. (2019)[234] and US EIA. (A black-and-white version of this figure will appear in some formats. For the colour version, refer to the plate section.)

once-through relative to 26–39 m³/MWh for recirculating cooling systems[118]. In contrast, the intensity of water consumption (i.e., the actual loss of water during cooling coal plants) in recirculation (closed-loop) cooling systems is typically higher than that of an open system. Macknick et al. (2012)[47] have estimated that the water consumption intensity for once-through cooling systems in the USA is 0.95 m³/MWh relative to recirculation cooling systems with water intensity of 2.6 m³/MWh for cooling towers and 2.1 m³/MWh for ponds (Table 3.4). Kondash et al. (2019)[118] have estimated higher water consumption intensity of 1.3–2.4 m³/MWh for once-through and 1.3–2.4 m³/MWh for recirculating cooling systems[118] (Table 3.4). While the water intensity values from data reported from China are similar to those of the USA, the water withdrawal intensity values are lower. The integrated water withdrawal intensity for China, including seawater, is estimated as 37 m³/MWh, while considering only the fresh water fraction of the water withdrawal results in a lower intensity of 22 m³/MWh (Table 3.4). The summary of the water intensity values are reported in both m³/MWh and L/GJ units for the different combination of open-loop and closed-loop cooling systems and the overall water intensity for cooling coal plants in the USA and China are presented in Table 3.4.

While traditional coal plants with open cooling systems withdraw a large volume of water but consume only a small volume, modern coal plants with closed-loop wet towers withdraw much lower volume, between 20 to 80 times, but consume higher volumes of water. In the USA, the Department of Energy's (DOE) evaluation in 2006 suggested a ~1:1 ratio of open- to closed-loop systems, with 43% of thermal electric generating capacity using once-through cooling, 42% wet recirculating cooling towers, 14% cooling ponds, and 1% dry cooling[238]. In 2015, the open-loop systems were 40%, while recirculating systems consisted of 60% of the cooling systems for coal plants in the USA[118]. In contrast, open-loop cooling systems with high water intensity in China have increasingly been replaced with closed-loop and dry cooling systems: between 2000 and 2015, the amount of closed-loop cooling systems increased from 55% to 75% in Central China, 30% to 50% in the South, and 70% to 90% in the Northeast. Likewise, the relative proportion of dry cooling increased during this period, from 10% to 50% in the Northwest, 8% to 20% in the North, and 12% to 30% in the Northeast[233]. Unlike the USA, the proportion of electricity generated from coal-powered plants with open-loop systems in 2015 was much lower (23%) relative to the proportion (77%) of recirculating cooling systems in China[235]. Given this difference in configuration of the types of the cooling systems between the USA and China, the water footprint of power generation in China is lower than that of the USA. Data calculated from Zhang et al. (2016)[235] for the water use for power generation in China indicate overall water withdrawal and consumption intensity values of 37 and 1.3 m³/MWh, respectively, for power generation in 2015. In contrast, the overall water withdrawal and consumption values for cooling coal plants in the USA in 2016 were 70.6 and 1.4 m³/MWh, respectively[118].

The International Energy Agency's (IEA) assessment[45] highlights the higher cost of the closed-loop, dry, and hybrid systems, lower electricity production efficiency, and limited technological knowhow. New thermoelectric plants' designs also utilize dry cooling systems, in which air flows through a cooling tower to condense steam with minimal water intake. Yet the cost of these plants is higher, and the dry cooling reduces the plant electricity production up to 7%[45]. In 2012, air-cooled coal-fired thermal power plants in China accounted for 14% of China's thermal power generation capacity, and yet the water conservation benefit of air-cooled units results in lower thermal efficiency, and consequently, higher carbon emission intensity[171]. Another option is combined-cycle gas turbines (CCGTs), which are more efficient and, combined with dry cooling, can further reduce the water withdrawal[118]. Some of the coal plants in the USA and China[233] are in transition to installing the water-conservative cooling systems while elsewhere this modernization is moving even more slowly.

3.4.2 Discharge of Cooling Water – Thermal Pollution

As much as the volume of water withdrawal for cooling thermoelectric plants is large, the volume of wastewater that is discharged from a coal plant is equally significant. We estimate that a volume of ~500 BCM (see Chapter 6 for evaluation of global water withdrawal) is globally discharged to waterways from thermo-electric plants using coal and natural gas every year. As shown earlier, the shift from open- to closed-loop cooling systems has drastically reduced the withdrawal as well as the volume of the coal plant effluents. Much of the impact of the wastewater outflow of coal plants that is *not* associated with the processing and treatment of coal combustion residuals (CCRs) (see Section 3.5) is related to the relatively high temperature of the wastewater relative to that of ambient natural water, known as *thermal pollution*[240–243]. Thermal pollution describes a combination of ecological effects induced by the discharge of high- temperature water into streams, river, or lakes causing (1) a decrease of dissolved oxygen (DO) in higher-temperature water that causes anaerobic conditions that can enhance anoxic conditions from fish kill; (2) an increase in algae blooms that in turn increase the eutrophication process and further enhance anaerobic conditions; (3) the loss of biodiversity for only thermotolerant organisms that can adapt to warmer waters; (4) acute effects such as thermal shock that can result in mass killings of fish, insects, plants, or amphibians in the aquatic system; (5) chronic effects such as a significant halt in the reproduction of wildlife, release of immature eggs, or prevention of normal development of eggs; and (6) an increase of the metabolic

Table 3.4. *Water intensity values in m³/MWh and L/GJ for different coal types and cooling systems from the USA and China as reported in Macknick et al. (2012)[47], Kondash et al. (2019)[234], and Zhang et al. (2016)[235]*

Coal plant type	Cooling system	Withdrawal (m³/MWh)	Consumption (m³/MWh)	Source
USA				
Generic	Tower	3.8	2.6	Macknick et al. (2012)
Subcritical	Tower	2.2	1.8	Macknick et al. (2012)
Supercritical	Tower	2.4	1.9	Macknick et al. (2012)
IGCC	Tower	1.5	1.4	Macknick et al. (2012)
Generic	Once-through	137.6	0.95	Macknick et al. (2012)
Subcritical	Once-through	102.5	0.42	Macknick et al. (2012)
Supercritical	Once-through	85.5	0.39	Macknick et al. (2012)
Generic	Pond	46.3	2.1	Macknick et al. (2012)
Subcritical	Pond	67.8	2.9	Macknick et al. (2012)
Supercritical	Pond	56.9	0.16	Macknick et al. (2012)
Subcritical	Once-through	173	1.4	Kondash et al. (2019)
Supercritical	Once-through	153.3	1.0	Kondash et al. (2019)
Subcritical	Recirculating	5.8	2.5	Kondash et al. (2019)
Supercritical	Recirculating	6.5	2.2	Kondash et al. (2019)
IGCC	Recirculating	5.2	1.9	Kondash et al. (2019)
Ultrasupercritical	Recirculating	5.2	2.2	Kondash et al. (2019)
Integrated US water use for cooling coal plants	*Once-through (40%), recirculating (60%)*	*70.6*	*1.4*	*Kondash et al. (2019)*
China				
Subcritical	Once-through	103.1	0.34	Zhang et al. (2016)
Supercritical	Once-through	100.6	0.28	Zhang et al. (2016)
Ultrasupercritical	Once-through	82.8	0.23	Zhang et al. (2016)
Subcritical	Recirculating	2.4	1.9	Zhang et al. (2016)
Supercritical	Recirculating	2.1	1.7	Zhang et al. (2016)
Ultrasupercritical	Recirculating	2.1	1.7	Zhang et al. (2016)
Integrated China (fresh water only) for 2015	*Once-through (23%), recirculating (77%)*	*22*	*1.5*	*Zhang et al. (2016)*
Integrated China (including seawater) for 2015	*Combined technologies*	*37*	*1.3*	*Zhang et al. (2016)*

Coal plant type	Cooling system	Withdrawal (L/GJ)	Consumption (L/GJ)	Source
USA				
Generic	Tower	1,056	722	Macknick et al. (2012)
Subcritical	Tower	611	500	Macknick et al. (2012)

Table 3.4. (*cont.*)

Coal plant type	Cooling system	Withdrawal (L/GJ)	Consumption (L/GJ)	Source
Supercritical	Tower	667	528	Macknick et al. (2012)
IGCC	Tower	417	389	Macknick et al. (2012)
Generic	Once-through	38,225	264	Macknick et al. (2012)
Subcritical	Once-through	28,475	117	Macknick et al. (2012)
Supercritical	Once-through	23,752	108	Macknick et al. (2012)
Generic	Pond	12,862	583	Macknick et al. (2012)
Subcritical	Pond	18,835	806	Macknick et al. (2012)
Supercritical	Pond	15,807	44	Macknick et al. (2012)
Subcritical	Once-through	161,791	667	Kondash et al. (2019)
Supercritical	Once-through	42,448	361	Kondash et al. (2019)
Subcritical	Recirculating	7,417	667	Kondash et al. (2019)
Supercritical	Recirculating	11,029	528	Kondash et al. (2019)
IGCC	Recirculating	528	361	Kondash et al. (2019)
Ultrasupercritical	Recirculating	528	444	Kondash et al. (2019)
Integrated US water use for cooling coal plants	*Once-through (40%), recirculating (60%)*	*19,613*	*389*	*Kondash et al. (2019)*
China				
Subcritical	Once-through	28,641	94	Zhang et al. (2016)
Supercritical	Once-through	27,947	78	Zhang et al. (2016)
Ultrasupercritical	Once-through	23,002	64	Zhang et al. (2016)
Subcritical	Recirculating	667	528	Zhang et al. (2016)
Supercritical	Recirculating	583	472	Zhang et al. (2016)
Ultrasupercritical	Recirculating	583	472	Zhang et al. (2016)
Integrated China (fresh water only)	*Once-through (23%), recirculating (77%)*	*6,111*	*417*	*Zhang et al. (2016)*
Integrated China (fresh water only)	*Once-through (23%), recirculating (77%)*	*10,279*	*361*	*Zhang et al. (2016)*

rate of organisms to trigger consumption of more food than what is normally required, and consequently, damaging the stability of the food chain and altering the balance of species composition in the aquatic system[243, 244]. The actual thermal impact depends on the proximity of the thermoelectric plant to the river/lake, temperature (heat) of the wastewater, and the mixing relationships between the wastewater and receiving streams; the higher the mix ratio of wastewater to natural water, the larger the impact of the thermal pollution[240, 245]. A global analysis has shown that coal plants generate the largest thermal emission rates, with 46.3% of global heat emission, particularly in the eastern USA, Northern China, Northern India, and Europe. The USA was ranked having the highest global thermal emission rate (26.1%), followed by Europe (17.8%) and China (16.4%). It was found that the Mississippi River receives the highest total amount of heat emission, with 62% derived from coal-fired power plants[245]. While it is challenging to quantify the global volume of the impaired water generated from thermal pollution, one can envision that even a small percentage of global discharge of the vast volume of 500 BCM of thermal wastewater from coal plants can result in a large impact on the water quality of the receiving water resources.

3.5 Coal Combustion Residuals

Coal combustion residuals (CCRs) are the by-product of coal combustion, comprised of fly ash, bottom ash, and flue gas desulfurization (FGD) (Figure 3.15). Fly ash is fine powdery material generated from the burning of coal, carried up from the furnace with the flue gas into the smoke stack, and entrapped in particulate collection devices, either electrostatic precipitators (ESPs) or bag-houses, designed to remove nearly all of the fly ash from the flue gas and prevent atmospheric emission. Fly ash is the largest solid waste of burning coal, and in the USA consists of 46% of the total CRR mass[246]. Dry fly ash collected in an ESP or

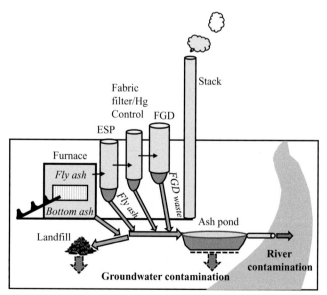

Figure 3.15 Schematic illustration of different types of CCRs generated in different stages following coal combustion in a coal-fire plant and their disposal in landfills and coal ash ponds. Modified from US EPA[254]. (A black-and-white version of this figure will appear in some formats. For the colour version, refer to the plate section.)

baghouse is then transferred to a hopper or storage silo, or mixed with water and transported to on-site impoundments. Fly ash commonly occurs as nano-size, round spheres (small balls) and is composed mainly of amorphous or glassy aluminosilicates. Bottom ash is made of coarse coal ash particles that are too large to be carried up with flue gas into the smoke stacks and accumulate in the bottom of the coal furnaces. Flue gas desulfurization (FGD) material is the residual material from the addition of lime or calcium oxide to reduce sulfur dioxide emissions from coal combustion, composed mostly of calcium sulfate (gypsum)[247]. In the USA FGD constituted 30% of the total CCR mass[246]. The mineralogy, chemical composition, and carbon contents of the parent coals determine the physical appearance, mineralogy, and chemistry of CCRs, in addition to pre- and post-treatment processes (e.g., FGD and halogens addition to reduce mercury emission) and combustion conditions. Low unburned carbon content in CCRs, typically associated with combustion of low-grade lignite and/or sub-bituminous coals, typically generates light grey to brown color ash. In contrast, CCR originating from high carbon bituminous and anthracite coals are dark grey with smaller-size particles (<0.075 nm). Based on their coal origin and chemical composition, CCRs are classified as (1) Class C-CCRs originated from lignite and sub-bituminous coals with relatively higher calcium contents ($>11.5\%$); and (2) Class F-CCRs originated from bituminous and anthracite coals with lower calcium content ($<11.5\%$) and a higher proportion of silica and aluminum. Overall, CCRs are composed of alumina, iron, calcium, metallic oxides, sulfates, phosphates, and residual unburned carbon[248, 249].

The global generation of CCRs is estimated at 1,222 million metric tons per year[250]. The volume of CCRs depends on the amount of coal mined and combusted as well as the ash content (yield) of the parent coal. The ash content can vary between 7% and 11%, like most of the US coals, and up to 40%, like coals in India and some coal deposits in China. Low-grade brown (lignite) and sub-bituminous coals would have higher ash content than bituminous and anthracite. The highest CCR production is in China, and in 2015–2018 the annual production of fly ash in China was 550–600 million metric tons[251, 252]. India and the USA generate about 200 and 100 million tons, respectively[249]. The high concentrations of toxic elements in CCRs (see below) requires adequate disposal without environmental impact and/or reuse for the industry. In China, about 65% of CCRs is reused, mostly for the cement (25%), low-end building materials (18%), and concrete (10%) industries[251, 252, 253], while in India and the USA the fraction of CCRs that are utilized is much lower (38% and 40%, respectively[247, 249]), largely for structure fill, mining applications, bricks, cement, and concrete (Figure 3.16). In China, most of the stored CCRs are located in the northern part of the country near coal plants, while the demand for CCRs as construction material is mostly in

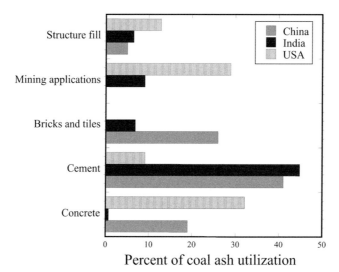

Figure 3.16 Major utilization and reuse of coal ash in China, India, and the USA[247, 249, 251].

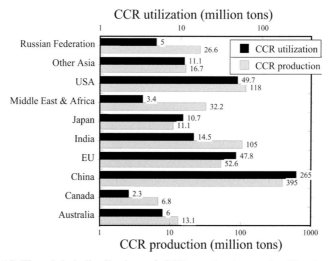

Figure 3.17 The global distribution of CCR production and utilization (million metric tons). Data from the report on production and use of CCRs in the USA[246].

the southeastern parts of China, resulting in a mismatch between CCR availability and utilization[252]. Globally, 53% of 777 million metric tons of CCRs that are annually generated is reused for different industries (Figure 3.17)[246]. The other fraction of CCRs (estimated as 412 million metric tons) is disposed primarily on on-site coal ash impoundments (in wet form) and landfills (dry form). In the USA, 60% of CCRs are stored in surface water impoundments and land fills[254]. The higher risk to the environment and human health is from the leaking of the coal ash ponds to nearby surface water and underlying groundwater, as well as from the formation of airborne particles of nano-size that can become a dust source for populations located downwind from coal-fire plants and disposal sites. Processing CCRs requires water; Grubert and Sanders (2018)[29] estimate a water intensity of 260 L/GJ for the use of 66 million cubic m per year for coal ash handling and water intensity of 27 L/GJ for the use of 300 million cubic m per year for sulfur scrubbing as part of the desulfurization (FGD) process. In China, the volume of water use for coal ash washing is 980 million cubic m per year[119], which is equivalent to a water intensity of 125 L/GJ, assuming that the volume of coal ash generated from coal combustion is 15% of the annual produced coal.

3.5.1 What's in Coal Combustion Residuals and Impact on Water Resources

The combustion of coal in a furnace operating at temperatures of over 1,400°C causes the vaporization of volatile elements, thermal decomposition, fusion, and disintegration of the coal particles and formation of spherical ash particles, known as fly ash. The partitioning of trace elements between the burning coal, flue gas, and fly ash depends on their chemical characterization, such as volatility. Trace elements in CCRs have been classified into three general groups:

(1) *Group 1* – low volatile elements that are concentrated in coarse residues or equally partitioned between coarse residues and finer fly coal ash particles and the concentrations of elements from this group (e.g., thorium, uranium, radium) in CCRs reflect the ash contents of the parent coals (i.e., enrichment as a function of carbon removal during coal combustion);

(2) *Group 2* – elements that are volatilized in the furnace but condense downstream; as the flue gas and ash particles flow from the furnace they begin to cool and the volatilized trace elements begin to condensate on the surface of the nano-size fly ash particles, with a very high ratio of surface area to volume. The enrichment of elements from that group on fly ash (e.g., arsenic, selenium, boron, lead) is typically higher than the enrichment induced from carbon removal, and along with smaller size fly ash it would retain higher mass and thus concentrations of trace elements of this group. In contrast, bottom ash particles would be depleted in these trace elements.

(3) *Group 3* – highly volatile elements with low boiling points that are not fully condensate and reattached into fly ash particles and may be emitted to the atmosphere in the gas phase from the stack. The retention of these highly volatile elements (e.g., mercury, fluoride) requires special scrubber devices (e.g., oxidation of mercury, see below)[255–262].

Non-volatile elements of Class 1 become enriched in the fly ash and bottom ash in proportion to the ash content of the parent ash. Therefore, low-ash coals would generate relatively high metals enrichment (e.g., US coals with ~10% coal ash content would generate CCRs with ~tenfold metal enrichment relative to the original coal), whereas high-ash coals would generate relatively low metals enrichment (e.g., India with 40% coal ash content would induce only 2.5-fold enrichment and yet much larger volume). The enrichment of volatile elements of Class 2 would be higher based on the elements' volatility and coal ash size. Consequently, CCRs are overall characterized by relatively high contents of metals. Figure 3.18 illustrates the relative enrichment of trace metals in fly ash relative to the global average sedimentary rocks, with some elements showing conspicuously high ratios such as beryllium, boron, arsenic, selenium, molybdenum, and mercury.

Radioactive elements that are defined as naturally occurring radioactive materials (NORMs) are also enriched in CCRs, in magnitudes that reflect the NORM contents and ash properties of the parent coals[168, 263]. In the USA, the radium activities ($^{226}Ra+^{228}Ra$ ~200–300 Bq/Kg) in CCRs were found to be about five times higher relative to the concentrations of average soil, reflecting the uranium content of common coals in the USA (2–3 ppm)[263]. Yet some coal deposits in China are characterized by much higher uranium concentrations (up to 376 ppm in Moxinpo deposits) that have much higher radium levels in associated CCRs (4600 Bq/Kg). Based on empirical data it has been shown that CCRs originated from coal with uranium contents >10 ppm exceed the standards for radiation in building materials, and therefore, such radium-rich CCRs should not be utilized for residential building[168].

The attachment of metals and metalloids onto fly ash and FGD solids during the cooling of flue gas and FGD processes generates unique reactive materials. Numerous leaching experiments reported in the literature have demonstrated that reacting CCRs with liquid causes mobilization of many of the soluble elements into the liquid phase[264–269]. Several studies have examined the mobilization of metals and metalloids from CCRs through leaching experiments under controlled laboratory conditions. The various experimental designs include factors such as different CCR sources, liquid to solid ratio (L/S), type of leaching medium (e.g., water, acid), as well as ambient temperature, pH, and redox state (i.e., aerobic versus anaerobic) conditions. Some of these leaching experiments were constructed to mimic specific scenarios of CCR interactions in the environment, and some others were aimed at obtaining predictions on available concentrations and used as a proxy for long-term releases of CCR contaminants[270]. The most common types of leaching experiments are (1) batch leaching using deionized water without pH control in which the pH conditions of the leachates are controlled

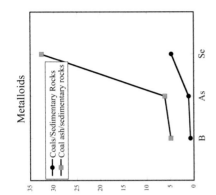

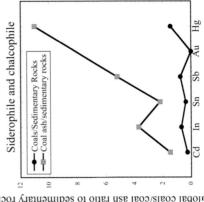

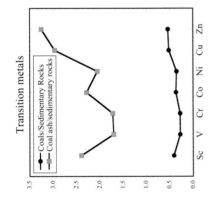

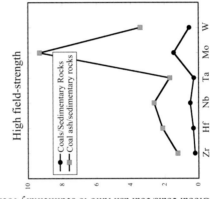

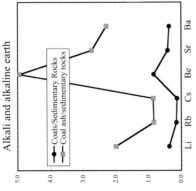

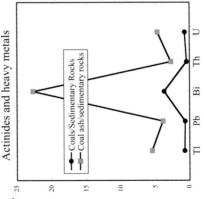

Figure 3.18 The ratios of trace elements sorted by their chemical properties in world coals and coal ash (Ketris and Yudovich 2009[139]) to the concentrations in average sedimentary rocks[139]. Most trace elements in coal ash are enriched (i.e. ratio >1) relative to the average composition of sedimentary rocks with some elements (boron, arsenic, selenium, antimony, bismuth, molybdenum, and mercury) having distinctive high ratios over common sedimentary rocks.

by the coal ash buffering capacity; (2) batch leaching using deionized water with pH control (using acid or base for pH adjustment); and (3) Toxicity Characteristic Leaching Procedure (TCLP), designed by the US EPA to characterize the toxicity of solid waste[271]. TCLP leaching experiments represent acid conditions (pH 2.9–4.9) that are commonly not occurring for many environmental settings and underestimate the mobilization of oxyanions like arsenic and selenium from CCRs[272, 273]. The US EPA in collaboration with Vanderbilt University developed the Leaching Environmental Assessment Framework (LEAF) as an alternative leaching assessment. The LEAF leaching methods include (1) the liquid–solid partitioning (LSP) of CCRs as a function of eluate pH (EPA Method: 1313); (2) the LSP of solid materials as a function of liquid-to-solid ratio (L/S) using either an up-flow percolation column (EPA Method: 1314) or parallel batch extractions (EPA Method: 1316); and (3) the rate of contaminants' mass transport from monolithic or compacted granular materials (EPA Method: 1315)[272, 274, 275].

In a number of controlled leaching experiments using the different leaching methods summarized above, several soluble elements like boron, strontium, arsenic, selenium, and lithium were found to be enriched in the leachates[258–260, 270–272, 274–276]. Therefore, it is important to distinguish between elements that are generally enriched in CCRs but insoluble relative to elements that are both soluble and easily leach out and mobilize into the aquatic phase[269].

Consistently, case studies that have monitored disposal of CCR wastewater into waterways, leaking of coal ash ponds to surface and groundwater, and contamination of water associated with spilled CCRs, like in the 2008 Tennessee Valley Authority (TVA) coal ash spill in Kingstone, Tennessee, have shown conspicuously high concentrations of boron, strontium, arsenic, selenium, and lithium with geochemical and isotope fingerprints that mimic the composition of coal ash. The conditions prevailing during the exposure of CCRs in the environment would control the mobilization of contaminants from solid CCRs into the ambient aquatic system. In particular, the redox state would control redox-sensitive elements like As, which under reducing conditions would preferentially mobilize and become enriched in water associated with CCRs. In contrast, under oxidizing conditions, elements like Se are preferentially mobilized[266–268, 276]. Other factors such as the pH would determine the leachability of cationic metals (low-pH) relative to metalloids (high-pH). The oxidation-reduction potential (ORP) and pH would therefore determine the speciation and mobility of elements from CCRs[266–268]. The selective enrichment and mobilization of some of the key CCR elements that are known to have major ecological toxicity effects (e.g., selenium) and human health effects (e.g., arsenic) make CCRs potential environmental hazards. In all cases where solid CCRs were released to the environment such as the 2008 TVA coal ash spill in Kingston, Tennessee[264, 265, 276–279], and unmonitored coal ash solids spills in Sutton Lake in

North Carolina[269], associated water, mostly in the form of pore water entrapped within the upper part of bottom sediments of the impacted river (Kingston) or lake (Sutton) had elevated levels of CCR contaminants. Likewise, discharge of effluents from coal ash impoundments caused massive contamination of the impacted lakes and rivers as demonstrated in many cases across North Carolina[280]. Given that many of the coal ash impoundments in the USA were built decades ago without any lining, and in some cases at a depth below the regional groundwater table, contaminants that were leached out from CCRs have migrated with CCR effluents and caused contamination of the underlying groundwater, in addition to seepages and contamination of surface water adjacent to the coal ash ponds[281]. An Earthjustice survey throughout the USA that included 265 coal plants and 550 individual coal ash ponds and landfills, which represents roughly three-quarters of the coal power plants across the USA, analyzed over 4,600 groundwater monitoring wells and found that the majority (91%) of the groundwater had elevated levels of CCR contaminants, from which about half had elevated arsenic[282]. Likewise, a Greenpeace study[283] reports several cases of ash dam failures in China such as 300,000 tons of ash slurry spilled directly into the Tuozhang River in 2006, polluting both the Tuozhang and Beipan Rivers, in addition to chronic leakage of CCR effluents from coal ash impoundments across China and associated groundwater contamination (Figure 3.18)[283].

3.5.2 Tracing Coal Ash Contaminants in the Environment

One of the major challenges in evaluating the impact of CCRs (or any other anthropogenic contamination source) on the environment, and particularly on water resources, is the ability to distinguish contaminants derived directly from the spillage or leaking of CCR effluents from naturally occurring (geogenic) contaminants that may characterize the presumably impacted water resources. In many cases, the option to measure the baseline water quality prior to the impact is impossible, and thus the ability to truly identify CCR contaminants in potentially impacted water requires objective scientific tools that can unequivocally identify CCR contaminants in disputed waters. In many cases, the discovery of one or more contaminants in either surface or groundwater near coal ash disposal sites raises concerns for an overall contamination into drinking water resources. One example is the detection of hexavalent chromium in drinking water wells near coal ash ponds in North Carolina that has been attributed to leaking from nearby coal ash ponds[284]. The public reaction to the elevated hexavalent chromium caused state officials to order a stop to using the well water and supply bottled water to residents with water supply wells within 1,000 feet of each of the utility's 14 coal-fired electrical generating facilities in North Carolina[284]. And yet systematic analysis of the distribution of hexavalent chromium in groundwater across North Carolina showed the widespread prevalence of hexavalent chromium in groundwater and the dependence on aquifer lithology, rather than proximity to coal ash impoundment. Furthermore, the geochemistry of groundwater with elevated hexavalent chromium was different from the expected composition of contaminated groundwater derived from CCR contaminants[285–287].

Numerous studies have shown that during the leaching of CCRs, multiple elements are leached out. Yet the magnitude of the leaching of the different elements can vary based on the elements' solubility and the conditions in which CCR contaminants are transformed to water sources. For example, it has been shown that the acidity (pH) and redox (oxidation-reduction) conditions in the ambient aquatic system can control and affect the differential leachability of different elements such as metals (e.g., lead, cadmium, copper) and metalloids (e.g., arsenic, selenium) from coal ash. Under low-pH conditions, metals from CCRs may mobilize more strongly, while under alkaline (high-pH) conditions, oxyanion-forming elements, like arsenic selenium, and hexavalent chromium are likely to mobilize in larger magnitude from CCRs. In contrast, under reducing conditions, elements like arsenic would preferentially mobilize and become soluble in the impacted water associated with CCRs, while under oxidizing conditions, elements like selenium would be preferentially mobilized[266–268, 276]. In cases of surface water contamination from CCRs, seepages, and/or coal ash pond wastewater, which are commonly under oxidizing conditions, elements like selenium with high bioaccumulation have potential to damage the aquatic system. In contrast, seepage from the bottom of a coal ash pond is likely to occur under reducing conditions, where arsenic is expected to be preferentially mobilized and enriched in the impacted underlying groundwater.

Unlike arsenic and selenium, other elements such as boron and strontium, which are highly enriched in CCRs (Figure 3.18), are not as much affected by the redox and pH conditions and can be detected in almost any CCR-impacted water[88, 265]. Boron is classified as a "Class III" element, which includes highly volatile elements with minimal condensation on ash-residue particles[257, 261]. It was estimated that 70% of boron present in coal could be released to the atmosphere during combustion[288]. However, the installation of high-efficiency ESPs, fabric filters, and FGD to mitigate atmospheric emission in many countries has caused the retention of over 95% of gaseous boron (50% by ESP and 45% by wet FGD) into CCRs[256, 257]. The high efficiency of boron retention through ESP and FGD results in high boron concentration in CCRs[256, 257, 261]. Boron data from the US Geological Survey archive (n = 6,981 [289]) show that the median values are 23 ppm, 72 ppm, and 60 ppm for boron concentrations in Eastern (Appalachian), Central, and Western coals, respectively[289] (see locations in Figure 3.3). Given the coal ash contents of US coals (9.7%, 10.8%, and 7.5%, respectively), the boron concentrations in CCRs is expected to be 9- to 13-fold higher, which is equivalent to 240 ppm, 670 ppm, and 800 ppm in CCRs derived from these coal deposits, respectively. These concentrations are consistent with literature on boron data in CCRs for both US and international CCRs[256, 257, 261].

Since boron and strontium concentrations are high in CCRs and are easily mobilized when CCRs interact with water, their concentrations in water have been shown to be the most sensitive tracers for the impact of CCRs on water resources, such as for the TVA coal ash spill in Kingstone, Tennessee[264, 276], the disposal of coal ash pond wastewater into waterways in North Carolina[280], leaking of CCR effluents into surface water and groundwater[281], and accumulation of CCR contaminants in pore water in bottom sediments from lakes impacted by CCR contaminants[269, 276, 280]. The boron isotope ratios of coals and CCRs are commonly different from those of meteoric water, naturally occurring boron in groundwater, as well as many other sources, making boron isotopes an ideal geochemical tracer for detecting CCRs in water resources[265, 290–292]. Likewise, the Sr isotope ratios of coals and CCRs can be distinguished from the composition of the local aquifer rocks or watershed, and in such circumstances, strontium isotopes can also serve as a sensitive tool to distinguish between CCR-impact vs. naturally occurring contamination[197, 265, 293–300]. For the case of the "false alarm" of finding hexavalent chromium in drinking water wells near coal ash ponds in North Carolina, as described above, the hexavalent chromium-rich groundwater had low boron commonly found in background groundwater, and the strontium isotope ratio was not consistent with the ratio expected for coals and CCRs. Instead, the strontium isotope ratios mimic the composition of the aquifer rocks and thus ruled out that the contamination is derived from CCR impact[285]. In conclusion, one can distinguish between elements enriched in CCRs, particularly "Class II" and "Class III" volatile elements that are captured in fly ash and elements that are likely to be soluble in water and thus would occur in CCR-impacted water, such as boron, strontium, arsenic, and selenium[265, 281].

3.5.3 The Impaired Water Intensity from Disposal of Coal Combustion Residuals

In the USA the total water use for coal ash handling is 66 million cubic m, and sulfur scrubbing is 300 million cubic m[29]; in China the annual water use for coal ash processing is 980 million cubic m[119], which is equivalent to the water intensity of 4–10 L/GJ and 42 L/GJ, respectively (applies for both water withdrawals and consumption).

   The wastewater that is generated from coal ash impoundments as well as DFG wastewater poses major environmental and human health risks (see Section 3.5.1). The US EPA has analyzed the chemistry of FGD wastewater from 150 sites in the USA and reported extremely high concentrations of toxic and carcinogenic elements[301] (Table 3.5). The concentrations of several toxic elements in wastewater generated in FGD, fly ash, and bottom ash wastewater, including arsenic, cadmium, chromium, lead, mercury, nickel, selenium, thallium, and antimony, exceed the drinking water standard (the US EPA Maximum Contaminant Level, MCL) and the Criterion Continuous Concentration (CCC) ecological standard for protection of aquatic life[302] (Table 3.5). FGD wastewater is also characterized by high salt content, in addition to extremely elevated levels of toxic elements even after treatment (e.g., selenium of 1170 µg/L). Likewise, Ruhl et al. (2012)[280] showed that effluents discharged from coal ash impoundments in North Carolina contained elevated levels of arsenic, boron, and selenium. Moreover, the discharge of these effluents from regulated outfall sites through the National Pollution Discharge Elimination System (NPDES) also causes contamination of the receiving water. The magnitude of this contamination depends on the concentrations of toxic elements in the wastewater, the volume of the wastewater, and the mixing relationship between the wastewater and the receiving stream/river/lake. Small lakes or low-flow streams are likely to have the largest impact induced from CCR wastewater discharge[280].

   According to the US EPA[254], in 2012, 470 coal-fired plants combusted 800 million metric tons of coal, which generated 110 tons of CCRs. About 60% of the CCRs were transferred to coal ash impoundment and landfills, with an estimated number of 735 surface coal ash impoundments. Based on a US EPA survey on 113 coal plants, each coal ash plant generates an average of 4.8 and 2.8 million cubic m of fly ash and bottom ash wastewater, respectively[301]. In 2009, the volume of wastewater discharged to CCR impoundments in the USA was 2.5 BCM, which included fly ash, bottom ash, and FGD wastewaters (885, 1,197, and 435 million cubic m, respectively) (Table 3.6). The wastewater discharge from the coal ash impoundments to the environment was estimated as 543 million cubic m for fly ash wastewater (from 113 plants) and 802 million cubic m per year for bottom ash (283 plants), with a total of 1.3 BCM of CCR wastewater[301]. Therefore, in the USA about half of the generated CCR wastewater that was stored in impoundments has discharged to waterways. In North Carolina, the volume of wastewater discharge to waterways from the coal ash ponds was reported to be on a range of 1–20 million cubic m per year with a total wastewater discharge of 86 million cubic m per year[280]. Since only 60% of the CCRs in the USA are transferred to coal ash impoundments, one can assume that the water intensity of the combined CCR wastewater that is released into the environment of 1.3 BCM divided by 60% of the coal production in the USA in 2009 (975 million tons; 21.67 exajoules) is equivalent to 96 L/GJ for combined CCR wastewater intensity (Table 3.6).

Table 3.5. *Concentrations of contaminants (mean values from 150 sites) in FGD, fly ash, and bottom coal ash wastewaters as measured in coal ash impoundments in the USA*[301]

Element	FGD wastewater (untreated)	FGD wastewater (treated)	Fly ash wastewater (treated)	Bottom ash wastewater (treated)	CCR standard	MCL standard
Chloride (mg/L)	7180	7120	13	28	230	250
Sulfate (mg/L)	13,300	1240	409	350	N/A	250
Boron (mg/L)	242	243	6.6	0.54	N/A	N/A
Iron (mg/L)	566	1.5	0.85	1.4	1	N/A
Arsenic	507	7.6	36.4	17.4	150	10
Cadmium	127	113	7.6	2.2	0.72	5
Chromium	1270	17.8	27.4	5.6	11	100
Lead	315	4.7	13.7	12.1	3.2	15
Mercury	289	7.8	0.8	0.6	0.8	2
Nickel	1490	878	30.5	16.5	52	N/A
Selenium	3130	1170	15.4	11.8	3	50
Thallium	22.1	13.7	10.3	89.4	N/A	2
Antimony	28.9	12.9	N/A	28.2	N/A	6
Beryllium	17.5	1.9	3.4	N/A	N/A	4
Barium	2750	303	121	110	N/A	2000

Table 3.6. *Volumes of CCR wastewater generated and discharged to the environment from US coal plants in 2009. Wastewater data were retrieved from US EPA report (2015)*[301]

Wastewater source	Volume per plant (10^6 m³/year)	Number of plants	Annual volume (10^6 m³/year)
Wastewater			
FGD wastewater	2.9	150	435
Fly ash wastewater	5.9	150	885
Bottom ash wastewater	3.4	348	1,197
Total CCR wastewater			*2,517*
Discharged wastewater			
Fly ash wastewater	4.8	113	543
Bottom ash wastewater	2.8	283	802
Total CCR discharged wastewater			*1340*

Wastewater generated in coal ash ponds and FGD treatment sites contains high concentrations of several CCR contaminants, including selenium, with concentrations of up to 1170 µg/L (Table 3.5). The US EPA conducted a survey of the volume and contaminants in wastewater discharge from coal ash ponds and showed a mean selenium concentration of 15 µg/L in CCR wastewater[301]. Selenium is known to cause severe ecological effects and therefore the US EPA has established limiting concentrations for effluents discharge to standing water (lakes, reservoirs) and flowing water of 1.5 µg/L and 3.1 µg/L, respectively[303]. Consequently, discharge of CCR wastewater with a concentration of 15 µg/L would require a dilution of at least fivefold in order to reduce the wastewater volume to get to a concentration below this harmful level. This suggests that the impaired water volume from the discharge of CCR wastewater in the USA is the annual CCR wastewater discharge rate × 5, which equals ~6.5 BCM per year with water impaired intensity of 481 L/GJ. This is a conservative estimate since it does not include the leaking of coal ash impoundment to the subsurface and contamination of underlying groundwater that have been detected in numerous sites across the USA[281], and also does not address the impact of leaking leachates generated from leaching CCR solids in unlined landfills that also cause contamination of underlying groundwater[282].

3.6 Integration: The Combined Water Footprint of Coal Utilization

Figure 3.19 and Table 3.7 present the summary of the water intensity values of water withdrawal and consumption for the different life cycles of coal mining,

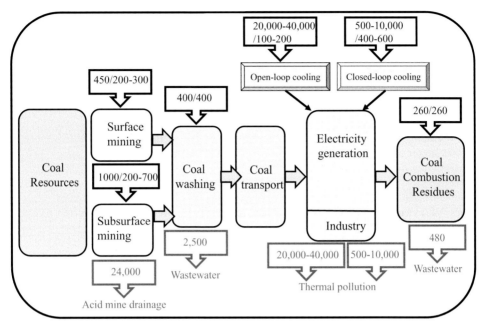

Figure 3.19 Schematic illustration of the different stages of coal generation and the water intensity values (L/GJ) for water withdrawal/consumption (in blue) and water impaired (red) at the different stages of coal production, including coal mining; coal washing and processing; coal transport, power generation, and processing of coal combustion residuals (CCRs). (A black-and-white version of this figure will appear in some formats. For the colour version, refer to the plate section.)

processing, combustion, and waste disposal. The estimated impaired water intensity evaluated in this book is also reported for the different life cycles of coal use. The data show that the impact of coal mining, combustion, and waste disposal on water quality, and thus on the volume of water that can be impacted, is comparable and even higher than direct water withdrawals and consumption. We used the example of the US energy sector to illustrate the magnitude of direct water use for coal production. Data extracted from Grubert and Sanders (2018)[29] for the 2014 US dataset suggest that the total water withdrawal for all stages of coal extraction was 111.7 BCM. The water withdrawal for cooling coal-fired plants was predominant (98.5%) by several factors relative to other stages of coal production (Figure 3.20). The total was consumption for all stages of coal generation in the USA in 2014 was 3 BCM, in which coal mining (19.8%), coal processing and washing (2%), and ash processing (12.1%) contributed to water use, in addition to water cooling (66.1%; Figure 3.20)[29]. In China, the water withdrawal and consumption specifically for cooling thermoelectric plans in 2015 were 124 and 4.9 BCM, respectively[233], while in India, the water consumption for power generation was estimated at 2.1 BCM[237].

Table 3.7. *Summary of the water withdrawal, water consumption, and impaired water intensities values (L/GJ) evaluated in this book**

Stage	Water withdrawal	Water consumption	Impaired water
Surface coal mining	450	200–300	2,500
Subsurface coal mining	1000	200–700	24,000
Coal washing	100–400	200–400	2,500 (1)
Groundwater impact			140 (1)
Coal mining legacy – Acid Mine Drainage			24,000
Coal combustion – open-loop cooling	20,000–40,000	100–200	(2)
Coal combustion – closed-loop cooling	500–10,000	400–600	(2)
Coal ash disposal	260	260	480 (3)

* (1) = Based on data from China; (2) = Thermal pollution from discharge of cooling wastewater; (3) = Based on data from the USA.

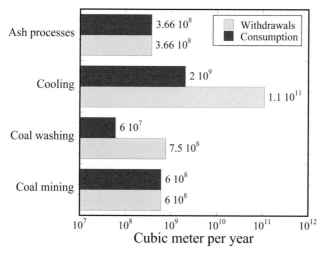

Figure 3.20 Water withdrawal and consumption volumes for different stages of coal production in the USA in 2014. Data from Grubert and Sanders (2018)[29].

3.7 Environmental Regulations and Coal

Environmental regulations concerning coal have largely focused on the reduction of atmospheric emissions associated with coal combustion, especially in the USA. In particular, the US EPA has promulgated several regulations designed to reduce contaminant emissions under the authority of the 1990 Clean Air Act

Amendments. Since the early 1980s, US EPA regulations have targeted reducing SO_2 and NO_2 emissions as a means to mitigate the widespread occurrence of acid rain. These mitigations included installation of FGD scrubbers in coal plants and utilization of low-sulfur coals[304, 305]. The emission prevention included installation of (1) ESPs to prevent fly ash/particulate matter emission; (2) FGD to reduce SO_2 emission; (3) selective catalytic reduction (SCR) to reduce nitrogen oxides emission; and (4) flue gas mercury controls (FGMCs) to reduce mercury emission[306]. The switch from high- to low-sulfur coals also reduced the concentrations of many toxic trace elements that were used for coal combustion, and consequently reduced their concentrations in CCRs. Yet the installation of FGD and other scrubbers to reduce atmospheric emission has enhanced the incorporation of trace elements into CCRs, and consequently has increased their transport to coal plant wastewater and coal ash ponds[304, 305, 307–309]. To further address SO_2 and NO_2 emissions, the US EPA finalized the Cross-State Air Pollution Rule (CSAPR) on July 6, 2011, which required a number of states in the eastern USA to reduce emissions that cross state lines from power plants – that is, to curtail coal consumption[310].

During the Obama Administration, the US EPA introduced a series of environmental regulations to lessen the negative impacts of coal production, including some to reduce mercury emissions. On February 16, 2012, the EPA issued the Mercury and Air Toxic Standards (MATS) rule[311]. The rule sets standards for all hazardous air pollutants (HAPs) emitted by coal- and oil-fired electric generating units (EGUs) with a capacity of 25 megawatts or greater[312]. Mercury-removal technologies were developed in parallel with regulations to reduce mercury emission from coal combustion. The technology involves "coal cleaning" with halogen additives, in particular bromine, that are used to increase the oxidation of mercury in the flue gas that is generated from coal combustion and improve the effectiveness of mercury removal by activated carbon[313–323]. As shown earlier (Table 3.1) coals contain typically high concentrations of mercury, which is highly volatile, and its emission from coal combustion poses a major global environmental and human health risk. The principle methods for mercury removal from coal-fired plants have thus entailed (1) separation of solids (like fly ash) from which mercury has been adsorbed on; and (2) absorption of the oxidized gaseous mercury species in aqueous media. A typical mercury control strategy utilizes one or both of these techniques, either by removing particulate-bound Hg using the particulate collection device or by removing gaseous Hg^{2+} in an FGD scrubber[313, 323]. Yet coal combustion generates elemental mercury (Hg^0) that is not reactive, and therefore oxidation of Hg^0 to Hg^{2+} is the condition for efficient removal of mercury. The addition of halogens like bromine is found to be effective for the oxidation and retention of mercury in combustion flue gas[313]. However, the addition of bromine to mitigate mercury emissions further degrades the quality of the wastewater that is generated from coal ash ponds with elevated bromide concentrations[324–327]. Elevated halogens such as bromide in wastewater can induce the formation of disinfection byproducts (DBPs), in particular the highly toxic brominated DBPs, in downstream water used for drinking water treated through chlorination[324–327]. Evidence for elevated bromide in waterways and spikes in DBPs in drinking water from Pennsylvania have been linked to wastewater from coal plants[324–327]. Higher bromide concentrations were reported particularly in wet FGD wastewater, and therefore the potential impact of bromide-rich FGD wastewater discharges from coal-fired power plants is high in scenarios where downstream water from power plants is used by large populations[327]. The projected increase of bromide to mitigate mercury emission and generate "clean" or "refined" coal is expected to cause further degradation of the quality of CCR wastewater.

In spite of decades of coal combustion, most regulations have, however, focused on preventing atmospheric pollution rather than addressing the impact of the quality of CCRs and wastewater. The 2008 TVA coal ash spill in Kingston, Tennessee triggered extensive public debate due to the numerous reports on the environmental impact of CCRs on water resources, including the US EPA's own investigation[306]. Consequently, the US EPA issued the first federal ruling concerning the fate of contaminant discharge in wastewater from CCRs storage facilities – *Steam Electric Power Generating Effluent Guidelines – 2015 Final Rule*[328]. These rules set, for the first time in the USA, limitation on the levels of contaminants allowed to be in CCR wastewater discharged to the environment. The 2015 EPA rule, promulgated under the Clean Water Act, aims to protect public health and the environment from toxic metals and other harmful pollutants by strengthening the technology-based effluent limitations guidelines and standards (ELGs) for the steam electric power-generating industry. The EPA suggests that the implementation of the rule would reduce the annual discharge of contaminants by 635,000 tons and water withdrawals by 12.3 BCM[328]. The rule distinguishes between existing sources like coal ash ponds that discharge effluents directly to surface water and new sources; while distinction was made for the types of existing facilities (fly ash versus FGD wastewaters), for new facilities, a zero discharge standard for all pollutants in all types of wastewaters was established[328].

A second US EPA ruling addressed the overall disposal of CCRs and established technical requirements for CCR landfills and surface impoundments under subtitle D of the Resource Conservation and Recovery Act (RCRA)[254]. In spite of increasing public requests to define CCRs as "hazard materials" and thus have full federal environmental protection regulations, the US EPA was reluctant to do so. Instead, the 2015 CCR rule required (1) any existing unlined CCR surface impoundment that is contaminating groundwater above a regulated constituent's groundwater protection standard to stop receiving CCR and either retrofit or close; (2) the closure of any CCR landfill or CCR surface impoundment that cannot meet the applicable performance criteria for location restrictions or structural integrity; (3) CCR surface impoundments that do not receive CCR after the effective date of the rule, but still contain water and CCR to be subject to all applicable regulatory requirements, unless the owner or operator of the facility dewaters and installs a final cover system on these inactive units no later than three years from publication of the rule; (4) establishment of new liner design criteria to help prevent contaminants in CCRs from leaching from the CCR units and contaminating groundwater; and (5) requirement that energy and CCR utilities owners will conduct groundwater monitoring and corrective action, including installation of a system of monitoring wells and public publication of the water quality data of groundwater underlying CCR landfills and impoundments. The latter component of this ruling forced energy utilities to conduct groundwater quality monitoring, which resulted in the establishment of a large dataset and discovery of large-scale groundwater quality deterioration under numerous CCR disposal sites throughout the USA[282].

One of the consequences of these CCR rulings is that in 2019 some energy utilities in the USA were in the process of phasing out CCR impoundments, raising a new debate on the actual removal (i.e., dragging CCRs out to protected and designated landfills vs. continued storage of CCRs on sites and cover by soil). The Trump Administration upon entering office in January 2017 actively sought to weaken environmental regulations regarding coal as a means to revive what was then a declining industry. Almost immediately, in February 2017, the US Congress took steps to revoke prior protections introduced during the Obama Administration. In particular, the US Congress sought to reverse the Stream Protection Rule to allow coal mining near streams. By keeping coal mining debris away from nearby streams, the Rule was designed to protect waterways from the practice of surface mining[329]. Yet despite the more than one hundred rollbacks during the Trump Administration, declining demand for coal in the face of cheap natural gas and growing use of renewables has led to the closure of approximately 15% of coal-fired electricity plants in the USA between 2017 and 2020[330].

3.8 Coal Transitions

Given the direct impact that fossil fuel consumption has had on the global climate, countries increasingly have sought to reduce carbon emissions through both international cooperation (e.g., the Paris Agreement) and domestic policies; in some cases, this has meant embarking on an energy transition away from fossil fuels to more renewable sources of energy. While reducing the number of coal-fired power plants and reengineering a country's fuel mix have often been central to these energy transitions, they also require rethinking the socio-technical systems that underpin them, which include a country's business model for energy provision[331, 332] as well as subsidies to support fossil fuel production and consumption[333].

Despite the global movement to reduce coal consumption as a means to address climate change, the Trump Administration instead through its actions to prop up the coal industry and its constituents in coal mining communities sought to undercut the Clean Power Plan that was introduced during the Obama Administration to reduce carbon dioxide emissions. Again, despite weakening environmental protections, coal consumption in the USA used for electricity generation in 2015 fell 29% from its peak in 2007. Most of this decline was in the Midwest and Southeast[334]. Part of the reason for this transition away from coal in the USA, despite government policy to bolster coal consumption, has to do with market factors. Simply put, the price and availability of other fuels have made coal less viable. The availability of cheap natural gas in the USA has led to investments in natural gas-fired power generation. With the use of horizontal drilling and hydraulic fracturing, shale gas and oil have transformed US energy markets since the mid-2000s. The use of natural gas and its impacts on water resources will be discussed in Chapter 5. In addition, consumers and industry have been turning to renewable sources of energy, including wind and solar. In April 2019, for the first time ever, renewables (hydro, biomass, wind, solar, and geothermal) outperformed coal for generating electricity in the USA[335]. The result has been the decline of coal prices and coal mining jobs over the last few decades – from about 800,000 in the 1920s to 130,000 in 2011 – as well as three of the four largest US coal companies filing for bankruptcy by the end of 2015[113].

What is clear is that the coal industry is declining in many industrialized countries, especially in Europe. Germany, for example, committed to an energy transition as part of its efforts to reduce greenhouse gas emissions. Much of this energy transition has depended upon the closing of coal mines and shifting away from coal to other renewable fuels. Whereby hard coal helped spur West Germany's economic growth and industrialization after World War II[336], the last of its hard coal mines were shuttered in December 2018 in the Ruhr region. Other policies that have facilitated energy transitions in Germany and elsewhere in Europe (as in the UK) have sought to reform coal subsidies accompanied by financial support for regional economic development and social assistance to communities affected by mine closures[337].

China is also seeking to reduce demand for coal despite its heavy reliance on coal over the last few decades for fueling its economic growth; owing to the changing structure of its economy that has been shifting away from heavy industry to manufacturing and the service sector, and a wealthier population that is demanding better air quality in its large urban centers, the government has taken steps to limit coal consumption. China's 13th Five-Year Plan (2016–2020) pledged to reduce the amount of coal consumption and shift instead toward renewables as well as shale gas as part of a broader economic transition away from heavy industry to promoting the service sector.

Where we are likely to still see a growing demand for coal, however, is in India, Pakistan, and other emerging economies. India, in particular, has embarked upon an ambitious program to expand electricity to its population through Prime Minister Modi's initiative, "24×7 Power for All," much of which will require increasing dependence upon fossil fuels, including coal[338]. Coal is India's main fossil fuel, and India planned to increase its coal production from 600 million metric tons in 2013 to 1.5 billion metric tons by 2020[339]. Continued reliance upon coal is likely to hamper any commitments from the Indian government to meet its Paris Agreement commitments and curtail CO_2 emissions; furthermore, coal use will continue to elicit concern about its environmental impacts on India's national forests and waterways[338].

Chapter 3 Take-Home Messages

- In spite of a temporary reduction in global coal production in the mid-2010s, the rates of coal mining and combustion increased by the end of the second decade of the 2000s, and coal comprised the second most popular source of global energy after crude oil. About half of the world's coal production is in China, and changes in coal use in China have worldwide implications for carbon emissions and water utilization.
- In China and India, coal is the predominant energy source for power generation. In the USA coal combustion has declined during the second decade of the 2000s, but since 2016 coal exports have increased (up to 15% of total production), implying that the actual carbon emission from US coal is higher than officially recorded on US soil.
- Water use for subsurface coal mining is higher than that of surface mining; most of global coal extraction is through subsurface coal mining. The formation and discharge of large volumes of metal-rich AMD from subsurface coal mining is a major environmental problem, implying that the legacy of coal mining has large water quality implications, beyond the direct use of water for coal mining.
- Coal cleaning and processing have a large water footprint that increases the overall water intensity of coal extraction.

- Most of the water used in the coal industry is for cooling coal-fired plants and electricity generation. The type of cooling systems determines the water use; once-through (open-system) water cooling systems withdraw a large volume of water that is mostly returned to the hydrosphere and thus has high water withdrawal and low water consumption. In contrast, recirculating (closed-system) water cooling withdraws less but consumes more water. The relative proportions between coal plants with once-through relative to recirculating cooling systems determines the overall water utilization. Air (dry) cooling reduces the water footprint but is more expensive to operate and requires more coal and thus carbon emissions to compensate for lower efficiency.
- The discharge of large volumes of water from thermoelectric plants causes thermal pollution of receiving water resources, particularly in cold regions.
- The disposal of CCRs into surface impoundments and landfills induces a major water quality impact due to the high leachability of highly toxic elements from coal ash into the aquatic system.
- The highly impaired water intensity of AMD and disposed coal combustion residues imply that the legacy of coal mining and combustion will remain for decades to come, even if the actual extraction of coal will reduce over time.
- Due to the relatively high water intensity of coal, the rise of unconventional shale gas and the transition from coal to natural gas have reduced the water footprint of power generation in the USA. Further replacement by renewable solar and wind resources would further reduce the water toll of energy production.
- Governments have largely sought to regulate atmospheric pollution generated by coal consumption with less attention to the water impacts. Despite attempts by the Trump Administration to bolster the coal industry by weakening existing environmental regulations, coal use is declining as natural gas and renewable energy are on the rise.

4

Conventional Crude Oil and Tight Oil–Water Nexus

4.1 Introduction

At the same time that coal helped fuel the Industrial Revolution in England during the eighteenth to nineteenth centuries, worlds apart, oil was about to take center stage in the mid-nineteenth century as another important fossil fuel. The world's first successful oil well was drilled in the Absheron Peninsula in Azerbaijan in 1848[340] followed by the discovery a decade later of commercial oil in Titusville, Pennsylvania in 1859. While the world's oil industry had its roots in Pennsylvania in the USA at the end of the nineteenth century, over the course of the twentieth century, oil exploration and production would spread across the globe, starting with new discoveries in Iran (then Persia) in 1908, Venezuela in the 1914, Iraq in 1927, Bahrain in 1932, Saudi Arabia in 1938, and then in Nigeria in 1956.

The oil and gas industry, however, is much more concentrated than the coal industry owing to it being capital-intensive, which creates greater barriers to entry. During the early years of oil and gas exploration in the first part of the twentieth century, a small number of oil and gas companies known as the Seven Sisters (i.e., Standard Oil of New Jersey (Exxon), Royal Dutch Shell, Socony-Vacuum (Mobil), Texaco, Gulf Oil, Standard Oil of California (Chevron), and British Petroleum) possessed the necessary capital to explore and produce new reserves and hence ended up dominating the international petroleum market[341].

In the 1960s and 1970s, following the period of decolonization in the 1950s and 1960s and the constitution of new nation-states across the globe, a wave of nationalizations also ensued as states sought to gain permanent sovereignty over their natural resources[342]; this led to the consolidation of state-owned oil companies across the globe[91, 343]. Most oil and gas production over the course of the twentieth century have thus taken place in countries with state oil and gas companies, as has been the case in Algeria, Iran, Venezuela, and the Soviet Union, among others. Some countries, such as Argentina and Mexico, nationalized their oil companies earlier: Argentina created Yacimientos Proliferos Fiscales (YPF) in 1922 and Mexico formed Petróleos Mexicanos (Pemex) in 1938.

For many oil- and gas-rich countries, oil, or what has often been referred to as "black gold," has been the main source of income for the ruling elites. Yet, countries that have depended on oil and gas for their primary source of revenue have also experienced the boom-and-bust cycles associated with the volatility in oil prices. Most notably, during the late 1970s, oil prices doubled in response to the Iranian Revolution and the start of the Iran–Iraq war; the 2000s witnessed even greater fluctuations in pricing from $25.00/barrel to $145.00/barrel in 2008[344]. Steep swings in the price of oil (especially downward) can wreak havoc on a country's economy and social services: without a steady stream of revenue, it becomes increasingly difficult to maintain, for example, universal health care or subsidized fuel. The collapse of oil markets in 2020 that ensued because of the COVID-19 pandemic raised new challenges for countries that depend predominantly on the export of oil and gas for their revenue base, including many in the Arabian/Persian Gulf.

That petroleum has had an unparalleled impact on building the world's transportation sector and fueling economic growth has led oil and gas companies and governments to scour the globe searching for new petroleum reserves so as to ensure access to reliable, plentiful, and cheap flows of oil and gas. Yet, in doing so, at times companies and governments have taken actions (often because of weak regulations or human error) that have resulted in large-scale environmental calamities, including those affecting water resources. When it comes to the energy–water quality nexus, media attention has often centered on large oil spills. Indeed, there are no shortage of large environmental disasters in the oil sector with water impacts. Just within the USA, the 1969 Santa Barbara oil spill, the 1989 Valdez oil spill, and the 2010 Deepwater Horizon oil spill stand out. Pictures that emerged of birds, fish, and sea mammals covered in oil in 1969 following a gigantic blowout from a drilling well off the coast of Santa Barbara ignited what became the environmental movement of the 1970s, and prompted the first set of new environmental regulations in the USA, including the 1969 National Environmental Policy Act that requires environmental impact statements[345].

Worldwide, the transport of oil, via tankers, barges, and pipelines, puts seawater, fresh water resources, and soil resources at peril. Between 1989 and 2018, there were 613 reported oil spills of 7 metric tons or more from tankers and barges, including the largest tanker collision in 1979 off the coast of Tobago, when the *Atlantic Empress* hit another tanker, spilling 287,000 tons of crude oil[346]. The most infamous tanker incident in the USA concerned the *Exxon Valdez* Spill in 1989. As the largest domestic oil spill at its time, causing widespread hardship to fishing communities and contamination, it served as a focusing event to trigger the US Congress to pass a law in 1990 requiring oil tankers in US waters to install double hulls[347]. Yet, tanker accidents, despite their decline, have not fully ceased, with the worst accident occurring in 2018 off the coast of China when an Iranian oil tanker, the *Sanchi*, collided with a cargo ship. In addition, international concern has mounted in 2020 as news broke about a leak in the engine room of an abandoned oil storage tanker (FSO SAFER) off Yemen's west coast, posing a threat to the Red Sea's ecosystem if a spill were to occur.

Oil exploration and production have also continued to threaten water resources even along the coast of the USA. On April 20, 2010, the Deepwater Horizon drilling rig exploded, killing 11 workers and sending more than 4 million gallons of oil into the Gulf of Mexico, causing damage to the marine environment and tourism along the Gulf Coast[348]. At the same time that these large accidents have made the news, there is no shortage of smaller accidents where oil has leaked from pipelines, as was the case in 2006, when 267,000 gallons of oil leaked from BP's network of pipelines in Prudhoe Bay, Alaska[349].

While countries like the USA have a larger tool kit at their disposal for ostensibly regulating oil spills, elsewhere in the world (the production of oil and gas in Nigeria and Ecuador, for example) spills have left a long trail of environmental externalities, particularly relating to water quality. In particular, the export of oil and gas has been central to Nigeria's economy since its independence in 1960. Despite the government receiving $350 billion in oil revenues between 1970 and 2000, the population plunged into poverty[350]. According to the World Poverty Clock, in 2018, Nigeria had overtaken India as the country with the largest number of people living in extreme poverty – that is approximately 87 million people (or about half the population) living on less than $1.90/day[351]. Furthermore, the populations living in the oil- and gas-producing regions, especially in the Niger Delta, have seen little of the economic benefits and rather have had to bear many of the environmental externalities generated from oil and gas exploration and development. Oil spills in the Niger Delta, including from ruptured pipelines, have caused damage to farmlands, wetlands, and the fishing industry[352]. Reports note that as much as 546 million gallons (2.1 million cubic m) of oil have spilled into the Niger Delta over five decades[353].

Even with greater knowledge about the environmental impacts of oil production on water as well as the links to climate change (discussed in Chapter 6), demand for oil has grown over the twentieth century and into the twenty-first century, particularly in emerging economies in Asia. Despite striking for energy self-sufficiency, oil demand outpaced domestic production in 1993 in China. Over the last few decades, China's increasing appetite for oil and gas has made it the world's second-largest consumer of oil and moved it from the second-largest net

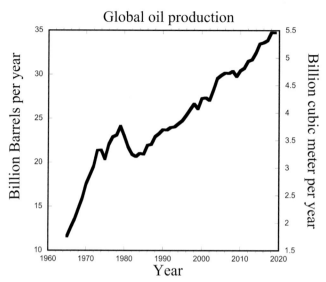

Figure 4.1 Global petroleum production over five decades. Since the mid-1980s, global oil production has increased at a rate of 0.34 billion barrels per year. Data retrieved from BP global energy dataset[10].

importer of oil to the largest in 2014[354]. After China, the USA, Japan, and India are the world's largest net importers of oil, respectively, although the rise of unconventional tight oil production in the USA has reduced the importation of oil to the USA[355]. Only with the COVID-19 pandemic in 2020 have oil markets ground to a halt, as oil prices plunged with the abrupt halt in global transportation and commerce.

The global production of petroleum since the mid-1960s is presented in Figure 4.1 (data from BP global data[10]). In spite of fluctuations in petroleum production in different counties, global petroleum production steadily increased up until 2020; since the early 1980s, global petroleum production has increased at a rate of 0.34 billion barrels per year (Figure 4.1). This constant rise in production has reflected growing demand, regardless of fluctuations in oil prices, internal political instability in oil-producing countries, and temporal availability. The continual rise in production also reflects the major transition away from conventional oil to unconventional oil (defined also as "tight oil") production, in addition to smaller contributions from oil sand and oil shale. This is noticeable in the changes of the relative proportion of US production in global oil since 2009; whereas during the 1960s and 1970s the USA was a major petroleum producer, with the nationalizations that took place in the 1970s, other countries outstripped the USA in oil production, and by the early 2000s, Saudi Arabia and Russia had

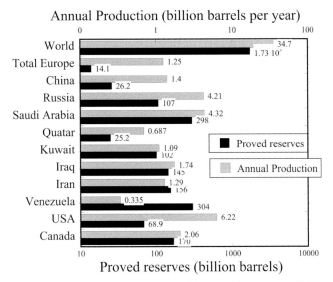

Figure 4.2 Distribution of global crude oil proved reserves (billion barrels) and annual crude oil production (billion barrels per year) among the major oil producing counties. Data from BP global energy dataset[11].

become the world's largest oil producers. Yet with the rise of unconventional tight oil exploration, US oil production started to rebound in 2009, allowing it to become the largest worldwide oil producer (Figure 2.3). In 2019, the USA produced 6.2 billion barrels per year, out of a global production of 34.7 billion barrels per year, which consists of 2% of the global proved oil reserves of 1,734 billion barrels (see distribution of global proved oil reserves and production in Figure 4.2). In 2019, global crude oil production consisted of 32% of total global energy production[355], which was the largest share of energy source (Figure 2.2).

While much media and public attention has focused on oil spills as the major component of the energy–water nexus, in this chapter we highlight the amount of water that is required for exploration (drilling and hydraulic fracturing) as well as the produced water that is co-produced with oil as critical components of the oil–water nexus. The rest of the chapter proceeds as follows. We first describe the essential differences between conventional and unconventional oil exploration and the implications for water utilization and impact. We evaluate the water footprint of the different life cycles of petroleum utilization that includes exploration (drilling, oilfield enhancement, and hydraulic fracturing), the generation of produced water and flowback water with petroleum, the environmental implications for oil wastewater disposal and management, the water footprint of oil sand, and the water footprint of oil refinement (Figure 4.3).

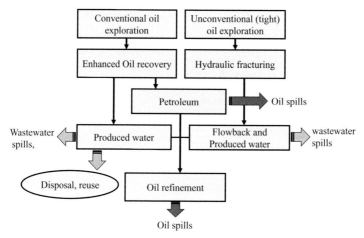

Figure 4.3 Major stages of the petroleum exploration cycle discussed in Chapter 4.

4.2 The Distinction between Conventional and Unconventional (Tight) Oil Exploration

Crude oils contain different types of hydrocarbons and heteroatoms (i.e., other organic and inorganic components such as sulfur, nitrogen, oxygen, iron, vanadium, nickel, and chromium). The density of oil depends on the proportions between hydrocarbons and heteroatoms, with higher fractions of heteroatoms resulting in higher molecular weight, known as heavy oils. For example, Venezuelan heavy crude oil contains less than 50% hydrocarbons and a higher proportion of organic and inorganic substances containing heteroatoms. A common unit to express oil density is the American Petroleum Institute gravity (API gravity), in which oils with API above 10 reflect oils with density lower than water, while oils with API less than 10 are heavier. Common oil has an average API gravity of 38, medium oil API gravity of 22, heavy oil API gravity of 16.3, and bitumen API gravity of 5. While the light hydrocarbons in crude oil are highly volatile and toxic (e.g., benzene), the heavy fractions and heteroatoms contain a higher concentration of other contaminants such as vanadium. During the oil refining process, many of the heteroatoms' components are removed to generate different petroleum products such as gasoline (mixtures of hydrocarbons that contain 4-to-12 carbon atoms), kerosene for jet fuel (hydrocarbons with 10-to-16 carbon atoms), and diesel fuel (hydrocarbons with higher numbers of carbon atoms)[356]. The chemical properties of crude oil define its grade and the common benchmarks for oil pricing, including the intermediate–low density (API gravity of 41) and low sulfur (0.24%), the so-called light and sweet West Texas Intermediate (WTI) crude oil that is produced in Texas and southern Oklahoma, and the intermediate sulfur (0.37%) and light (API gravity of 38) Brent Crude from the Northern Sea. The major crude oil types that are extracted in the USA and Canada

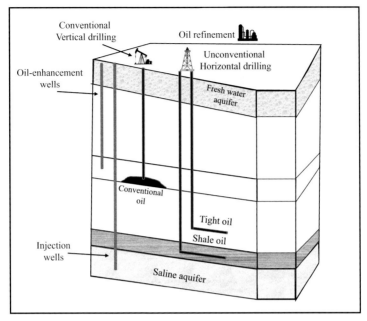

Figure 4.4 Illustration of the geological structure of conventional oil and unconventional tight and shale oil exploration through vertical and horizontal wells. (A black-and-white version of this figure will appear in some formats. For the colour version, refer to the plate section.)

are characterized by a wide range of gravity and sulfur contents, including crude oils from the Bakken (API of 43.3, Sulfur = 0.07%), Permian (API of 30.2, Sulfur = 1.5%), San Joaquim Valley California (API of 13, Sulfur = 1.19%), Eagle Ford (API of 45.7, Sulfur = 0.24%), and Cold Lake (oil shale) in Canada (API of 22.7, Sulfur = 4.1%)[357].

Since the beginning of conventional crude oil exploration, oil extraction was directly linked to the geology and permeability of the formations from which oil is extracted, in addition to hydrocarbon sources. The maturation of kerogen to crude oil occurs in organic-rich formations under a specific temperature and pressure window; the hydrocarbons then flow from the source rocks through permeable formations (e.g., sandstone) to a geological trap, where the formation boundary (e.g., upper layer) is no longer permeable, which is a common target of conventional oil exploration (Figure 4.4). The reservoir productivity typically depends on the size and permeability of the host geological formations that allow the flow of hydrocarbons to the production well (Figure 4.4). Since the geological formations also contain formation water, produced water is co-extracted with oil. The pumping of oil and produced water is defined as primary recovery. Yet, the oil production rates commonly decrease over time, and in order to increase oil production, secondary oil recovery technology is often applied, mostly through water injection into the oil field, which can enhance oil extraction by 30% to 60%. The injection of water through separated injection wells, defined as "water flooding" or "enhanced recovery," commonly utilizes recycled produced water blended with "make up" external water sources. This process involves a large volume of water: an average ratio of 8.6 gallons of water is injected to recover one gallon of crude oil[358]. With time, the effect of secondary oil recovery starts to decline and enhanced oil recovery (EOR) is applied to maintain oil production. EOR utilizes several techniques, including the injection of water, carbon dioxide, steam (thermal EOR), and synthetic (polymer) materials. In all of the EOR methods, water is used for injection, with high ratios of water to crude oil (e.g., steam injection of 5.4, carbon dioxide injection of 13[359]). In 2008, secondary oil recovery constituted 75% of total water use for oil production in the USA and thus presents the largest water footprint associated with the production of conventional crude oil.

In contrast, unconventional oil or "tight oil" exploration involves tapping hydrocarbons directly into the source rocks, typically composed of shale formations, or overlying rocks with relatively low hydraulic permeability in which the natural flow of fluids is limited. Through hydraulic fracturing technology, in which water is mixed with sand and fracking chemicals and then injected under high pressure to generate small fractures along vertical and, more commonly, horizontal wells, hydrocarbons, which are entrapped within the shale and/or tight oil formations, are released and flow to the well. In addition to vertical wells, hydraulic fracturing is conducted along horizontal wells with a lateral length of up to 2 km to maximize the areas of fractures. Together with the hydrocarbons, the fluids that are returned to the surface (defined as "flowback water" during early stages of production) are composed of the injected water blended with the formation water. Over time, the relative proportion of the formation water in the flowback water increases and the water that is generated from the unconventional oil wells is defined as "flowback and produced water" (FP water)[360]. The major water use for unconventional oil extraction is through the water that is injected as part of the hydraulic fracturing process.

The mid- to late 2000s mark the transition from the predominance of conventional to unconventional oil exploration in the USA. The development of the hydraulic fracturing method and the successful outcome of shale gas exploration (Chapter 5) have sparked a rapid growth in the magnitude of unconventional crude oil production. Figure 4.5 shows crude oil production in the USA during the last 100 years and the rise of production since 2008 due to increasing tight oil exploration. In 2019, about 2.81 billion barrels (449.4 million cubic m) of crude oil were produced directly from tight oil resources, which comprise 63% of total crude oil production in the USA (4.46 billion barrels; 709.8 million cubic m)[361]. The rapid rise of tight oil production relative to the steady and

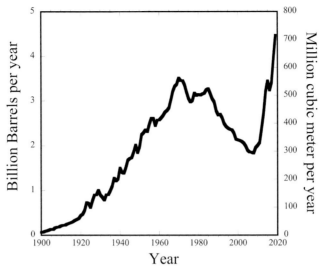

Figure 4.5 Crude oil production in the USA during the last 100 years (reported in billions of barrels per year, left, and millions of cubic m per year, right). Since 2008, the rise of unconventional oil production has boosted oil production to the highest levels ever in US oil production history. Data from US EIA crude oil production[361].

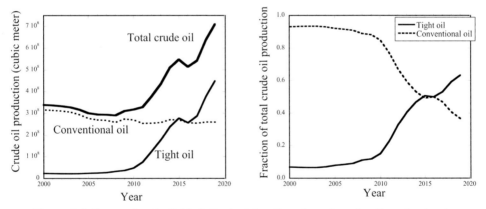

Figure 4.6 Conventional oil (dash line), tight oil, and total crude oil production in the USA since 2000. Since 2008, the rise of unconventional oil production has boosted the tight oil production that in 2015 exceeded conventional oil production. The bottom figure presents the relative proportions of conventional vs. tight oil of the total crude oil production in the USA. Data from US EIA crude oil and tight oil production[361].

even decline of conventional oil production resulted in higher extraction of tight oil relative to conventional oil in the USA (Figure 4.6). The transition from predominantly conventional to unconventional oil exploitation in the USA occurred in 2015. The extraction of unconventional oil has taken place in several

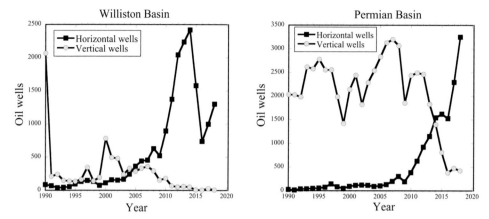

Figure 4.7 Number of vertical and horizontal oil wells installed in the Permian and Williston basins in the USA. The shift from vertical to horizontal wells reflects the rise of unconventional oil exploration. Data from US Geological Survey National Produced Water Database[362].

major basins in the USA, specifically in the Permian, Bakken, and Eagle Ford basins. The discovery of the Permian Basin in western Texas as the largest oil production field has further increased unconventional oil exploration in the USA. The transition from conventional to unconventional oil exploration in the Permian and Williston basins, the largest oil basins in the USA, is illustrated by the shift in the number of vertical to horizontal oil wells in the basins since 2008–2009 (Figure 4.7). Likewise, Canada has tapped unconventional oil, particularly from the Bakken/Three Forks Shale Play in Saskatchewan and Cardium, Davernay, and Montney shales in Alberta. Beyond the USA and Canada, large global tight oil reserves have been identified in China, Russia, Argentina, and Australia (Figure 4.8).

4.3 Water Use for Enhanced Oil Recovery and Hydraulic Fracturing

While drilling for conventional oil wells has a relatively low water intensity of 5 L/GJ[365], the water flooding and EOR for further oil extraction have a much larger water intensity with estimates between 5 L/GJ to 81 L/GJ[359]. In 2005, a volume of 4.4 million cubic m was used to extract 3.5 million barrels in the USA, which gives a water-to-oil ratio of 8 (i.e., water intensity of 208 L/GJ)[359]. In 2014, the majority (94%) of on-shore conventional crude oil exploration was conducted by water flooding[359, 365], with an overall water intensity of water-to-oil volume ratio of 15.7[365], which is equivalent to 408 L/GJ. In many cases, produced waters that are co-extracted with oil are used for oil recovery and thus the net fresh water use to crude oil production ratio ranges from 0 (i.e., 100% reuse of produced water) to 50% for the major oil production regions in the USA, with an average net fresh water intensity of 117 L/GJ[365] (Table 4.1). It was estimated that in 2014,

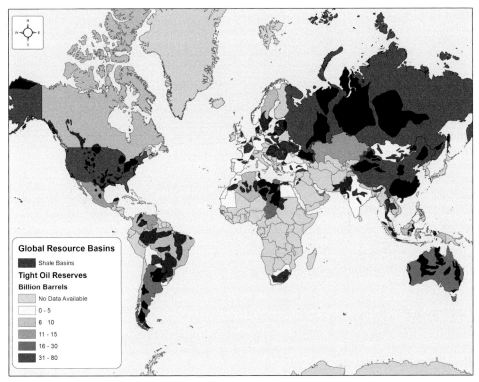

Figure 4.8 Global tight oil resources and shale basins sorted by potential production. Map was prepared by A.J. Kondash (Duke University) based on US EIA report (2013)[363] and EIA oil and gas maps website[364]. (A black-and-white version of this figure will appear in some formats. For the colour version, refer to the plate section.)

about half of the volume of the water used for conventional oil enhancement across the USA was oil produced water, yet in some regions, such as in West Texas, lower production rates of produced water requires a higher supplement of fresh water or alternative water sources[365]. Because the water that is used for drilling and oil enhancement is not recovered, the water withdrawal is identical to water consumption for the upstream conventional oil exploration.

For unconventional oil extraction, much of the water intensity is directed to the hydraulic fracturing process. Chapter 5 describes the details of the hydraulic fracturing process. During the early stages of hydraulic fracturing development in the USA (2011–2013), the water used for unconventional oil extraction varied from 1,000–15,000 cubic m per well (median values for shale play), which is equivalent to 2–7 L/GJ[23]. This water footprint estimate was lower than the water intensity of conventional oil recovery (Table 4.1). Yet between 2012 and 2016, unconventional oil exploration intensified with longer lateral wells (increased by 6% to 40%) and higher volumes of water used for hydraulic fracturing (by 14% to 770%)[24].

Table 4.1. *Summary of the water intensity of water withdrawal and consumption for different types of oil extraction*

Source	Water withdrawal intensity (L/GJ)	Water consumption intensity (L/GJ)	Reference
Conventional oil extraction			
Drilling	5	5	Wu et al. (2018)[365]
Flooding for oil recovery	233	233	Bush and Helander (1968)[358]
Enhanced oil recovery	140–338	140–338	Wu et al. (2009)[359]
Enhanced oil recovery (2014)		408	Wu et al. (2018)[365]
Enhanced oil recovery (2014) – net fresh water		117	Wu et al. (2018)[365]
Overall conventional oil	73–171	73–171	Wu et al. (2009)[359]
Water and steam flooding	13	13	Grubert and Sanders (2018)[29]
Produced water (2012)		239	Veil (2015)[309]
Produced water (2014)	67	67	Grubert and Sanders (2018)[29]
Unconventional (tight) oil exaction			
Hydraulic fracturing, USA (2011–2013)	2–7	2–7	Kondash and Vengosh (2015)[23]
Hydraulic fracturing, USA (2014)	12.9	12.9	Grubert and Sanders (2018)[29]
Hydraulic fracturing, USA (2016)	13	13	Kondash et al. (2018)[24]
Hydraulic fracturing, USA (first 12 months, 2016)	37	37	Kondash et al. (2018)[24]
Flowback and produced water (first 12 months)	18.1–39.1	18.1–39.1	Kondash et al. (2018)[24]
Oil sand extraction			
In situ steam injection	78	78	Allen (2008)[368]
Oil sands mining	104	104	Wu et al. (2008)[359]
Oil sands mining	234	234	Allen (2008)[368]
Oil sands life cycle (extraction, upgrading, refining)	75–134	54–109	Ali and Kumar (2017)[369]
Oil processing			
Oil refinement	39	39	Sun et al. (2018)[370]
Gasoline production	18.1–21.1	18.1–21.1	Sun et al. (2018)[370]

Table 4.1. (*cont.*)

Source	Water withdrawal intensity (L/GJ)	Water consumption intensity (L/GJ)	Reference
Diesel production	5.7–11.4	5.7–11.4	Sun et al. (2018)[370]
Jet fuel production	5.7–11.4	2.6	Sun et al. (2018)[370]
Oil combustion			
Cooling oil-fired power plants	76	1	Grubert and Sanders (2018)[29]

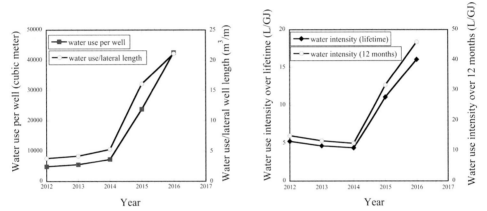

Figure 4.9 Changes in total water use (cubic m per well), water use per length of lateral well (cubic m per m), water use intensity over the lifetime oil production (L/GJ), and water use intensity over the lifetime oil production (L/GJ) of unconventional oil exploration from the Permian Basin, USA (data from Kondash et al., 2018).

In the Permian region, for example, the median water use for hydraulic fracturing of oil wells increased by 770%, from 4,600 cubic m per well in 2012 to 42,560 cubic m per well in 2016. Likewise, the water use normalized to the length of the lateral oil well increased during this time, from 3.8 cubic m per m to 21.1 cubic m per m (Figure 4.9)[24]. The water use intensity that was measured for the first 12 months of oil production increased from 15 L/GJ in 2012 to 50 L/GJ in 2016, while estimates for the water use intensity upon the lifetime of oil production increased from 5.3 L/GJ in 2012 to 16.1 L/GJ in 2016[24] (Figure 4.9). The water intensity across the unconventional oil basin in the USA in 2016 was 37 L/GJ, considering oil production during the first 12 months, and 13 L/GJ upon estimating oil production during the lifetime of unconventional oil production (i.e., longer time and higher oil production and thus a lower water intensity value) (Table 4.1). It is interesting to note that in spite of the public perception of the water intensity of unconventional oil exploration[19, 26, 366], and the intensification of the water use for unconventional oil exploration[24], the water intensity of hydraulic fracturing for unconventional oil is lower than that of conventional oil due to the large volume of water that is used for conventional oil enhancement (Table 4.1).

The water source that is used for conventional oil enhancement or for hydraulic fracturing is the key for evaluation of the actual blue water footprint of conventional and unconventional oil exploration. One of the alternative sources for fresh water is the wastewater that is co-produced with oil; produced water for conventional oil and flowback and produced water for tight oil (see Section 4.3). In the USA, the majority (71%) of produced water that is co-generated with conventional oil has been used for oil enhancement[359], and thus the net (i.e., fresh) water intensity of conventional oil, at least in the USA, is lower than the absolute water intensity values shown in Table 4.1. While the reuse of flowback and produced water for hydraulic fracturing is possible, the high salinity and chemistry of the oil and gas wastewater is a critical limiting factor for its reuse[26]. Several studies have suggested that seawater salinity is the upper limit of water that can be used for hydraulic fracturing, and because the salinity of flowback and produced waters from the major unconventional oil produced basins (Bakken and Permian basins) is much higher, the reuse option is limited. Other options to remediate the water footprint of hydraulic fracturing include using alternative water resources such as domestic wastewater and brackish groundwater, but again, the water quality of these alternative water sources and possible interactions with fracking chemicals limit this option (see further discussion on shale gas in Chapter 5).

Overall, while the water use for hydraulic fracturing/unconventional (tight) oil extraction is higher than the drilling of conventional oil, the water use for enhanced conventional oil recovery is much larger (Table 4.1, Figure 4.10). Similarly, it was demonstrated that unconventional water-to-oil ratios (0.2–1.4) are within the lower range of those for US conventional oil production (a range of 0.1–5)[23, 367]. While the intensification of unconventional oil exploration and associated water use has increased the overall water use for oil production in the USA[24], the water use per energy unity (i.e., water intensity) of conventional oil production is larger[367]. For decades, when conventional oil extraction dominated the oil sector, less attention was paid to the high water footprint of oil exploration and production; instead public attention focused primarily on environmental externalities such as oil spills. Yet, with the rise of unconventional oil exploration, public attention has pivoted toward the water footprint of hydraulic fracturing in spite of the fact that it has a lower water footprint than conventional oil extraction. The perception of the high water intensity of hydraulic fracturing has overshadowed the ongoing reality of the much higher water footprint of conventional oil extraction.

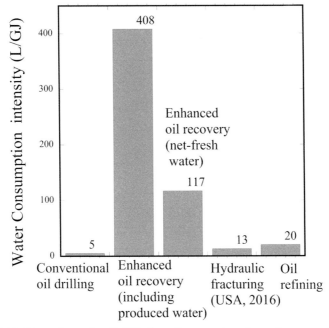

Figure 4.10 Water intensity (L/GJ) for oil extraction through conventional and unconventional technologies. The data show that the water intensity of enhanced oil recovery conducted to simulate conventional oil extraction is higher than the water use for hydraulic fracturing for extraction of tight oil, in contrast to the conceptual paradigm of the high water footprint for hydraulic fracturing.

4.4 Flowback and Produced Waters

4.4.1 The Water Intensity of Flowback and Produced Waters

Oil extraction also involves production of wastewater known as "produced water" for conventional oil exploration and "flowback and produced water" for unconventional tight oil extraction. In conventional oil settings, the extraction of hydrocarbons in permeable formations is accompanied by produced water originated from formation water in the geological formations. Over the lifetime of a conventional oil well, the ratio of formation water to oil increases (Figure 4.11), until the relative proportion of water is too high to make the water–oil blend profitable. The volume of the formation water in the mix depends on the geological condition and hydraulic permeability that allows the formation water to flow to the well-pumping zone. The global production of oil and gas produced water in 1999 was estimated as 12.2 BCM per year[371]. In 2012, the volume of combined oil and gas wastewater in the USA from on-shore and off-shore operations was 3.37 BCM per year, with the majority (3.27 BCM) from on-shore operations[372]. While the majority of the oil and gas from on-shore operations

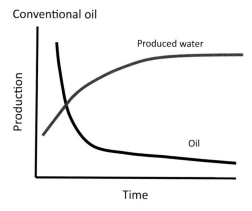

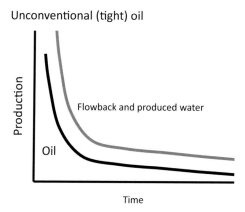

Figure 4.11 The relationships between produced water and oil over time for conventional and unconventional (tight) oil extractions. Note the rise of produced water from conventional oil wells over time relative to the parallel decrease of flowback and produced water with tight oil over time.

is discharged to the subsurface through injection into deep wells or reuse for oil enhancement (see below), oil and gas produced water from off-shore oil and gas operations is discharged to the ocean with only minimal water treatment (oil removal). The volume of OPW that was discharged to the ocean from off-shore facilities in the USA is estimated as 107 million cubic m per year[372, 373] and globally as 800 million cubic m per year[374].

The overall national US weighted average produced water-to-oil volume ratio in 2012 was 9.2[372, 373], which is equivalent to a wastewater intensity of 239 L/GJ (Table 4.1). Some states with large oil production have had a much higher ratio. In particular, in California, 197.7 million barrels oil and 3.7 billion barrels (366 million cubic m) of produced water were generated in 2012, corresponding to a produced water-to-oil volume ratio of 15.5 (water intensity of 402 L/GJ)[375]. Grubert and Sanders (2018)[29] estimated that in 2014 the annual produced water

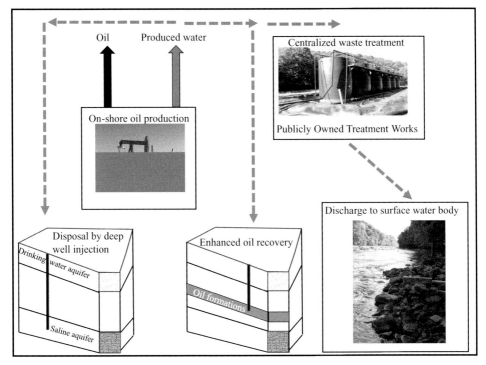

Figure 4.12 A schematic illustration of the transport and fate of oil produced water generated on-shore in the USA. (A black-and-white version of this figure will appear in some formats. For the colour version, refer to the plate section.)

volume from conventional oil wells in the USA was 2.3 BCM (67 L/GJ; Table 4.1). In the USA, the produced water is managed through disposal in deep-well injection, used for enhanced oil recovery, transferred to water treatment centers, and discharged to surface water (Figure 4.12). In 2012, out of 3.17 BCM of produced water that was generated from on-shore operations in the USA, 1.47 BCM (46%) was injected for enhanced recovery of oil , 1.26 BCM (40%) was disposed through injection into deep Class II wells, 110 million cubic m (3.5%) was lost by evaporation, 218 million cubic m (6.8%) was disposed offsite through commercial operations, 20 million cubic m (0.6%) was allocated for beneficial use, and 96.2 million cubic m (3%) was discharge to surface water[375] (Figure 4.13). Direct release of produced water to waterways causes direct and indirect water contamination and accumulation of radioactive elements in the sediments/soil.

For unconventional (tight) oil, the water that is returned after hydraulic fracturing is defined as flowback and produced (FP) water. Several studies in the USA[360] and China[16] have shown that the production of the FP water follows the production of oil; during the first 3 to 6 months after hydraulic fracturing, the oil

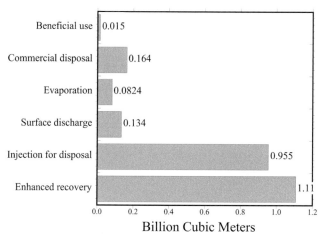

Figure 4.13 Management of the US-produced water generated from conventional oil and gas operations in 2012. Data from Veil (2015)[375]. Note that a volume of 134 million cubic m, which is 5.4% of the produced water generated in the entire USA, has been annually discharged into surface water.

and FP water flows are high, followed by a gradual decrease, down by an order of a magnitude, through the lifetime of a well (Figure 4.11). In most shale basins, the accumulated volume of the FP water is systematically lower than that of the volume of water injected as part of hydraulic fracturing[360] (see Chapter 5). Yet parallel to the intensification of the water use for hydraulic fracturing, the volume of FP water co-extracted with tight oil also increased during the time period of 2008 to 2016 in the USA. In the major tight oil production basins (Bakken, Eagle Ford, and Permian basins) the volume of FP water increased from 2,300–9,200 m^3 per well in 2011 to 15,100–36,2000 m^3 per well in 2016 (up to 550%; Figure 4.14). The FP water intensity increased during this period from 3.6–35.4 L/GJ to 18.1–39.1 L/GJ (Figure 4.14; Table 4.1)[24]. Unlike conventional oil extraction, the ratio of FP water to unconventional oil remains constant over time as both unconventional oil and FP water production decrease over time after hydraulic fracturing (Fig 4.11). Based on oil and FP water production rates during the first 12 months after hydraulic fracturing reported by Kondash et al. (2018)[24], the FP water-to-oil ratio in the major unconventional oil basins in the USA varies between 0.9 (Bakken and Eagle Ford basins) and 1.9 (Permian Basin), which is equivalent to a water intensity of 18.1–49.4 L/GJ. These ratios are significantly lower than the produced water-to-conventional oil ratios (an average ratio of 9.2[372, 373] or 239 L/GJ in the USA; Table 4.1). Therefore, the rise of unconventional oil exploration in the USA has increased both the oil production and the overall volume of oil and gas wastewater that is generated with unconventional oil, but in lower intensity due to the

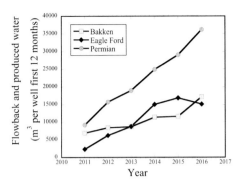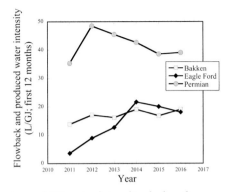

Figure 4.14 Changes in the volume of FP water and FP water intensity during the intensification of hydraulic fracturing for tight oil extraction between 2011 and 2016 in the USA. Data from Kondash et al. (2018)[24].

lower FP water-to-oil ratios. We calculated the expected volume of conventional produced water (assuming that on-shore conventional oil constitutes only 78% of the total crude oil production in the USA[375]) and unconventional FPW using the produced water-to-oil ratio of 9.2 and FPW-to-oil ratio of 1. The production of conventional produced water and unconventional FPW in the USA over time is illustrated in Figure 4.15. Our estimates show that due to the relatively lower FP water-to-oil ratio, the volume of the combined produced water and FP water has only marginally changed since 2000 in spite of the rapid rise of tight oil production. In 2019, a volume of 2.32 BCM of combined conventional water and FPW was generated from both conventional and unconventional crude oil production in the USA.

The rise of the overall volume of wastewater derived from increasing production of unconventional oil (as well as natural gas, see Chapter 5) from horizontal wells has increased the volume of the water injected into Class II injection wells (Figure 4.12). The injection of increasing volumes of injected oil and gas wastewater has resulted in the increasing frequency of earthquakes in some areas associated with deep-well wastewater disposal in the USA, suggesting a clear association between the volume of oil and gas wastewater and the induced siesmicity[376–388]. Given the expected rise of unconventional oil exploration, particularly in the Permian Basin in New Mexico and Texas, the projected high FP water is expected to further increase the risk of induced seismicity. While the magnitude of previous earthquakes associated with wastewater disposal was not high, a potential large earthquake could stop the practice of injection that could result in disposal and contamination of surface water (see Section 4.4.4).

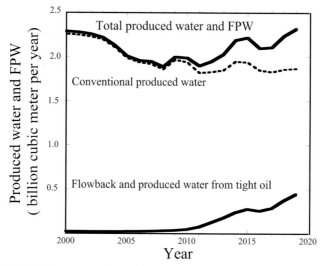

Figure 4.15 Variations over time of the volumes of conventional produced water (dashed line), unconventional FP water, and combined produced water and FP water co-generated with crude oil production in the USA. Conventional produced water volume was calculated based on the weight water-to-oil ratio of 9.2 and the assumption that on-shore conventional oil consists of 78% of total crude oil production in the USA[375]. Unconventional FP water production was calculated based on the FPW-to-oil ratio of 1 demonstrated in tight oil wells during the first 12 months of production[24].

4.4.2 The Origin of Flowback and Produced Waters

In many conventional oilfields, oil is co-extracted with produced water, originated from "formation water" or "oilfield brine" in the geological formations. In many cases, the formation water is derived from connate water (i.e., fossil water that was entrapped within the formation) mostly but not exclusively from a marine origin. The principle seawater source of formation water implies high salinity relative to meteoric water. The salinity of the formation water determines other water quality parameters and is the key for understanding the origin of the formation water and the ability to manage and/or reuse the produced water or FP water in the case of unconventional oil exploration. In many basins, the salinity of the formation water is controlled by (1) the source of the original fluids, (i.e., seawater with TDS of 35,000 mg/L); (2) the degree of evaporation of the original seawater typically after being entrapped in a closed open-water basin (e.g., lagoon) before infiltrating into the subsurface and migration into the geological formation; and (3) magnitude of interactions of the fluids with the formation rocks, which results in mobilization of different elements from the host formation rocks into the formation water. Combined, these factors control the salinity and the overall chemistry of the formation water[88, 389–402].

Since the chemistry of seawater is uniform, and the seawater evaporation process can be predicted, researchers have utilized the chemistry of formation water to reconstruct its geochemical evolution. For example, seawater is characterized by a uniform Br/Cl ratio. During early stages of seawater evaporation (up to degree of evaporation of 10), Br/Cl ratio in the residual evaporated seawater is not modified since both bromide and chloride remain in the solution (defined as "conservative elements"). Evaporation of seawater to concentrations in which mineral halite (NaCl, table salt) starts to precipitate, typically at degree of evaporation of 10, results in incorporation of chloride and bromide into the solid phase. Yet the incorporation of the two elements is not equal as chloride is incorporated into halite while most of the bromide remains in the solution, resulting in an increase of Br/Cl ratio in the residual evaporated seawater with further evaporation and halite precipitation. In contrast, dissolution of bromide-depleted halite salts would generate saline water with a relatively low Br/Cl ratio that would be lower than seawater or evaporated seawater. The salinity of the evaporated seawater is determined by the degree of evaporation of the original seawater as well as the dilution effect, as formation water can be mixed with low-saline meteoric water. In such a case, the Br/Cl ratio of the evaporated seawater would not be affected given the relatively negligible concentrations of chloride and bromide in meteoric water. Consequently, Br/Cl variations in formation water can provide useful information for the origin of the connate water, including marine vs. non-marine origin, the degree of evaporation of the original seawater, and the dilution effect of the original evaporated seawater[88].

The salinity of produced water varies across oil basins and depends on the geological history of the connate water. In the eastern USA, for example, formation water from different geological formations in the Appalachian Basin is characterized by relative high salinity (chloride of up to 200,000 mg/L) and high Br/Cl ratio (higher than the ratio in seawater), reflecting a high degree of evaporation of the original seawater and extensive water–rock interactions. Likewise, formation water associated with the Bakken Basin in North Dakota has typically high salinity (chloride of up to 300,000 mg/L) and high Br/Cl ratio. In contrast, formation water from the western USA has much lower salinity and Br/Cl ratios reflecting much less evaporation of seawater, as shown in oilfield waters in California and Colorado (Figure 4.16). In some cases, such as the Permian Basin, the original Ca-chloride connate brine was mixed with lower salinity Na-Cl water that was recharged into the basin and caused dissolution of evaporite deposits[403]. In addition to the tracing of the sources of the original fluids, the geochemistry of the formation water can be used to detect its migration history and types of rocks with which the fluids have interacted[88]. Several studies have developed diagnostic and forensic tracers that are capable of detecting the types of the source fluids

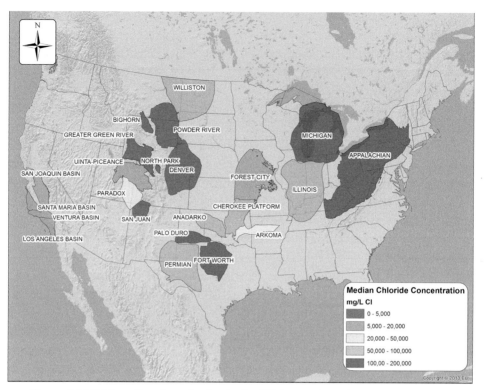

Figure 4.16 Major oil and gas basins in the USA sorted by the chloride concentrations in the formation waters. Data for map from the US Geological Survey archive of the chemistry of produced water[362]. Map prepared by A.J. Kondash (Duke University) based on US EIA oil and gas maps website[364]. (A black-and-white version of this figure will appear in some formats. For the colour version, refer to the plate section.)

(e.g., Br/Cl ratios) and water–rock interactions. Different isotope tracers such as strontium, boron, radium, and lithium isotopes have been utilized to evaluate the sources of the oil brines and modes of water–rock interactions[295, 404–413].

While the chemistry of produced water generated from conventional oil wells reflects the geological history of the formation water, the salinity and chemistry of the FP water from unconventional oil and shale gas wells (see Chapter 5) represent the mixing between the formation water and the frac water that is injected during hydraulic fracturing and return to the surface. The relative proportion of the injected water is typically reduced, however, after the initial hydraulic fracturing, and the proportion of the formation water in the blend increases over time and becomes predominant after only a few months of operations. Over time, the salinity of FP water from unconventional oil wells becomes similar to the salinity of produced water from conventional oil wells. A comparison between the salinity of conventional and unconventional OPW shows typical higher salinity in

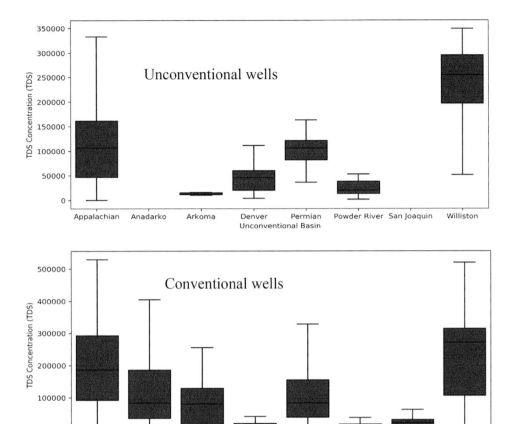

Figure 4.17 Variations of Total Dissolved Salts (TDS) in produced water and flowback and produced water from major oil and gas basins in the USA. Data from US Geological Survey National Produced Water Database[362].

conventional relative to unconventional OPW from the same basin, except for the Permian Basin, which shows an opposite trend (Figure 4.17). The distribution of total dissolved salts (TDS) in formation water from the major oil basins in the USA is presented in Figure 4.17. The mixture between the injected hydraulic fracturing water and the formation water controls the chemistry of the early stages of FP water generated from shale formation (see more discussion in Chapter 5), yet many of the unconventional oil wells are drilled into low-permeability geological formation overlying the shale rocks, and therefore could introduce higher flow rates and different salinity trends that characterize shale formations. This explains the high volume of FP water from unconventional oil wells in the Permian Basin in western Texas[24]. Likewise, in California the salinity variations of the FP water from the Monterey Formation show an inverse pattern where the first flow of FP water is more saline than later FP water[414].

4.4.3 The Water Quality of Oil Produced Water

The chemical constituents in common oil produced water (OPW) are divided into different groups (Figure 4.18), including (1) organic constituents; (2) salts and halides; (3) metals and metalloids; (4) nutrients; (5) radioactive elements known as Naturally Occurring Radioactive Material (NORM); and (6) gases[374, 395, 415, 416]. Organic constituents, composed of dispersed components like residual oil, can have potentially toxic effects and increase biological oxygen demand in impacted water[417] upon releasing of OPW to the environment. Organic constituents in OPW are also composed of dissolved hydrocarbons, including organic acids, polycyclic aromatic hydrocarbons (PAHs), phenols, and volatiles. Generally, the concentration of organic compounds in OPW increases as the molecular weight of the compound decreases, resulting in higher concentrations of lighter-weight, highly volatile compounds such as benzene, toluene, ethylbenzene and xylenes (known as BTEX), and naphthalene[416]. The high solubility of many of the organic constituents results in high concentrations of dissolved organic matter (DOC) that can reach concentrations of hundreds to thousands of mg/L (e.g., Gulf of Mexico 6,400 mg/L) in OPW[418 374, 419]. PAHs and alkylphenols are the most toxic components in OPW and can occur as both dissolved components and as part of the dispersed oil content of OPW[420, 421]. Some of the toxic organic constituents (e.g., PAHs) are highly persistent to biodegradation and thus their presence in OPW and impacted water increases the toxicity effects of OPW.

The second type of chemical constituent in OPW is salts. Since many of the OPWs are derived from evaporated seawater, the concentrations of elements

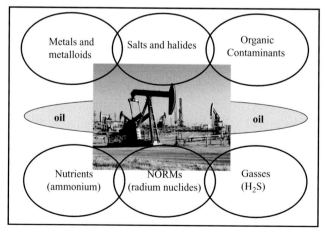

Figure 4.18 Major types of contaminants in oil produced water. (A black-and-white version of this figure will appear in some formats. For the colour version, refer to the plate section.)

associated with evaporated seawater such as chloride, sodium, bromide, calcium, and magnesium are high. Figure 4.17 shows the wide salinity distribution in the major oil fields in the USA; while TDS levels in OPW from the Appalachian (e.g., Marcellus Shale[411–413, 422]) and Bakken[409, 423–426] basins are high (up to 300,000 mg/L), the TDS content in OPW from other basins like the Monterey Basin in California are lower (up to 20,000 mg/L[427–430]). The high salinity in OPW is commonly associated with high bromide (e.g., up to 2,000 mg/L in the Appalachian OPW[411]), reflecting the marine origin of OPW. Yet OPW from non-marine geological settings has much lower concentrations of salts (e.g., chloride concentrations of up to 9,000 mg/L in Nigeria OPW[417]). In addition to chloride and bromide, OPW are commonly enriched in iodide[416] (up to 40 mg/L in the Appalachian OPW[411]). The combination of high DOC and halide concentrations (like bromide and iodide) in OPW poses high risks upon the release of OPW to surface water given the potential of formation of disinfection byproducts in downstream chlorinated river water[411, 431].

The third type of chemical constituents in common OPW is metals and metalloids. Metals typically found in OPW include zinc, lead, manganese, iron, barium, lithium, strontium, and boron[415, 416]. In many cases the concentrations of barium, strontium, boron, and lithium are higher than what we would expect from evaporation of seawater, which implies mobilization of the metals from the host formation rocks[405, 412, 415, 422, 432]. The high concentrations of toxic elements like barium in many of the OPW (e.g., 4,000 mg/L in OPW from the Appalachian Basin in western Pennsylvania[433]) relative to drinking water regulations (2 mg/L for barium) imply high potential for water contamination upon the release of OPW to the environment. Boron and lithium are known to be enriched in OPW[415, 434–437] and have been used to detect OPW in the environment[88, 405], as well as reconstruction of the mineralogical modification of clay minerals upon increasing temperature and pressure in buried sedimentary basins[438–442].

The fourth type of contaminants in OPW is nutrients, in particular ammonium cation. High ammonium concentrations have been recorded in OPW from numerous oil fields (e.g., OPW from the Lower Silurian Oneida Formation in the Appalachian Basin with NH_4^+ concentrations up to 432 mg/L[411]). Given the high toxicity of ammonium, even a small volume of OPW discharge into aquatic systems could pose a major ecological risk[411].

The fifth type of contaminants in OPW include radioactive elements that are defined as Naturally Occurring Radioactive Material (NORMs), in particular radium (^{226}Ra and ^{228}Ra) nuclides. Radium-226 is a radioactive element that is part of the uranium-238 decay series, with a half-life of 1,600 years. Radium-228 has a shorter half-life (5.6 years) and is part of the thorium-232 decay series. The decay of uranium and thorium nuclides results in accumulation of radium nuclides in the geological formations. Radium in the rocks mobilizes into the co-existing formation water, particularly under conditions of high salinity, low pH, and reduced conditions that are common for oil formation water[443–447]. Consequently, OPW commonly contains high concentrations of radium nuclides, which could pose human health and environmental risks upon the release of OPW to the environment[404, 409, 448].

The sixth type of chemical constituents common in OPW is gases, in particular hydrogen sulfide (H_2S) gas. Due to the highly reduced conditions in oil fields, H_2S can be formed from decomposition of organic matter and oil, as well as from sulfate-reducing bacteria. The occurrence and emission of H_2S in OPW is problematic since sulfide is a flammable, irritating, and highly toxic gas[449]. In addition, it causes sulfide-stress-corrosion cracking of pipes, which can result in leaks and spills (see below). The extremely high toxicity of H_2S poses major risks to the residents living near oil operations, as well as to the workers in the oil industry.

The combination of multiple toxic organic and inorganic contaminants in OPW (Table 4.2) presents a major risk upon the release of OPW to the environment. In addition to the direct contamination from toxic elements, the retention of some contaminants like radium nuclides into soil or stream sediments at sites of OPW spills and/or discharge causes accumulation of NORMs on the solids in these sites, generating a radioactivity legacy as demonstrated in spills and in oil and gas wastewater discharge sites in North Dakota and Pennsylvania[404, 409, 411, 448, 450]. The concentrations of both organic and inorganic contaminants are commonly associated with the salinity of the OPW, and therefore hypersaline OPW contains higher levels of contaminants relative to low-saline OPW.

The water quality of OPW from the major unconventional oil basins in the USA (the Bakken[409], Permian[451], and Central Valley in California[414]; see locations in Figure 4.16) are presented in Table 4.2. The wide range in salinity of OPW that characterize conventional oil basins across the USA (Figure 4.17) is also shown in the salinity variations of OPW from unconventional (tight) oil production. In addition to the major inorganic constituents that also characterize OPW from conventional oil (e.g., high halides, ammonium, radium nuclides), OPW from unconventional oil wells contains stimulation "frac" chemicals that are used for hydraulic fracturing. In hydraulic fracturing water in California, the frac chemicals include biocides, gelling agent, acid/base, solvent, crosslinker, clay control, scale inhibitor, corrosion inhibitor, and surfactants with large variations in the proportions of these chemicals in frac water used by different operators[414].

Table 4.2. *Mean values of major and trace elements as well as NORMs in oil produced water from the major unconventional oil basins in the USA as compared to the US EPA ecological, secondary drinking water, and drinking water (Maximum Contaminant Level) standards* *

Component	Bakken Formation Williston Basin	Wolfcamp Shale Permian Basin	Monterey Formation, California	CCR/ secondary standard	MCL standard
Major elements (mg/L)					
Total Dissolved Salts (TDS)	144,000	105,400	70,000	500	N/A
Dissolved Organic Carbon (DOC)	381	80	N/A	N/A	N/A
Alkalinity	353	412	2,000	N/A	N/A
Boron	165	32.4	90	N/A	N/A
Barium	13.5	N/A	8	N/A	2
Bromide	199	493	110	N/A	N/A
Calcium	9,298	1,463	7,000	N/A	N/A
Chloride	94,317	63,050	30,000	250	250
Iron	15.5	19.7	40	N/A	N/A
Iodide	N/A	62.8	N/A	N/A	N/A
Potassium	N/A	388	N/A	N/A	N/A
Lithium	23.9	25.8	30	N/A	N/A
Magnesium	933	222	500	N/A	N/A
Manganese	11.5	N/A	2	0.05	N/A
Ammonium	1,281	633	N/A	1.7 (chronic)	N/A
Sodium	38,644	38,101	9,000	N/A	N/A
Silica	12	11.1	N/A	N/A	N/A
Sulfate	182	363	90	250	N/A
Strontium	561	316	100	N/A	N/A
Trace elements (µg/L)					
Lead	847	N/A	30	N/A	15
Thallium	134	N/A	N/A	N/A	2
Cadmium	19	N/A	0.02	N/A	5
Zinc	6,711	N/A	300	5000	N/A
NORMs (Bq/L)					
Combined radium-226 and radium-228	43.4	N/A	2.2	N/A	0.185

* Data for the Bakken Basin from Lauer et al. (2016)[409], the Permian Basin from Engle et al. (2016)[451], and Central Valley in California from Stringfellow and Camarillo (2018)[414].

4.4.4 The Impact of Oil Produced Water on the Environment: The Impaired Water Intensity

OPW is being released into the environment through three major processes: (1) release to streams and waterways typically after some treatment as part of the wastewater management plan; (2) uncontrolled spills and leaking; and (3) subsurface leaking into groundwater systems (Figure 4.19). These processes can affect the quality of water and soil in the impacted areas, which would result in increased water footprint, defined in this book as *impaired water*. Given the high concentrations of organic matter, salts, metals, nutrients, and radioactive elements in OPW (see Section 4.4.3), the release of OPW into water resources could have multiple effects. First, it causes direct salinization and contamination of impacted water resources to levels that would make the water unusable for domestic use and/ or induce ecological risks. Second, the release of OPW containing high concentrations of halides (chloride, bromide, and iodide) to river water in which chlorination treatment is applied to the downstream river water (i.e., a common practice for using river water for drinking water) could trigger the formation of highly toxic disinfection byproducts in the treated drinking water[411, 431]. Experimental work has demonstrated that even a small addition (~1%) of OPW to river water containing naturally occurring organic matter could trigger the formation of highly toxic bromine- and iodine-disinfection byproducts (like halomethanes)[431]. Third, the release of OPW containing high radium levels into low-saline rivers and/or on soil would result in retention and accumulation of the radium nuclides on the riverbed sediments and/or soil. Highly radioactivity has been observed in river sediments from disposal sites[404, 448, 452–455], as well as in soil impacted by OPW spills[409, 450].

Out of 3.17 BCM produced water that was generated on-shore in the USA in 2012, 96 million cubic m (3%) were discharged into surface water[309] (Figure 4.13). Only a few treatment centers in the USA, however, can adequately remediate the quality of OPW, whereas the treatment of most sites is inadequate and does not remove the high salts, metals, and NORMs in OPW (see also discussion in Chapter 5). Consequently, the discharge of inadequately treated OPW to streams

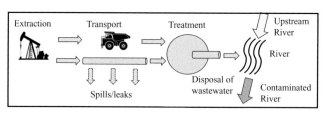

Figure 4.19 Illustration of the possible impact of OPW on water and soil quality as part of routine operation of oil wastewater in the USA.

Table 4.3. *Estimated water intensity and impaired water intensity values evaluated in this book*

Impact	Water intensity (L/GJ)	Impaired water intensity (L/GJ)
Conventional OPW discharge to surface water (California case)	7.9	1106
Conventional and unconventional OPW spills (North Dakota case)	0.02–0.06	1.7–5
Conventional OPW spreading on roads (Pennsylvania case)	2.3	794
Oil spills		25–49

and rivers in the USA has resulted in water contamination, halide accumulation in the river water, leading to the formation of disinfection byproducts in downstream river water used for drinking water, and radium accumulation in the river sediments[26, 404, 448].

In California, the volume of the annual oil wastewater discharge to the environment in 2012 was 9.6 million cubic m, whereas the total oil production was 198 million bbl[372]. This implies a wastewater-to-oil volume ratio of 0.3, which is equivalent to the water intensity of 7.9 L/GJ. Given that the salinity of the OPW from the Central Valley in California is 70,000 mg/L (Table 4.2), the discharge of the saline OPW would require a dilution factor of 140 with fresh water in order to compensate the contamination of the impacted water and reduce the salinity to 500 mg/L, which is the common US EPA standard for discharge of effluents into waterways. Therefore, the impaired water intensity from the discharge of the saline OPW is 1,106 L/GJ (i.e., 140 × 7.9; Table 4.3). This is a conservative estimate since OPW contains even more toxic constituents that would require much larger dilution to balance the concentrations in the impacted water resources. In addition to California, other states have discharged oil wastewater into surface water, including Alabama (3.7 million cubic m), Montana (3 million cubic m), and particularly Texas (59 million cubic m for both gas and oil wastewater)[372].

The magnitude of the impaired water intensity varies given the large diversity in operation practices, and that the majority of both conventional and unconventional oil and gas wastewater is discharged through injection into deep wells. Yet the increasing volume of oil wastewater injection has triggered induced seismicity and earthquakes in areas without known seismic history across the USA[376–388], and therefore some basins, like the Permian Basin in western Texas, could reach a stage where deep-well injection disposal practice would no longer be possible, which would result in increasing disposal of oil wastewater into surface waters, and thus increase the impaired water intensity. It is also interesting to note that in many areas, like in Pennsylvania, where decades of disposal of oil and gas wastewater into surface water did not trigger public concerns, the generation of waste water from unconventional oil and gas operations has elicited major public apprehension, largely owing to films like Gasland (also see Chapter 5 for broader discussion of natural gas). While most of this attention has focused on the gas sector in the Marcellus Basin, it is important to note that unconventional oil is also extracted from the Marcellus Basin, and therefore the wastewater that is discharged in Pennsylvania contains both oil and gas wastewater. Furthermore, while surface water disposal of conventional oil and gas is permitted in Pennsylvania, unconventional oil and gas wastewater is not allowed, and thus the impaired water intensity from disposal of unconventional OPW is lower.

The second mechanism that induces impaired water intensity from OPW is spills. In addition to oil spills (see Section 4.4), spills of OPW are often associated with the increase in activities of oil exploration. OPW spills associated with conventional oil operations have been recorded all over the world, including sites in the Niger Delta[456], Western Siberia[457], Peruvian Amazon[458], Oklahoma, USA[459], and eastern Gansu Province in China[460]. Yet the rise of unconventional oil exploration and development in the Appalachian Basin (Pennsylvania) and Williston Basin (North Dakota) in the USA has been associated with higher rates of OPW spills, and the frequency of spills is further linked to the density of unconventional oil and gas wells[26, 409]. Figure 4.20 illustrates that the rise of oil production from 4.5×10^7 bbl in 2007 to 4.3×10^8 bbl in 2015 (9.6-fold increase) was associated with an increased OPW spill volume of 2,600 m^3 per year to 17,700 m^3 per year (6.7-fold increase) in the Williston Basin (Bakken Shale) of North Dakota. Since 2016, the ratio between OPW spill volume and oil production decreased, but the rise in oil production in 2019 led to an increased volume of spilled OPW (7,450 m^3; Figure 4.20). The ratio between spilled OPW to the total oil production in the Williston Basin is low (8.6×10^{-5} in 2019 to 2.6×10^{-4} in 2016), which would imply a relative low water intensity of 2.2×10^{-3} to 6.7×10^{-3} L/GJ. Yet the spilled OPW has a major impact on waterways and soil in the spill sites. The organic contaminants levels[418] as well as total dissolved salts (TDS), chloride, ammonium, barium, manganese, lead, and radium concentrations of OPW from the Bakken Basin in North Dakota are higher by several orders of magnitude relative to the threshold values of drinking water and ecological standards (Table 4.2; Figure 4.21). Therefore, the release of OPW like the Bakken brine to waterways would require a large dilution of fresh water to balance the water loss of contamination. Figure 4.22, further, shows the range of the enrichment of different contaminants from the Bakken brines relative to drinking water and ecological standards, which ranges from 7 (for barium) up to 750 (for the

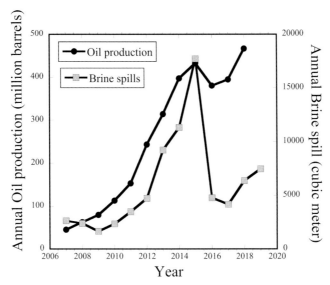

Figure 4.20 The parallel rise of oil production and volume of spilled OPW in the Williston Basin (Bakken Shale) of North Dakota. Data from North Dakota Oil and Gas Division[471].

highly toxic ammonium). Consequently, the impaired water footprint of OPW spills in North Dakota is not negligible and is estimated between 1.7 L/GJ to 5 L/GJ (Table 4.3).

In addition to direct water contamination, OPW spills result in accumulation of radium nuclides on the impacted soil, much higher above the background levels[404, 409, 448, 450, 456]. This means that even if the spilled water is removed from the site, the legacy of the environmental impact is not, and the impact is expected to remain for centuries (i.e., the half-life of ^{226}Ra is 1,600 years). Lauer and Vengosh (2016)[450] have shown that the decay of the ^{226}Ra and ^{228}Ra and the ingrowth of ^{210}Pb and ^{228}Th nuclides on soil impacted by OPW spills can be used to detect the timing and the source of the spill given the different decay and growth rates of the different nuclides. This method can be used to delineate between OPW spills that occurred from conventional oil and gas operation and relatively recent spill events from unconventional OPW[450].

The third process by which OPW can affect water resources is through subsurface leaking and/or flow through inactive orphaned oil and gas wells and contamination of fresh water aquifers. It has been shown that the legacy of conventional oil and gas operation in the northeastern USA has resulted in hundreds of abandoned wells in the Appalachian Basin[461, 462]. Corrosion and deterioration of the cement along the wells could provide direct pathways to deep formation water from the oil and gas to flow into overlying aquifers and,

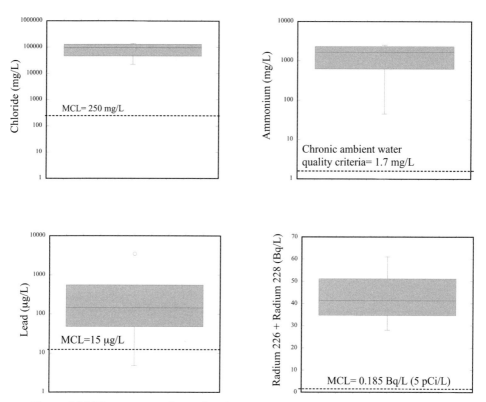

Figure 4.21 The concentrations of major contaminants in OPW from the Bakken Basin as compared to the US EPA Maximum Contaminant Level (MCL) for drinking water and ecological standard for chronic ambient water quality criteria. Data from Lauer et al. (2016)[409].

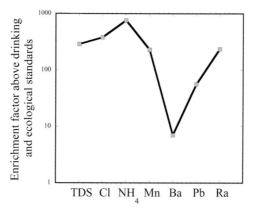

Figure 4.22 Enrichment factors of several major contaminants in OPW from the Bakken Basin relative to the US EPA MCL for drinking water and ecological standards. Data from Lauer et al. (2016)[409].

consequently, could contaminate groundwater from shallow drinking-water aquifers. Several studies in fact have shown direct contamination of shallow groundwater through direct leaking of abandoned conventional oil and gas wells such as in Pennsylvania[463, 464] and Garfield County in Colorado[465]. This is opposed to the natural flow of deep saline formation water that has caused the salinization of shallow drinking water wells in the Appalachian Basin, as witnessed in Pennsylvania[466, 467] and West Virginia[410]. The ability to distinguish between groundwater contamination from natural flow deep saline formation water and direct leaking of oil and gas wells is challenging given the similarity in the chemical composition[26], and several methods, such as the identification of boron and lithium isotopes, have been developed to identify flowback water from unconventional oil and gas vs. conventional produced water[405] (see more details and discussion in Chapter 5).

In addition to leaking OPW through abandoned conventional oil and gas wells, direct injection of OPW as part of enhanced oil recovery (EOC) can cause groundwater contamination due to the increasing pressure and flow potential in the deep formations relative to the overlying aquifers. Decades of injection of OPW for EOC in Martha oil field, located in Lawrence and Johnson counties, Kentucky, resulted in the upwelling of the highly pressurized injected OPW through breached well casings, uncemented annuli along the legacy oil wells, and improperly plugged and abandoned wells and caused widespread contamination of the major aquifer of drinking water. The chloride content of the contaminated aquifer increased from low background values of 13 mg/L to 3,950 mg/L[468, 469]. Likewise, the injection of OPW into hundreds of thousands of Class II injection wells across the USA has also been associated with overlying groundwater contamination, as demonstrated in multiple cases across the USA. In all of these cases it has been shown that abandoned, improperly plugged oil and gas wells without having been properly sealed off near an active injection well provided the pathway through which OPW can reach the overlying shallow aquifer. It has been estimated that already in late 1980s there were about 1.2 million abandoned wells across the USA, many of which may be improperly installed with potential leaking to fresh water aquifers[470].

4.4.5 Beneficial Use of Oil Produced Water: Is That Possible?

The intensification of unconventional oil exploration in the USA has increased the volume of OPW[24] and hence the need to manage this increasing volume of oil wastewater[376–388]. While OPW from conventional oil operations was either used for enhanced oil recovery (45% of the total wastewater in the USA in 2012) or deep well injection (39%), for unconventional OPW deep well injection is the ultimate option; yet, this increases the risk of induced seismicity. One of the alternative OPW management options is recycling and utilization for hydraulic fracturing. The reuse of OPW for hydraulic fracturing is, however, constrained by the high salinity and chemical constituents in OPW, specifically upon interaction with the man-made frac chemicals that are added to hydraulic fracturing water [472–476]. In addition to the salinity, the high concentration of DOC in OPW can affect the viscosity of the frac water[476].

Recycling of flowback and produced water would reduce both the water footprint of hydraulic fracturing and the economic cost of wastewater management (transport, disposal, treatment). In the Marcellus Shale, for example, the majority of FPW is recycled for hydraulic fracturing[477]. Yet one of the implications for reusing OPW for hydraulic fracturing is the reduction of natural gas. In a study in Sichuan Basin in China it has been shown that using FP water for hydraulic fracturing results in 20% lower natural gas production per well (during the first 12 months of operation) and an 18% increase in FP water production relative to wells from the same geological setting that were using fresh water for hydraulic fracturing[478]. This study argued that the tradeoff between the reduction of the water footprint for hydraulic fracturing (45,000 m^3 per well in Sichuan Basin) and reduction of the economic cost of wastewater treatment would outweigh the economic cost of lower natural gas production[478]. While the reduction of natural gas could be a limiting factor for the shale gas industry (see discussion in Chapter 5), for unconventional oil exploration, this reduction could be an advantage since natural gas that is co-generated with oil is commonly not captured but rather released into the atmosphere via flaring as conducted in the major unconventional oil productions of the Bakken and Permian basins (see Chapter 6). The reduction of co-produced natural gas could reduce the fugitive gas emissions from unconventional oil exploration, and at the same time would reduce the water footprint of hydraulic fracturing. In some waste-scarce areas like the Permian Basin, the use of an alternative water source for hydraulic fracturing could be vital for the continued exploration of unconventional oil[24].

Wastewater from conventional oil and gas operations has also been utilized for deicing, dust suppression, and land spreading on roads as well as for road maintenance across the USA, particularly in Pennsylvania, New York, and West Virginia[454]. The spreading of OPW that contains high levels of salts, toxic organic and inorganic chemicals, and NORMs has directly contaminated runoff in impacted areas. The spreading OPW derived from conventional oil and gas operations on roads has released over 4 times more radium to the environment than OPW treatment facilities, and 200 times more radium than spill events in Pennsylvania[454]. Between 2008 and 2016, the volume of oil and gas wastewater spread on roads in Pennsylvania was estimated between 20,000 m^3 to 60,000 m^3, which is equivalent up to 6% of the total conventional OPW generated in Pennsylvania[454]. Since the relative proportions between wastewater generated from oil wells relative to natural gas wells are not known, we use the relative ratio between the expected volume of produced water generated from conventional oil (estimated as 3.8 million cubic m) and the predicted volume of produced water generated from conventional natural gas in Pennsylvania (2.7 million cubic m). Since the oil producing water-to-natural gas producing water ratio is 1.4, we estimate that out of the total 60,000 m^3 produced water spread on roads, ~35,000 m^3 came directly from conventional oil production. Given the volume of conventional oil that was generated in Pennsylvania in 2012 (4.3 $\times$ 10^6 bbl [372]), this implies a spread water-to-oil ratio of 0.05 and water intensity of 1.3 L/GJ. Yet the median TDS concentration of the spread water in Pennsylvania was 293,000 mg/L[454], which is 586-fold greater than the common ecological threshold of 500 mg/L, implying an impaired water intensity of 794 L/GJ (Table 4.3).

An alternative beneficial use of the increasing volume of OPW is for the agricultural sector. The over-exploitation of natural water resources, coupled with reduction in precipitation in arid and semi-arid regions, has incentivized water users to explore alternative water resources. In the Central Valley in California, the large depletion of groundwater resources and increasing limitation of the ability to transfer surface water from northern to central and southern California have triggered the utilization of OPW to supplement the local water supply for irrigation in the Central Valley[479]. For over 25 years, OPW blended with surface water has been used for irrigation in the Cawelo Water District of Kern County in Central Valley of California. While typical OPW from Central Valley in California contains high levels of salts (mean TDS value of 70,000 mg/L[414]; Table 4.2), in Kern County, the salinity of the local OPW is much lower (chloride content between 64 and 92 mg/L), similar to the salinity of the local groundwater[480]. Consistent with the low salinity, the blended OPW from Kern County used for irrigation has low concentrations of DOC (8.9 mg/L), trace metals, and radium nuclides, with most parameters below drinking water standards and irrigation water guidelines. Yet, systematic evaluation of the soil irrigated by blended OPW relative to soil irrigated by groundwater showed relatively higher salts and boron contents, implying long-term salts and boron accumulation in the OPW-irrigated soils[480]. High boron in soil can pose long-term risks to plant toxicity, and therefore the high boron that characterizes many OPWs could be a limiting factor, even for the low-saline OPW from Kern County in Central Valley. In contrast, irrigation with OPW with high salinity and DOC, such as the OPW from Colorado (chloride ~10,000 mg/L; DOC~800 mg/L[481]), or even 50% diluted OPW, could have negative effects such as lower crop yields[482–484], weakening of the disease resistance of crops such as wheat[482, 485], as well as the accumulation of toxic metals in crops, thus posing human health risks[486]. Therefore, it seems that the utilization of common saline OPW for irrigation could induce major plant safety- and human health issues, and thus is not a sustainable solution for the beneficial use of OPW. Alternatively, the removal of salts, metals, and organic matter through treatment could make OPW suitable for irrigation, but would increase the economic cost and thus may reduce its viability. In addition, the development of treatment technologies for the adequate removal of the complex chemistry of OPW, including mechanical vapor compression (MVC), membrane distillation (MD), and forward osmosis (FO) technologies, could help address the desalination of OPW for possible beneficial utilization[487–489].

4.5 Oil Sands

4.5.1 The Water Intensity of Oil Sands

In certain geological terrains, unconsolidated sandstone rocks contain high concentrations of kerogen, the earlier stage of maturation of hydrocarbons known as *oil sands* or *tar sands*. The extraction of crude oil from oil sands presents an alternative method for conventional oil and tight oil extraction methods. The largest oil sand reserves are in Alberta, Canada with an estimated proven oil sand reserve of 165 billion barrels[490], about 70% of global potential oil sand production. High reserves of oil sands are also located in Kazakhstan, Russia, and Venezuela. Since 1984, crude oil production from the Canadian oil sands has increased rapidly, and in 2018 reached 1,059 million barrels per year, which accounted for 64% of Canada's oil production[490]. The oil sand deposits are located in three major basins in northern Alberta, covering over 140,000 km^2 and composed of shallow oil sand deposits (up to 75 m depth) and deeper oil sand deposits in which bitumen is extracted through *in situ* stream injection[368].

The high viscosity of the heavy hydrocarbons in the tar sands requires special extraction methods in which water is the principle medium for crude oil extraction, and therefore involves high water intensity. Crude oil extraction from tar sands involves two major operations: (1) *in situ* extraction with stream injection into subsurface oil sand deposits (> 75 m depth), which accounted in 2018 for 53% of total crude oil production; and (2) oil sand mining for surface oil sand deposits, which accounted for 47% of total oil production (Figure 4.23). Approximately 82% of Canada's oil sand reserves are recoverable via *in situ* technologies (a potential of 132 billion barrels), while the potential of surface oil mining is only 31.8 billion barrels (18%)[490]. In the in-situ extraction process, water is utilized to generate steam that is injected into subsurface deposits to generate an oil–water mixed slurry that is pumped to the surface for oil separation. The major

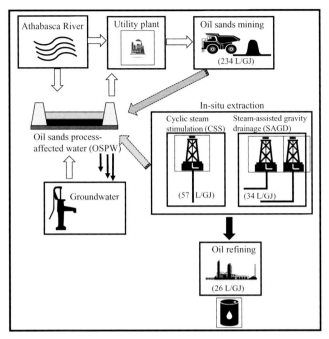

Figure 4.23 Illustration of the different stages in the production of crude oil from oil sand mining for shallow oil sand deposits and *in situ* extraction for deep oil sand formations in Alberta Canada. The values in brackets represent the water intensity of the different stages of oil extraction.

technologies for *in situ* injections are cyclic steam stimulation (CSS), where steam is injected and oil–water slurry is recovered through the same vertical well, and steam-assisted gravity drainage (SAGD), in which steam is injected via a horizontal injection well and the oil–water slurry is pumped through a vertical recovery well[359]. While CSS technology requires a 2.2 volume ratio of water to oil (57 L/GJ), SAGD requires 1.3 volume ratio of water to oil (34 L/GJ)[359]. Allen (2008)[368] estimated a higher volume ratio of 3 (78 L/GJ) and highlighted the high percentage (90%) of recycling of water for steam production, which results in net fresh water replenishment of only 0.3 water-to-oil volume ratio (water intensity of 8 L/GJ) (Table 4.1).

In contrast, surface oil sand mining requires a much larger volume of water per oil and thus is highly water intensive. Heated water is blended with oil sands to generate a slurry from which oil is separated (Figure 4.23). Wu et al. (2009) estimated a water-to-oil volume ratio of 4 (104 L/GJ), while others estimated a ratio of 9[490], which is equivalent to a water intensity of 234 L/GJ (Table 4.1). Yet, 80% of the used water is recycled from the process plant and only 20% of the water is extracted from the Athabasca River and/or local groundwater, which is equivalent to a fresh water replenishment intensity of 47 L/GJ. In 2005, the total allocation of the Athabasca River flow for oil sand producers was 360 million cubic m per year, which amounted to only 1.7% of the river's annual flow[368].

In addition to the water use for bitumen extraction from oil sands, water is a critical component of upgrading and refining to convert bitumen into oil (Figure 4.23). The water to refined oil ratios for water consumption and withdrawals during the total life cycle of the combined processes was estimated as of 2.1 to 4.2 (54–109 L/GJ) and 2.9 to 5.2 (75–134 L/GJ), respectively[369]. The relative contribution of the refining stage (water-to-refined oil ratio of 1.1) consists between 25% (for open mining) to 50% (for *in situ* extraction) of the total water footprint of crude oil production from oil sands[369].

4.5.2 Oil Sands Wastewater and Water Quality Impacts

Following the extraction of bitumen from oil sands, the residual water, known as oil sands process-affected water (OSPW), becomes brackish with TDS ~2,000 mg/L, high alkalinity, and high levels of toxic organic acids[491–494]. Over 840 million cubic m of OSPW have accumulated in the Canadian oil sand operations and are stored in tailings ponds that cover a total area of about 130 km² in a region north of the city of Fort McMurray, Alberta[495–497]. The OSPW from the tailings ponds is reused for further bitumen extraction in the Canadian oil sands[368]. Several studies have highlighted the potential environmental impacts of OSPW in the Athabasca oil sand region in northern Alberta[492, 497–501]. One of the major contaminants found in OSPW in tailings ponds is naphthenic acids[500, 502], which have a high toxicity effect on fish and other wildlife[503–511]. While direct disposal of OSPW to surface water or shallow groundwater is not allowed[368], seepage of OSPW from tailings ponds to underlying groundwater and subsequent migration to the nearby Athabasca River is a potential long-term risk in areas of oil sand operations in Alberta[495].

The oil sands wastewater is characterized by high chloride (~400 mg/ L), boron, and lithium concentrations relative to the Athabasca River, which is used for bitumen extraction. The isotope composition of boron, lithium, and strontium in OSPW reflects a mix of residual and saline formation water entrapped within the oil sands and fresh surface water that has been modified by interactions and the leaching of elements from the solid oil sands[491]. Boron, lithium, and strontium isotope fingerprints, as well as stable isotopes of oxygen and hydrogen[512], could be used as potential monitoring tools for detecting OSPW contamination of local fresh water resources[491]. In addition, the carbon isotope fingerprint of naphthenic acids generated during bitumen separation from oil sands was found to be different from that of natural background organic acids, and thus has been used to distinguish between industrial impact from naturally occurring organic acids that helped to delineate the source of organic compounds in the environment[502]. Although there are some indications for the leaking of OSPW into the underlying aquifers, no evidence has been reported for OSPW contamination of the nearby Athabasca River[513].

In addition to direct water contamination, atmospheric emissions of organic substances with potential toxicity to humans and the environment are a major concern for the industrial development associated with oil sand extraction. Polycyclic aromatic hydrocarbon (PAH) combined with nitrogen and sulfur were detected in particulate matter from air in the Athabasca Oil Sands Region in northeastern Alberta, Canada, indicating emission from industrial activities and combustion associated with oil sand operations[514, 515]. The indirect pathway of transporting PAHs to the aquatic systems from open pits via the atmosphere may be as significant a contributor of PAHs to aquatic systems[516]. Yet, Ahad et al. (2014)[517] also emphasized the role of oil sand mining, with evidence for co-occurrence of PAHs and metals in the environment, up to around 200 km away from the main area of mining activities[518–521]. In addition, PAHs in sediment cores from nearby lakes with carbon isotopic fingerprints indicating petroleum-derived PAHs were recorded over the last 30 years and originated from the deposition of bitumen in dust particles associated with airborne particles emitted from open pit mines[514, 522].

4.6 Refining Crude Oil

4.6.1 The Water Intensity of Refining Crude Oil

The conversion of crude oil to useable petroleum products is conducted through the oil refining process; when crude oil is heated, the oil breaks down and generates hydrocarbon gases. The gases are then cooled and condensate to generate different fuel liquids of different density such as gasoline, jet fuel, and diesel fuel. The liquids are then processed through cracking (breaking down large molecules of heavy oils), reforming (changing molecular structures), and isomerization (rearranging the atoms of the hydrocarbon molecules). The US EPA has identified several petroleum refining process, including: (1) topping – separating crude oil into hydrocarbon groups that include desalting and distillation; (2) thermal and catalytic cracking – breaking large into small hydrocarbons through coking and catalytic cracking; (3) combining/rearranging hydrocarbons to generated desired products; (4) removing impurities such as sulfur, nitrogen, and metals from products; and (5) generating final products through blending and manufacturing[523]. The heating and condensation processes are energy intensive, and water is also an important component of the oil refinery process. Water is mainly used for generating steam and cooling (~90% of the water use) as well as for removing salts and contaminants from the crude oil, generating hydrogen for some fuels, and for equipment cleaning and maintenance. Estimates of water consumption in the oil refining process in the USA suggest that the magnitude of water use depends on the quality/density of the crude oil and the energy allocation for the specific refinery products[370]. While the withdrawal water is estimated to be a volume ratio of 1.5 of water to crude oil, which is the equivalent of a water intensity of 39 L/GJ (Table 4.1), the water consumption depends on the refinery product. Gasoline production requires the highest volume of water with a volume ratio of 0.6–0.7 water to gasoline (equivalent to a water intensity of 18.1–21.1 L/GJ), followed by diesel (volume ratio of 0.2–0.4 water to diesel, 5.7–11.4 L/GJ), and jet fuel diesel (volume ratio of 0.09 water to jet fuel, 2.6 L/GJ). In the USA, the majority (72%) of water sources for oil refining comes from surface (lake, river) water, with municipal water as an alternative water source consisting of 18% of the total water use for oil refinement[370].

4.6.2 The Quality of Wastewater from Oil Refineries and Environmental Impacts

It is estimated that about 40% of the used water for oil refinery is discharged to the environment, and thus the water intensity of oil refinery wastewater has a volume ratio of 0.6 of water to crude oil, or 16.4 L/GJ. Since the oil refinery process involves the volatilization and condensation of crude oil, some of the chemicals in oil, particularly organic constituents and metals, are transported into oil refinery wastewater. Given the multiple processes that are involved with oil refineries, the US EPA identified 13 different types of wastewater that contain different types of contaminants. The crude oil desalter, for example, generates wastewater with high salts and metals, while the distillation, cracking, and coking processes generate wastewater with high concentrations of organic contaminants such as phenols and BTEX (benzene, toluene, ethylbenzene, xylenes). The contribution of the different waste sources generates a blend of wastewater with high concentrations of organic and inorganic contaminants[523]. The major common constituents in oil refinery wastewater are phenols, ammonia, H_2S, and combined highly toxic and volatile contaminants of BTEX[523]. The concentrations of organic matter expressed as Chemical Oxidation Demand (COD) vary between 400 and 1,000 mg/L, hydrocarbons up to 1,000 mg/L, phenols up to 200 mg/L, and benzene up to 100 mg/L[524–526]. Given that the US EPA MCL for benzene is 0.05 ppb, the high levels of the organic contaminants in oil refinery wastewater imply a risk to the ecosystem and human health. The enrichment of benzene in oil refinery wastewater is 200,000-fold relative to the drinking water standard. Therefore, the impaired water intensity of oil refinery wastewater is extremely high.

It has been shown that the direct discharge of oil refinery wastewater to the environment causes direct contamination of receiving waters and toxicity impact on the associated ecosystem[527]. The toxic effects of oil refinery wastewater was shown by the absence of all or most living species and reduction of the biomass close to the wastewater outfall[527]. Given the variations in the different types of refining processes and associated contaminants, remediation technologies, quality of the wastewater, and the ratio between wastewater volume and receiving water volume (i.e., dilution effect), generalization for the environmental impact of oil refinery wastewater is challenging[527]. The zero discharge policy for oil refinery wastewater is the only practical mechanism that can prevent the environmental effects of oil refinery wastewater[523].

4.7 Oil Spills

Large oil spills in the oceans are considered the major environmental issue associated with oil exploration, especially due to their direct effects on the marine ecology. Increasing global oil production and associated oil tanker transportation have resulted in a growing number of oil spills in the ocean. In 2002, the US National Research Council (NRC) estimated that about 800,000 m^3 of petroleum entered the sea each year from the extraction, transportation, and consumption of crude oil and the products refined from it, with an additional 680,000 m^3 derived from seepage[528]. Due to increasing awareness and enforcement, the number and volume of major tanker oil spills have decreased over time, from 150,000–750,000 tons per year in the 1970s to 20,000–420,000 tons per year in the 1980s and 1990s, and 10,000–70,000 tons per year in the 2000s[529]. The major oil spills on record are the 1989 Exxon Valdez oil spill in Alaska North Slope (40.8 million liters), the 2010 Deepwater Horizon on the ocean floor into the Gulf of Mexico (507 million liters), and the 2007 Hebei Spirit oil spill in Taean area of western Korea (13 million liters). Concentrations of total petroleum hydrocarbons (TPH) and PAH in most affected areas decreased to background levels within 7−18 months for seawater, surface sediments, and oyster tissues, although residual oils remained in the subsurface layers for longer periods in some protected areas. Yet the evaluation of the long-term ecological impacts of large-scale oil spills is still ongoing and shows that early assumptions about the degradation of hydrocarbons may not have been adequately evaluated[530].

The transition from conventional to unconventional oil exploration also marks a transition from off-shore to on-shore oil spills, as the rapid development of unconventional oil in the USA has been associated with the rise of on-shore oil

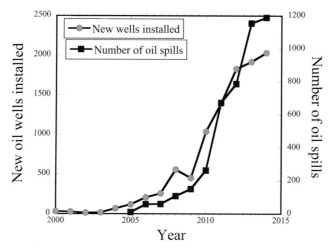

Figure 4.24 The parallel increase of new unconventional oil installations and number of oil spills in the Williston (Bakken) Basin of North Dakota. Data from Patterson et al. (2017)[531].

spills. Since the production of unconventional oil is reduced at about 3–6 months after the installation of a well, unconventional oil development is associated with rapid rates of drilling and hydraulic fracturing. For example, in North Dakota the number of new wells installed in the early 2000s was 15–30 wells per year, which exponentially increased to about 2,000 wells per year in 2014 (Figure 4.24)[531]. Yet the high rate of unconventional oil well installation has resulted in an increasing frequency of oil spills in the Williston Basin in North Dakota and the Permian Basin in New Mexico, and the majority (75–94%) of spills occurred within the first three years of a well life time after drilling, hydraulic fracturing, and production of the highest oil volumes[531]. The increased rate of new oil well installation was associated with an increased number of oil spills[531]. From 2008 to 2013, the annual oil spill volume also followed the rise of oil production in North Dakota, followed by a reduction of the oil spills volume from 2013 to 2019 (Figure 4.25). Consequently, during the unconventional oil development in North Dakota, the volume ratio between the annual oil spilled and oil produced in North Dakota varied between 0.8×10^{-4} to 3×10^{-4}.

The impact of oil spills on groundwater resources depends on the oil properties (e.g., density) as well as hydrogeology (e.g., aquifer lithology, permeability, and structure). The oil viscosity and density control the fate of the oil in the saturated zone of aquifers; heavy oil (i.e., density higher than that of fresh water) would sink to the aquifer bottom, generating a contaminated zone along the saturated zone underlying the spill zone, whereas light oil (i.e., density lower than that of groundwater) would float and occupy the upper part of the saturated zone. In both

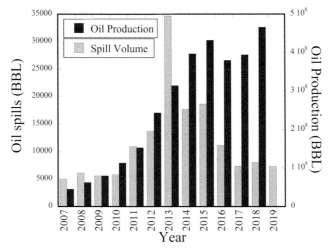

Figure 4.25 The volume (BBL) of annual oil spills and oil production in the Williston (Bakken) Basin of North Dakota. Data from North Dakota Oil and Gas Division[471].

cases, the unsaturated zone underlying the spill area is contaminated with residual oil droplets entrapped in the porous or fracture areas along the infiltration pathway. In addition to the physical mobilization of the spilled oil, its properties would determine its impact on the groundwater quality. While very light oil (e.g., jet fuels, gasoline) would evaporate, and because it contains high concentrations of organic contaminants, it would become water soluble; intermediate light crude oil (e.g., diesel) would instead be moderately volatile, and as such, it would leave oil residuals along the flowpath. Common crude oil with lower density has the potential to cause long-term contamination of water resources. In contrast, heavy crude oil (e.g., oil sand) with high density would not evaporate, and accordingly would cause even more severe and long-term water contamination, accumulating in the porous media of the aquifer[532].

One of the most common phenomena of crude oil spills is pipeline burst and leaking of the oil to the subsurface. The US Office of Pipeline Safety estimated that an average of 83 crude oil spills occurred per year between 1994 and 1996 in the USA, each spilling about 50,000 barrels of crude oil (an annual spill volume of 3.4 million barrels)[533, 534]. For over 30 years, the US Geological Survey has investigated the quality of groundwater impacted by the 1983 crude oil spill near Bemidji, Minnesota[533–537]. A volume of 1,700 cubic m of crude oil spill infiltrated the unsaturated zone (~11 m depth) and floated to the upper section of the groundwater table, at a distance of 40 m downgradient from the spill site (Figure 4.26). Yet the impact on the water quality of groundwater in the spill area far exceeded the distribution of the oil phase in the aquifer. The US Geological Survey study[533–536]

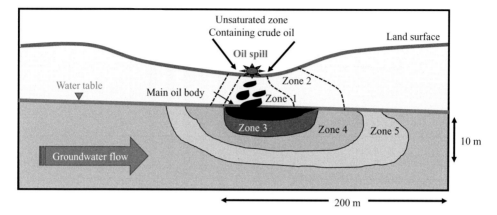

Figure 4.26 A cross-section of the crude oil spill site near Bemidji, Minnesota. Areas in the unsaturated and saturated zones were defined by changes in the groundwater quality as monitored in 1997 by the US Geological Survey[533–536]. The figure is based on figure 4 in Delin et al. (1998)[533]. (A black-and-white version of this figure will appear in some formats. For the colour version, refer to the plate section.)

has identified several impacted zones underlying the spill site (Figure 4.26): (1) Zone 1: the unsaturated zone underlying the spill site and above the main oil body, which is relatively anoxic and contains residual insoluble oil, high carbon dioxide (>10%), methane (>10%), and hydrocarbon (>1 ppm); (2) Zone 2: the transition unsaturated zone with lower concentrations of oxygen (10–20%), moderate carbon dioxide (0–10%), methane (0–10%), and hydrocarbon (<1 ppm); (3) Zone 3: the saturated zone beneath and immediately downgradient of the floating oil, which consists of an anoxic plume of groundwater containing high concentrations of hydrocarbons, dissolved manganese, iron, and methane; (4) Zone 4: the transition zone from anoxic to oxygenated conditions in the saturated zone, which contains groundwater with low concentrations of hydrocarbons following aerobic degradation; (5) Zone 5: oxygenated groundwater downgradient from the contamination plume that contains slightly higher concentrations of dissolved toxic constituents, such as BTEX. Overall, the impact on groundwater quality in areas of oil spills varies with time and space given the dynamic transport and transition of insoluble and soluble contaminants in the groundwater. In the anoxic parts of the aquifer, metals that are usually retained into the aquifer solids are released in the aquatic phase and further degrade the quality of the impacted groundwater. It has been demonstrated that naturally occurring arsenic, cobalt, chromium, and nickel that are commonly retained on iron oxides, can mobilize to groundwater upon contamination by BTEX and ethanol mixtures under iron- and nitrate-reducing conditions. Therefore, biodegradation that is part of the remediation process can trigger mobilization and contamination of trace metals in the impacted groundwater[538]. The transition to oxygenated conditions would trigger organic compounds such as BTEX to become soluble in the groundwater. More recent analysis has revealed that in addition to BTEX, other organic contaminants including *n*-alkanes and other aromatic compounds also control the quality of the impacted groundwater[537]. Overall, the volume of impacted groundwater far exceeds the volume of the spilled oil. It has been estimated that groundwater may be contaminated for several kilometers from an oil spill site in the direction of groundwater flow, and it may take at least a decade for the contaminated groundwater to reach its maximum contamination distance from the spill site[532].

In the case of the Bemidji, Minnesota oil spill, the volume of contaminated groundwater (about 200 m length, 50 m wide, and saturated depth of 10 m, porosity = 0.3; Figure 4.26) was estimated as 30,000 m^3, which is ~18-fold higher than the volume of the spilled oil on the site (1,700 m^3). Given the impact on the groundwater quality, the dilution factor that would be required to remediate the contaminated groundwater to pre-spill conditions was estimated to be between 140 and 750[532]. Assuming a middle value of 500-fold dilution factor, the ratio of the impacted groundwater volume to spilled oil volume is estimated as ~8,800. By using this ratio, we use both state (North Dakota) and national oil spill data to evaluate the impaired water intensity induced by oil spills in the USA. In 2013, North Dakota had a total in oil spills of 5,542 m^3, which would impact groundwater of a volume of 49 million cubic m, assuming the same ratios reported in the case study in Bemidji. Since the total oil production in North Dakota was 50 million cubic m in 2013, the impaired water intensity of oil spills in the USA is calculated as 25 L/GJ. Using the total spills in the USA, in 1994 the volume of oil spills was 0.54 million cubic m, which would impact groundwater volume of 4.8 BCM. Given the total oil production of 3.4 BCM during that year, the impaired water intensity of oil spills in the USA is calculated as 36 L/GJ (Table 4.3). These water intensity values suggest that the chronic on-shore oil spills associated with both conventional and unconventional oil spills have meaningful impact on fresh water resources. These estimates were based on the case study of Bemidji, Minnesota[533–536] that represents a common crude oil spill. However, spills of heavy oil would generate a larger impact on the groundwater quality, and therefore a larger impaired water intensity.

4.8 Oil Combustion and Fugitive Atmospheric Emission

Oil products used for combustion include natural gas liquids (NGL) and petroleum derivatives. Combusted products are categorized as liquefied petroleum gas (LPG), light distillates, middle distillates, and fuel oil. In 2019, 19 billion kWh of electricity was used out of petroleum combustion, which is only 0.5% of the total electricity production of 4.12 trillion kWh generated at electricity generation facilities in the USA (Figure 2.14)[112], indicating that oil contribution to the overall electricity sector in the USA is small. The water withdrawal and consumption for cooling oil-fired power plants in 2014 was 2.6 BCM and 36 million cubic m, implying a water intensity of 76 L/GJ and 1 L/GJ, respectively[29]. Two types of fuel oil are commonly burned by combustion sources: distillate oils and residual oils in the form of petroleum coke. Petroleum coke is carbon-rich residual material that is generated as part of the petroleum refining process and has been increasingly used as a combustion fuel in coal-fired power plants and cement kilns. In 2017, global petroleum coke production was 128 million tons, from which the USA generated 47%[539]. In addition to the direct combustion of petroleum, oil refining, which involves thermal processes, also causes fugitive atmospheric emission and indirect water pollution. Volatile components such as hydrocarbons (e.g., BTEX), sulfur, vanadium, mercury, and lead are emitted to the atmosphere during combustion processes.

The increasing use of bitumen from oil sands as well as heavy oils has resulted in a larger utilization of petroleum coke over time[355]. Because vanadium is highly enriched in heavy oil and bitumen, petroleum coke produced during distillation of bitumen from the Athabasca Oil Sands Region in Canada, for example, could have very high vanadium concentrations ($1,280 \pm 120$ mg/kg)[540]. The combustion of petroleum coke results in further enrichment of vanadium in the residual ash (up to 40,000 mg/kg), and thus in the absence of effective scrubber systems, the combustion of heavy oils and petroleum coke can dominate regional atmospheric vanadium emissions[541]. Schlesinger et al. (2017)[541] have, furthermore, estimated that the global annual emission of vanadium to the atmosphere from burning both conventional and heavy petroleum products is 283,000 tons per year, which exceeds the naturally occurring flux of vanadium (anthropogenic to natural ratio of ~1.7)[541].

Another example of contaminant emission from oil combustion is mercury. In 2004, the total mercury concentration of oil processed in the USA of both dissolved and suspended forms, expressed as a volume-weighted mean, was 3.5 ± 0.6 µg/kg[542]. The range of measured concentrations extended from below the analytical detection limit (0.5 µg/kg) to approximately 600 µg/kg. It has been estimated that ~10% of the mercury in the crude oil process in oil refineries is emitted to the atmosphere, which is equivalent to 0.12 mg-Hg per ton of petroleum product[543]. Given the global crude oil processing (100 million barrels per day, or 5 billion tons per year), this would entail a mercury atmospheric flux of 600 kg-Hg per year. While this flux consists of only a small fraction of the total anthropogenic mercury emission (5,000–8,000 tons per year[544]), it is still an important source of local mercury contamination. Since vanadium and nickel are the two most abundant metals in petroleum, with concentrations of up to 1,600 mg/kg and 340 mg/kg, respectively[545], oil refining also causes large vanadium emissions, as shown by elevated vanadium in aerosols near oil refinery sites[546, 547].

4.9 Integration: The Combined Water Footprint of Crude Oil Utilization

Figure 4.27 presents the different stages and associated water intensity values of oil and wastewater production during the different life cycles of drilling, hydraulic fracturing for tight oil, oil enhancement, oil refinement, oil combustion, and wastewater disposal (Table 4.1 and 4.3). Data collected for the USA[29] have shown that in 2014 the total water withdrawal for all stages of conventional and unconventional crude oil exploration was 3.5 BCM (Figure 4.28). This included water withdrawal for cooling oil-fired plants (73.6%), oil refining (18.7%), water flooding (for conventional oil, 3.7%), hydraulic fracturing (for tight oil, 3.4%), drilling (0.4%), and transport (0.2%). The total water consumption in the USA during 2014 was 744 million cubic m, which included water consumption for oil refining (59.2%), cooling oil-fired plants (4.8%), water flooding (17.5%), hydraulic fracturing (16.1%), drilling (2.3%), transport (8.1%), and processing (0.1%) (Figure 4.28). In 2014, conventional oil production (263 million cubic m) consisted of 52%, whereas unconventional tight oil (244 million cubic m) was 48% of the total annual crude oil production in the USA (507 million cubic m)[361].

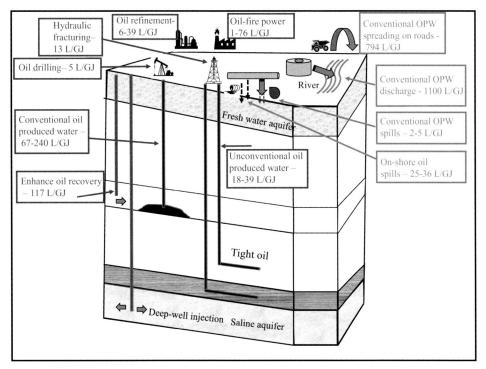

Figure 4.27 Schematic illustration of the different stages of oil exploration and processing as well as the associated water intensity values (L/GJ) for withdrawal/consumption (in blue) and water impaired (red). (A black-and-white version of this figure will appear in some formats. For the colour version, refer to the plate section.)

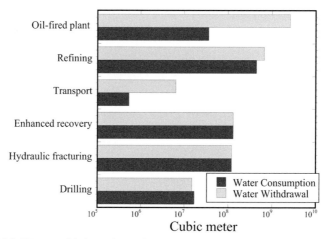

Figure 4.28 Water withdrawals and consumption volumes associated with oil exploration in the USA in 2014. Data from Grubert and Sanders (2018)[29].

Since the exploration of unconventional tight oil was almost equal to conventional crude oil extraction in 2014, a comparison of the water use for conventional and unconventional oil shows a much higher water withdrawal volume for conventional crude oil over unconventional tight oil (by a factor of 2.5). The water consumption of conventional crude oil was only slightly higher (1.08-fold) than that of unconventional tight oil (Figure 4.29). It is interesting to note that the high net addition of water consumption for hydraulic fracturing for unconventional oil (120 million cubic m) was almost balanced by the volume of water flooding for conventional oil (130 million cubic m), indicating that the rise in tight oil exploration in the USA did not induce a higher water footprint as opposed to the common conceptual idea. Nonetheless, the data reported for 2014[29] reflect relatively the early stages of tight oil exploration in the USA, as intensification of hydraulic fracturing has increased the water footprint of hydraulic fracturing and, therefore, more developed tight oil exploration such as shown in the Permian Basin in Texas is expected to increase the volume of water consumption for unconventional oil exploration[24].

In 2014, the total volume of produced water co-extracted with oil in the USA was 2.27 BCM, from which 2.2 BCM (97.2%) originated from conventional oil and gas wells, and only 63 million cubic m (3%) originated from tight oil wells (Figure 4.30) (data from Grubert and Sanders, 2018[29]). This overwhelmingly higher volume of produced water from conventional oil wells reflects the high ratios of conventional produced water to oil (8.4) relative to tight oil (0.3). These estimates are consistent with both the conventional OPW-to-on-shore oil ratios of 9.2 reported by Veil (2015)[375] and the unconventional FPW-to-tight oil ratio of

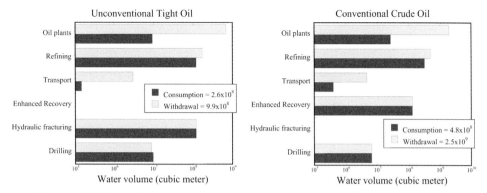

Figure 4.29 Water withdrawals and consumption associated with conventional and unconventional oil exploration in the USA in 2014. Data from Grubert and Sanders (2018)[29].

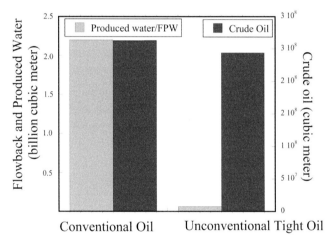

Figure 4.30 The volumes of produced water from conventional oil and FPW from unconventional oil as compared to oil production in the USA. Data from Grubert and Sanders (2018)[29] and US EIA[361].

0.3–1 reported by Kondash et al. (2018)[24]. The case studies presented in this chapter clearly indicate that the disposal of conventional produced water after some treatment to surface water could induce water contamination with a large water impaired intensity. Likewise, the development of unconventional tight oil is expected to induce large volumes of spills that would further increase the water footprint of unconventional tight oil beyond the direct water use. Figure 4.27 shows that the release of petroleum and oil wastewater to the hydrosphere and the contamination of water resources induces a large footprint of the impaired water.

The transition from conventional to unconventional oil exploration through hydraulic fracturing and installation of horizontal wells has increased water utilization for hydraulic fracturing. At the same time, there has been a reduction in the water used for oil enhancement that is part of conventional oil exploration. This has resulted in an overall lower water use and lower water intensity for oil exploration. This is mainly due to the high volume of water requires for oil enhancement, which can be mitigated by using OPW for oil enhancement instead of using fresh water sources. Without this factor, the water use intensity of hydraulic fracturing is higher. On the other hand, the transition to unconventional oil exploration has increased the production rates of both crude oil and OPW. In spite of the lower ratio of OPW to oil in unconventional wells relative to higher OPW-to-conventional oil ratios and thus high water intensity in conventional wells, the absolute volume of OPW generated from unconventional (and tight) oil has considerably increased, and so has the magnitude of seismicity events induced from OPW disposal through deep-well injection. In addition, the frequency of oil and OPW spills has also increased due to the increasing rates of new well installation and the overall intensification of the oil industry.

In addition to the USA, the rise of global on-shore unconventional oil exploration in other countries like Canada, China, and Saudi Arabia (Figures 4.2 and 4.7) is expected to have similar implications for water resources. Given the depletion and degradation of the quality of global water resources, most public attention has focused on the volume of water needed for hydraulic fracturing and the possible conflicts with other water users such as the agricultural sector. Yet, the overall life cycle of the different stages of oil production and wastewater disposal should be considered in protecting water resources. Any release of oil produced water and crude oil to the environment (e.g., disposal to surface water, leaks, spills) induces a large impaired water intensity that needs to be included in the water balance policy associated with oil exploration.

4.10 Policy and Regulations

This chapter has sought to survey the different types of conventional and unconventional oils, ranging from the Venezuelan heavy crude oil to the crude oil from the Bakken and the oil sands in Canada. What has changed over the twentieth century and into the twenty-first century is as some conventional oil fields have become less accessible and more expensive to exploit, the ability to access other deposits, such as unconventional oil fields, has become increasingly possible with the introduction of new technologies such as hydraulic fracturing and horizontal drilling.

Regulating the environmental impacts of conventional and unconventional oil production on water is, however, as complex as it is to trace all the different ways in which conventional and unconventional oil production affects water quality. Complicating the policy process is that no one authority exists for regulating oil exploration and production globally, as well as within many different countries. Few instances of international cooperation exist. At the international level, nation-states have come together to address the prevention of pollution of the marine environment. To prevent the dumping of oil at sea as well as tanker accidents, states signed onto the International Convention for the Prevention of Pollution from Ships (MARPOL) in 1973 and then the 1978 Protocol, which entered into force in 1983[548, 549]. Within the USA, the 1990 Oil Pollution Act (OPA) authorized the EPA to work to prevent and respond to catastrophic oil spills through publishing regulations for the aboveground storage of oil; the Coast Guard has responsibility for oil tankers and deep water ports[550].

Yet, preventing oil pollution on land or around inland bodies of water has proven difficult, owing to the fragmentation and distribution of actors that play a role in regulating the energy–water quality nexus. While often national governments are responsible for promulgating laws regarding the energy and water sectors, in many cases jurisdiction falls to subnational political entities, including local governments, which have responsibility for issuing permits. Furthermore, the companies involved in the production and refining of oil themselves carry tremendous political and economic clout and are able to influence the policies and regulations that vary across the countries in which they operate. For instance, within the USA, when the Commission on Fiscal Accountability passed the Oil and Gas Royalty Management Act in 1982, the way in which it was designed allowed the agency responsible for implementing the act to be captured by the companies that it was supposed to be regulating. Specifically, the Minerals Management Service (MMS) was charged with both (1) collecting royalties from the oil companies operating within the USA and (2) protecting US lands and waters from oil company operations[551]. Because the MMS was charged with ensuring that the USA had enough energy resources, it relied on industry to help design standards, and as such, some have argued that this regulatory structure was partially responsible for the BP oil spill in the Gulf of Mexico in which the MMS did not provide enough oversight regarding oil extraction technologies used in the Gulf of Mexico[551]. In the aftermath of the BP Blowout, the MMS was renamed the Bureau of Ocean Energy Management, Regulation and Enforcement in 2010 and then dissolved a year later.

Indeed, the importance of the states is distinctive of the regulatory culture in the USA; some of this has to do with the fact that much of oil and gas regulation is devolved to the states and the nature of private ownership in which oil and gas companies must acquire the lease to the mineral rights before they can drill. These leases lay out the rights and responsibilities of the different parties to the agreement. While oil production has been taking place in North Dakota since the

1950s, the use of horizontal drilling combined with hydraulic fracturing in the Bakken beginning in 2009 has turned North Dakota into one of the largest oil producers in the USA. Companies are required to comport with existing environmental regulations that protect water quality; yet, much policy depends upon voluntary action undertaken by companies to adhere to "best practices" that include making publicly available the list of chemicals used in the fracking process (http://fracfocus.org). In addition, if a spill happens, companies are required to report the release only to the appropriate local agencies – for example, the Department of Mineral Resources or Department of Health.

Outside of the USA, large companies increasingly have sought to influence the policy space through voluntary initiatives, or what is often referred to as private regulation[552]. Many of these voluntary initiatives, as an alternative to government regulation, fall under the rubric of Corporate Social Responsibility (CSR)[553]. Owing to concern over large oil spills, corporations have furthered CSR programs as a means to prevent oil spills in their global operations. Such programs often entail sustainability and transparency initiatives such as the UN Global Compact and the Extractives Industries Transparency Initiative. Companies, furthermore, will voluntarily report their environmental practices, including information about spills, as a mechanism for protecting and enhancing their reputations[553].

While in many other parts of the world where oil and gas is owned by the state, the oil and gas industry is regulated at the federal level, often with the national oil companies playing an enlarged role in managing oil and gas exploration and production along implementing environmental legislation. The USA is quite different in that oil and gas regulation takes place across many different agencies and implementation takes place at the state level. O'Rourke and Connelly highlighted the dispersed, fragmented, and at times overlapping nature of the regulatory system[554]. They found that the following statutes/laws have some form of jurisdiction for the US oil industry at the federal level: the Federal Land Policy and Management Act; the Federal Oil and Gas Leasing Reform Act; the Outer Continental Shelf Lands Act; the National Environmental Policy Act; the Oil Pollution Act; the Clean Air Act's National Emission Standards for Hazardous Air Pollutants, National Ambient Air Quality Standards, New Source Review (NSR), and New Source Performance Standards; the Clean Water Act's (CWA) National Pollutant Discharge Elimination System (NYDES) and Spill Prevention Control and Countermeasure Requirements; the Emergency Planning and Community Right-to-Know Act; and the Underground Injection Control (UIC) program of the Safe Drinking Water Act (SDWA). When it comes to protecting water resources within the USA, the CWA's National Pollutant Discharge Elimination System (NPDES) and the SDWA's Underground Injection Control program are of significant importance.

That the USA has such varied policy responses for dealing with the impacts on water resources from conventional and unconventional oil exploration and production can be seen again by how different states respond to regulating OPW as well as the chemicals used in the fracking process. As this chapter has shown, for decades, conventional oil production has generated large amounts of wastewater with a significant direct water footprint and impaired water intensity (Table 4.3, Figure 4.24). When it comes to the energy–water quality nexus and the focus on produced water, in the USA, produced water is considered to be an "'exempt' oil and gas waste stream," and as such not subject to the provisions of the Resource Conservation and Recovery Act (RCRA) that deals with hazardous waste[555].

Following the shale revolution, the recognition of toxic components in frac fluids increased public attention and concerns pertaining to the volume of water used for hydraulic fracturing, the fate of the wastewater from unconventional oil development, and the associated human health and ecological risks. For decades, conventional OPW in the USA has been discharged to surface waters through designated outfalls after minimum if not negligible treatment, causing massive water contamination and accumulation of radium nuclides on stream sediments[404, 409, 448]. Furthermore, localities in some US states (e.g., Pennsylvania) have allowed conventional OPW to be spread on roads for deicing and dust suppression in spite of the high level of contaminants known in conventional OPW[454, 555]. In contrast, soon after the beginning of the shale revolution, OPW from unconventional oil wells was not permitted to be discharged to surface waters like the conventional oil wastewater, but rather disposed of in deep injection wells[556]. Specifically, the EPA sets the regulations for how the NPDES regulates the discharge of wastewater into water bodies that fall under the CWA. Depending on the particular state, the EPA itself may issue permits for the disposal of wastewater or may allow states to issue discharge permits, including for produced water[555].

Ultimately, even within these two federal permitting programs, the states retain the greatest regulatory authority for how OPW is disposed: the NPDES program requires the state to issue permits to discharge OPW into surface water bodies, and the wells that are used for injection of produced water must also receive permits under the UIC program[555]. Yet, unlike in other parts of the world where there is greater standardization at the national level, that there are more than 30 oil and gas producing states in the USA has meant different states have sought different ways to manage their OPW. Thus, with time, states such as Oklahoma, Texas, and New Mexico have sought EPA permission to release the produced water into streams and rivers instead of relying upon deep injection wells.

States that have experienced significant drought or have experienced induced seismicity from underground injection of OPW have sought other alternatives to the disposal of OPW, including reuse. One particular form of beneficial reuse is crop irrigation. Consider California, which has been prone to drought and has a large agricultural sector. The California Water Board for the last 25 years has allowed for OPW to be blended with surface water and used for irrigation in places like the Cawelo Water District of Kern County[480].

Chapter 4 Take-Home Messages

- Since the early 1980s, global crude oil production and consumption have constantly increased to become the largest (33%) globally consumed energy source. Since the early 2010s, the rise of hydraulic fracturing and unconventional tight oil exploration has made the USA the largest worldwide oil producing country.
- Many of the worldwide tight oil basins are located in arid and semiarid areas, which raises the concerns of sustainable water allocation for hydraulic fracturing and competition with the agriculture and domestic sectors. The Permian Basin in western Texas is one of the world's largest unconventional oil reserves, and intensification of hydraulic fracturing and production has also increased water use in this mostly arid region.
- In spite of the public perception, the water use for conventional crude oil exploration through water injection for enhancement oil recovery is similar and even higher than the water volume use for hydraulic fracturing and tight oil exploration. In both technologies, the use of alternative water sources (e.g., reuse of oil produced water) can potentially mitigate the limit availability of potable water.
- Water use for oil includes oil recovery (conventional and unconventional methods) and oil refining.
- The volume of oil produced water from conventional oil wells increases over the lifetime of a well, whereas the production of flowback and produced water from tight oil wells deceases a few months after the initial hydraulic fracturing. Consequently, the ratio of conventional produced water to crude oil is far larger than the ratio of flowback and produced water to tight oil. Nonetheless, the rapid rise of unconventional tight oil in the USA has generated an overall large volume of wastewater that is mostly injected into deep wells, which in some cases induces earthquakes in areas with deep well injection.
- Most of the injected water during hydraulic fracturing is retained into the rock matrix and the flowback water, and produced water that is generated during most of the lifetime of a well is derived primarily from formation water, similar to produced water generated from conventional oil wells. Consequently, the chemistry and quality of the two wastewater sources are similar.
- Most of the formation waters were derived from relics of evaporated seawater that interacted with the host rock formations, resulting in enrichment in salts, nutrients,

heavy metals, and radioactive elements in produced water and flowback water. The release of the highly saline with toxic contaminants to natural waters poses major ecological and human health risks. In addition to direct contamination of impacted water resources, accumulation of radioactive elements on soil and stream sediments in impacted areas can generate long-term radiation and possibly bioaccumulation. Even a small contribution of oil wastewater to surface water would increase the probability of the formation of toxic disinfection byproducts in the downstream utilization of chlorinated surface water for drinking water.

- The high concentrations of toxic and radioactive elements in produced water require a large volume of clean water to mitigate the impact on water quality in cases of spills or permitted disposal to surface water. Consequently, the impaired water intensity upon the release of produced water and flowback water to the environment is high.
- In spite of similarity in water quality, produced waters from conventional oil wells are permitted to be discharged after some treatment to surface water or be used for deicing and dust suppression, while wastewater from unconventional oil operation must be injected into deep wells or reused for hydraulic fracturing in the USA.
- The rapid rise and high intensity of unconventional oil development have been associated with increasing cases and volumes of wastewater and oil spills in the USA.
- The infiltration of spilled oil into the subsurface and contamination of groundwater generates a large impacted area in the aquifer, which affects the quality of a large volume of groundwater and therefore induces a large impaired water intensity.
- Because of a dispersed and fragmented regulatory system in the USA, governing the water quality impacts of oil exploration and production is difficult and often devolved to the states.

5

Conventional and Unconventional Natural Gas–Water Nexus

5.1 Introduction

Natural gas has been around for thousands of years, frequently escaping and bubbling from the ground. At the end of the eighteenth century, new technology in Britain utilized gas in streetlights and household lamps, allowing working hours to be elongated with additional light. Before the technology spread across the Atlantic in the early nineteenth century and the invention of the incandescent lightbulb later that century, natural gas along with kerosene provided most households within the United States with artificial light[557].

Yet, for most of the twentieth century, natural gas played a secondary role in global energy markets, especially within the USA. By the late 1940s and early 1950s, natural gas was considered, as Daniel Yergin (p. 429)[557] writes in *The Prize*, to be "a useless, inconvenient by-product of oil production and thus burned off – since there was nothing else to do with it. Natural gas was the orphan of the oil industry." At that time, there were less uses for natural gas despite the large reserves that existed in the Southwest of the USA.

When it comes to natural gas, the twentieth century was largely the era of conventional gas whereas the twenty-first century has been defined by "unconventional gas" with major discoveries of natural gas[558, 559]. One of the first large reservoirs of natural gas was discovered in Kansas in 1922 – the Hugoton reservoir. Several decades later, gas fields were discovered in the Netherlands (Groningen in 1959) and in Western Siberia in the then Soviet Union (Urengoy in 1966)[559]. At present, the largest gas fields are concentrated in Russia, Iran, and Qatar with the latter two sharing the gigantic South Parse Field. Large gas fields have also been found in Algeria and Turkmenistan.

The main constraint on utilizing natural gas for homes and industry has concerned its transmission, which is highly dependent upon an infrastructure built of a network of pipelines. Whereas oil can easily be moved by tankers and even rail, gas transport requires pipelines where it can be compressed and pushed through the pipe. In 2013, approximately 89% of natural gas was transported by pipeline over land whereas another 10% was converted to liquefied natural gas (LNG)[559] and shipped by sea; a small amount (~1%) was also compressed and liquefied and sent by rail or truck in the USA. Bradshaw and Boersma[558] argue that the "materiality of natural gas" makes it fundamentally different from oil, resulting in most natural gas being consumed in the region in which it is produced; as such, natural gas is frequently organized around regional markets rather than global markets. Indeed, much of the pipeline trade has been concentrated in Europe. After the oil embargo in the 1970s, European countries, in particular, sought to increase their use of natural gas so as to diversify their energy markets and reduce their dependence on oil. A natural supplier was the Soviet Union with it massive Siberian gas fields[560]. The 1980s saw the construction of a large pipeline network that would bring gas from the Soviet Union to Western Europe, despite the objection of the United States that warned of increasing Soviet leverage over Europe[561]. A number of European countries are also highly dependent on pipelines that bring gas across the Mediterranean via Tunisia to Italy and via Morocco to Spain. Such regional pipeline networks, however, create forms of energy interdependence that make subsequent diversification difficult owing to the high investment costs upfront for constructing a pipeline that then locks in exporting and importing countries. As such, owing to the particular characteristics of natural gas, three large regional gas markets have emerged[558] – a North American market, a European market, and an Asia-Pacific market.

China's increasing demand for energy resources has also affected natural gas production and pipeline construction. Turkmenistan, in particular, signed a production sharing agreement with China's National Petroleum Company (CNPC) in 2007 to explore and develop gas fields in Turkmenistan; with the construction of the Central Asia–China Gas pipeline, Turkmenistan has shifted its gas exports away from Russia toward China[562] with the pipeline coming online in 2009. At the same time, Russia has also expanded its pipeline infrastructure to export gas to China; at the end of 2019, Russia began to deliver natural gas from its Siberian gas fields to China's northeast, running from Heilongjiang on the Chinese–Russian border to Jilin and Liaoning, which ostensibly would help China's transition away from coal[563].

Based on the British Petroleum (BP) global dataset[10], the global natural gas reserves are estimated as ~200 trillion cubic m, from which 2% (about 4 trillion cubic m) was extracted in 2019 (Figure 5.1). The global production of natural gas has continuously increased since the early 1970s (Figure 5.2; data from BP global dataset[10]), with an annual increase rate of 56.6 BCM per year. During the late 1980s and 1990s, Russia was the largest natural gas producer, but since 2009, the USA overtook Russia (Figure 2.5). In 2018, US natural gas production (21.5% of global natural gas production) was followed by Russia (17.3%), Iran (6.2%),

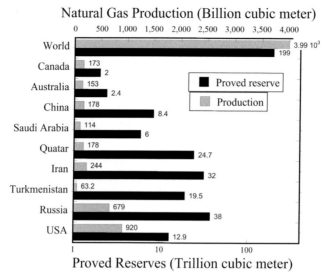

Figure 5.1 The distribution of global natural gas proves reserves (trillion cubic m) and annual production (BCM) in 2019. Data from BP global dataset[10].

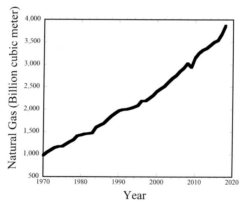

Figure 5.2 Global natural gas production over time. Data from BP global dataset[10].

Canada (4.8%), Qatar (4.5%), and China (4.2%) (data from BP global dataset[10]). In the 2000s, new gas reservoirs were discovered under the Eastern Mediterranean Sea; these reservoirs are also likely to transform the local economies and domestic energy markets of Israel, Lebanon, and Syria.

Yet, the biggest story of the twenty-first century has been the expansion of unconventional natural gas production, beginning in the USA around 2005. What is now known as the "shale revolution" has fundamentally altered US natural gas production as well as world global energy markets (see Section 5.2.3). The widespread use of hydraulic fracturing coupled with horizontal drilling has allowed US companies to extract gas from shale rock, resulting in the USA becoming a net exporter of natural gas in 2019[558]. The International Energy Agency (2011)[564] has called this period the "Golden Age of Gas" as demand for natural gas, especially from unconventional gas, coupled with lower gas prices, was expected to lead to an increase in natural gas use in North America, the Middle East, India, China, and Australia.

In 2018, global natural gas production (3.3 billion tons of oil equivalent) consisted of 23.9% of global energy production, followed by coal (27.2%) and oil (33.6%)[10]. The shale revolution has caused a major shift in the energy source for electricity in the USA; while coal has been the predominant source for the electricity sector during most of the twentieth century (~50% in 2000), the decline of coal and rise of shale gas during the first two decades of the twenty-first century have made natural gas the major source of the US electricity sector (36% of electricity production was from sources of natural gas in 2018). Much enthusiasm accompanied the development of natural gas, as it was heavily promoted as an alternative to dirtier coal and was considered to be a bridge fossil fuel for addressing the climate crisis; furthermore, natural gas was considered to be an important alternative fuel for electricity generation, especially for countries that had growing urban populations and increasing energy demand[564].

The US Energy Information Administration (EIA) estimates that in 2019, US dry shale gas production was about 740 BCM, and equal to about 75% of total US annual dry natural gas production. While the USA has been at the forefront of developing unconventional gas reserves, shale gas is also found across the globe, including in China, the UK, Mexico, Canada, France, Poland, Brazil, Argentina, and Algeria. Outside of the USA, beginning with its 12th Five-Year Plan (2011–2015), China began to prioritize shale gas development[565]. Indeed, China has one of the largest global shale gas resources; estimates suggest natural gas volumes from 12.8 trillion cubic m to 31.2 trillion cubic m[16, 104–106, 566]. In 2017, the proven shale gas reserves in China approached about 1 trillion cubic m, and shale gas production increased from 25 million cubic m in 2012 to more than 9 BCM in 2017, reflecting the rapid development of shale gas exploration mostly in the Sichuan Basin in in China[478]. In 2020, China produced 13–17 BCM, with annual production goals predicted of up to 100 BCM by 2030[567, 568].

Natural gas has been extracted from conventional oil and gas wells, shale gas wells, tight sand gas wells, and coalbed methane sites (Figure 5.3). In this chapter we evaluate the water footprint of the different life cycles of natural gas exploration and production that includes exploration (drilling, hydraulic fracturing, water extraction for coalbed methane), transport to gas processing plants, water use for electricity generation in natural gas plants, and the volume and chemistry of flowback and produced water generated from conventional and shale gas wells, as well as its impact on water resources (Figure 5.4). We conclude by discussing some of the regulatory and policy challenges for governing natural gas and the energy–water quality nexus.

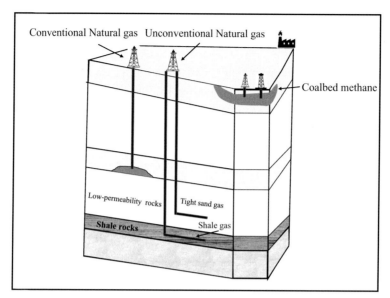

Figure 5.3 Different sources of natural gas exploration. Figure based on US EIA[15]. (A black-and-white version of this figure will appear in some formats. For the colour version, refer to the plate section.)

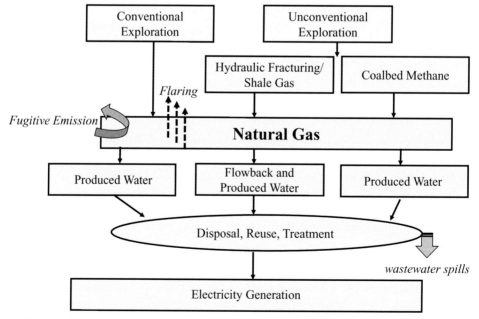

Figure 5.4 Major stages of natural gas exploration and impact on water discussed in this chapter.

5.2 The Origin of Natural Gas

5.2.1 Classifications

Natural gas is commonly defined as the mixture of gases and low-boiling liquid hydrocarbon derivatives, known as condensates[569–571]. The classification of commercial natural gas is commonly based on its origin, hence the distinction between *conventional gas* and *unconventional gas* (Figure 5.5)[569–571]. Conventional gas originates from the maturation of organic matter buried in sedimentary rocks, followed by the migration of the natural gas into permeable reservoir rocks (e.g., sandstone; Figure 5.3), which is then extracted through traditional vertical wells. In some basins, conventional natural gas is associated with crude oil and occurs as either free gas or dissolved gas in petroleum, while in other basins natural gas is solely extracted. Unconventional gas includes natural gas extracted from the source shale rocks (*shale gas*), extracted from low permeable formations (*tight gas*), associated with coal formations (*coalbed gas* or *coalbed methane*), and associated with deep-sea sediments and polar regions (*gas hydrates*)[569–571] (Figure 5.5). Natural gas is also classified by the mechanisms of its origin; natural gas that is generated from bacteriological activities under reducing environments such as shallow aquifers, deep-sea sediments, bottom-lake sediments, and landfills is commonly defined as *biogenic gas* or *bacterial gas* as opposed to *thermogenic gas* that originated in deep sedimentary basins from the maturation of organic matter. Natural gas can also be defined as *abiotic gas*, typically in geothermal and high-temperature geological settings[572, 573] (Figure 5.5). The distinction between the sources of natural gas is critical for evaluating the possible mechanisms by which natural gas may occur in groundwater resources and for evaluating the

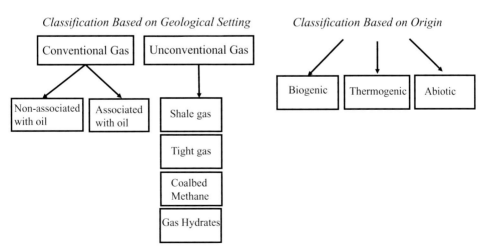

Figure 5.5 Classification of natural gas based on its geological settings and origin. Modified from Faramawy et al. (2016)[571].

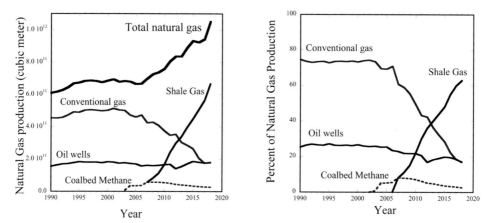

Figure 5.6 Natural gas production from different sources in the USA over time, expressed as annual production rates (cubic m) and percentage of different sources relative to gross natural gas production in the USA. Data retrieved from US EIA[574].

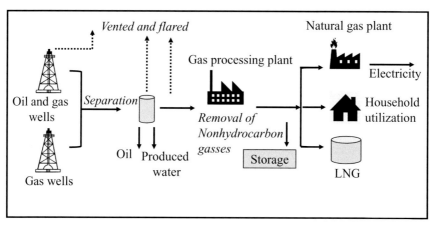

Figure 5.7 Major stages of natural gas production from gas wells, oil and gas wells, and delivery to potential users. Modified from US EIA[15].

source of the fugitive emission of methane into the atmosphere, and thus the impact of natural gas exploration on the rise of atmospheric methane.

In the USA, natural gas is extracted from conventional natural gas, conventional oil wells, shale gas wells, and coalbed methane. The relative proportions of natural gas production from these sources have changed over time; during the twentieth century and early 2000s, gas wells (~75%) and oil wells (~25%) were the predominate sources for natural gas production in the USA, with steady production rates over time (Figure 5.6). Yet since 2005, the rapid rise of shale gas production was associated with the decline of conventional natural gas production, but nonetheless resulted in an overall increase in gross natural gas production in the USA (Figure 5.6). The different stages of natural gas production are illustrated in Figure 5.7; these include extraction from natural gas wells, gas separation (from oil, water, and heavy hydrocarbons), transport, storage, and delivery to users.

5.2.2 The Chemistry of Natural Gas

Natural gas contains hydrocarbon gases (alkanes), including methane (C1-CH_4), ethane (C2-C_2H_6), propane (C3-C_3H_8), butane (C4-C_4H_{10}), pentane (C5- C_5H_{12}), and hexane (C6-C_6H_{14}), as well as nonhydrocarbons gases such as sulfur, helium, nitrogen, hydrogen, carbon dioxide, and water. The abundances of methane relative to other hydrocarbon gases defines the type of natural gas: *dry natural gas* is defined by a high abundance ($>85\%$) of methane, whereas *wet natural gas* contains high abundances of ethane, propane, butanes, and pentanes, which are typically removed from natural gas before supply to users (Figure 5.7). Gas wetness is defined by the ratio of the "wet" components (C2–C6) to the total alkanes, expressed in percentage, or by the ratio of methane to other alkane components (e.g., C1/[C2+C3]). Wet gas is typically associated with crude oil ("Associated with oil"; see Figure 5.5), whereas dry gas with the predominance of methane is typically not associated with oil. Another important characteristic of natural gas is the isotope composition of carbon ($^{13}C/^{12}C$ expressed as $\delta^{13}C$) and hydrogen ($^2H/^1H$ expressed as δ^2H), commonly in methane, but also in the other hydrocarbons. The combined proportions of hydrocarbons and the isotope compositions have been used to determine the origin and evolution of natural gas[575–582].

The origin of natural gas is derived from the burial and evolution of organic matter. Organic matter can be accumulated in low-oxygen conditions in both marine (deep sea) and terrestrial (e.g., swamps) environments and be preserved over geological time. During early stages of organic matter maturation, bacterial processes produce exclusively dry natural gas with a predominance of methane (C1) that has a distinctively low (negative) $\delta^{13}C$-CH_4 isotope fingerprint. The bacterial methanogenesis process is predominantly through two major process of CO_2 reduction generated from biodegradation and fermentation. At that stage of evolution, natural gas is defined as *biogenic methane* or *bacterial methane* and is characterized by a high ratio of methane to other alkanes and low $\delta^{13}C$-CH_4 induced from the selective removal of ^{12}C from CO_2 during the methanogenesis (with higher carbon isotope fractionation during the CO_2 reduction process)[575–581]. Figure 5.8 illustrates the evolution of natural gas with organic matter burial and the relationships between the gas wetness and carbon isotope variations along the different stages of organic matter maturation. The bacterial methane stage is restricted to burial up to ~80°C.

At higher depth, pressure, and temperature, the buried organic matter that is preserved in both marine and terrestrial sedimentary rocks is decomposed and converted to kerogen. The thermal degradation of kerogen generates hydrocarbons, including crude oil and wet natural gas that includes methane and the other alkanes (C2-C5). The thermal degradation of kerogen also generates hydrocarbons with a

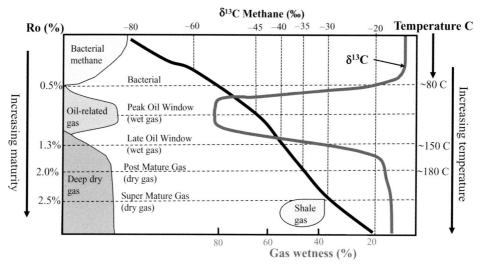

Figure 5.8 Evolution paths of natural gas formation from organic matter buried in sedimentary rocks. The rise of temperature with depth increases the maturity of the organic matter (defined by Ro percent for kerogen maturity) and is associated with a gradual increase of the $\delta^{13}C$ of methane that is generated (black line) and also changes the gas wetness (gray line) that reflects the relative proportions of methane relative to other hydrocarbons (C1/[C2+C3]). The evolution with depth and temperature results in the formation of different types of natural gas: shallow bacterial gas with low $\delta^{13}C$ and high C1/[C2+C3], wet gas associated with peak oil with higher $\delta^{13}C$ and lower C1/[C2+C3], and mature gas with a higher (positive) $\delta^{13}C$ and lower C1/[C2+C3] ratio. Figure was modified from Schoell (1980)[575] and Schoell (1983)[582].

relatively high (positive) $\delta^{13}C$-CH$_4$ isotope fingerprint. Consequently, the thermogenic gas that is generated under these conditions of the "peak oil window" (Figure 5.8) will have higher $\delta^{13}C$-CH$_4$ values (>−60‰) and lower C1/[C2+C3] ratios that are different from the composition of bacterial methane. The relationship between the gas wetness and stable carbon isotope ratios is one of the tools used to identify the origin of natural gas, known as the Bernard Plot (Figure 5.9), which was based on Bernard et al. (1977)[583]. Increasing depth, pressure, and temperature causes further thermal maturation of kerogen, which progressively generates a larger fraction of methane with higher (positive) $\delta^{13}C$-CH$_4$ values (Figure 5.8). Consequently, mature thermogenic gas evolves into dry gas (i.e., high C1/[C2+C3] ratios) with a high (positive) δ ^{13}C-CH$_4$ isotope fingerprint (Figure 5.9). Based on over 20,000 natural gas analyses, Milkov and Etiope (2018)[579] have shown that $\delta^{13}C$-CH$_4$ values of biogenic gas vary between −80‰ to −60‰ (peak at −70‰), whereas $\delta^{13}C$-CH$_4$ values of thermogenic gas vary between −60‰ to −20‰ (peak at −40‰[579]; Figure 5.9). Shale gas extracted from shale and other low-permeability rocks could have this distinctive mature gas composition, as shown

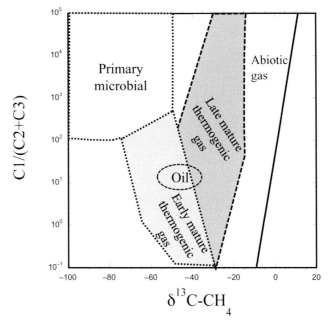

Figure 5.9 The composition of different natural gases in the framework of the ratio of methane to ethane and propane (C1/[C2+C3]) vs. the stable carbon isotope ratio in methane (δ^{13}C-CH$_4$), commonly defined in the literature as the Bernhard Plot after Bernhard et al. (1977)[583]. The separation of fields was based on revaluation of the natural gas data provided by Milkov and Etiope (2018)[579].

for shale basins with large gas production such as the Marcellus and Haynesville shale formations that contain distinctive late–mature thermogenic gas[584]. Coalbed gas (also known as coalbed methane) is found in coal seams and is generated during the coalification process (see Chapter 3). Biogenic coalbed gas can be generated during early stages of coal maturation, while further coalification under higher temperature and pressure would generate thermogenic coalbed methane. Methane, ethane, carbon dioxide, and nitrogen are the main constituents of coalbed gases, with a general lack of heavier hydrocarbon species (i.e., dry gas) in gases produced from shallow levels and more mature coals[585].

The original composition of natural gas can be modified due to secondary processes, including (1) gas migration from the original source rock to the geological trap reservoirs, which is common for most conventional gas resources; (2) migration of natural gas to shallow geological formations overlying the original source rocks; (3) mixing between bacterial and thermogenic sources that would change the original chemical composition; and (4) degradation of organic matter inducing secondary methanogenesis with a different geochemical fingerprint and oxidation of natural gas. Several studies have demonstrated that intrusion of meteoric water into deep geological formations will induce methane oxidation and formation of CO_2 enriched in ^{13}C that is converted to bicarbonate to generate formation water with high bicarbonate (alkalinity) with distinctively high $\delta^{13}C$[578–580]. Likewise, degradation of crude oil will generate secondary methane associated with ^{13}C-rich CO_2, and consequently ^{13}C-rich bicarbonate in co-existing formation water[586]. Overall, the modification processes might mask the original composition of natural gas (Figure 5.9) and introduce more challenges in the evaluation of its source and migration. Delineating the original source of natural gas becomes critical in evaluating the possible methane contamination of groundwater resources and the ability to distinguish between natural occurring methane flow in groundwater and anthropogenic contamination from the leaking of shale gas or conventional natural gas wells to the adjacent aquifer. The role of methane contamination on water resources is discussed in Section 5.6.

5.2.3 Shale Gas and Hydraulic Fracturing

While conventional gas extracts natural gas trapped in permeable geological formations using vertical wells, unconventional shale gas and tight gas explorations tap gas entrapped in low-permeable rock formations, such as organic-rich shale rocks. The development of horizontal drilling through the impermeable rocks combined with hydraulic fracturing technology has resulted in the production of natural gas from deep and low-permeable rocks. The hydraulic fracturing technology involves high-pressure injection (e.g., 9,500 psi in the Marcellus Shale[587]) of water and frac chemicals into organic-rich shale rocks and the formation of micro-fractures along the horizontal well. The reduction of pressure releases natural gas and flowback water from the shale matrix back to the well and then to the surface (Figure 5.10). The total organic contents (TOCs) in shale gas reservoirs require concentrations above 2% to become economically viable.

Since 2008, these technological developments have resulted in exponential growth of natural gas well drilling in the USA, particularly in basins with abundant natural gas resources such as the Barnett, Haynesville, Fayetteville, Woodford, Utica, and Marcellus shale formations (Figure 5.11). The rise of shale gas production in the USA has changed the landscape of the energy sector in the USA, with a continuous increase of shale gas production (Figure 5.6). From zero production prior to 2005, shale gas production increased to 716 BCM (25.3 trillion cubic feet) by 2019, which was equivalent to 75% of total US natural gas production (data from EIA[574]). While the rise of shale gas was associated with the general trend in the decline of conventional natural gas in the USA (Figure 5.6), it increased overall natural gas production by 150% as compared to the early 2000s, prior to unconventional shale gas development. In early 2019, major shale gas

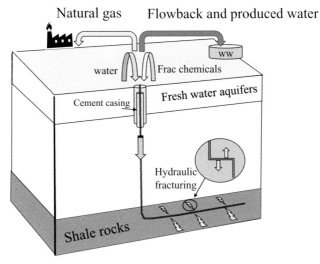

Figure 5.10 Schematic illustration of the configuration of shale gas wells and the process of hydraulic fracturing. The injection of high-pressure water and frac chemicals into separated sections along the horizontal sections of the well generates micro-fracturing in the low-permeable shale rocks. Imbibition of the injected water releases entrapped natural gas and water within the shale matrix and upon reduction of the well pressure the natural gas and water are tapped to the surface. (A black-and-white version of this figure will appear in some formats. For the colour version, refer to the plate section.)

production in the USA occurred in the Marcellus (34%), Permian (13.7%), Haynesville (12.7%), and Utica (11%) basins (Figure 5.12). The intensification of the hydraulic fracturing process was accompanied by the longer length of horizontal drilling. For example, in the Marcellus Play, the average horizontal well length increased from 1,200 m per well in 2011 to 1,700 m per well in 2016. The increasing of the horizontal well length resulted in higher volumes of water used for hydraulic fracturing[24]. The water use intensification associated with the rapid rise of shale gas development and hydraulic fracturing is further discussed in this chapter (see Section 5.3).

The EIA estimated that the proven technically recoverable US dry natural gas resources in 2018 was 13.4 trillion cubic m. The relative proportion of shale gas reserves increased over time to 9.7 trillion cubic m, which is equivalent to 72% of the potential recoverable dry natural gas in the USA (Figure 5.13)[588].

Organic-rich shale formations are not restricted to the USA, and globally, numerous shale gas basins hold a large potential for shale gas exploration. Figure 5.14 presents the global distribution of shale plays. The EIA (2013) report[363] estimated that the global technically recoverable shale gas resources are 5.9 trillion cubic m (207 trillion cubic feet), from which China (15.3% of

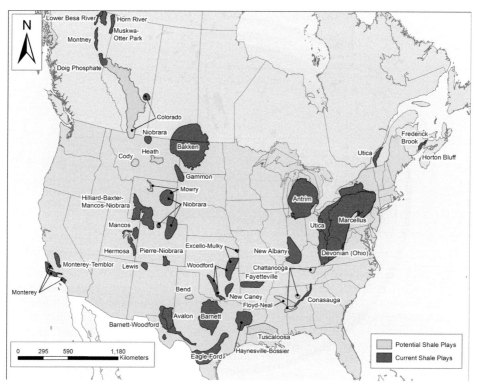

Figure 5.11 Distribution of shale gas plays across the USA. Map was generated by A.J. Kondash (Duke University) based on US EIA oil and gas maps website[364]. (A black-and-white version of this figure will appear in some formats. For the colour version, refer to the plate section.)

worldwide potential shale gas reserves), Argentina (11%), Algeria (10%), the USA (9%), Canada (8%), Mexico (7%), Australia (6%), South Africa (5%), Russia (4%), and Brazil (4%), are the top 10 countries with the largest shale gas potential (Figure 5.15). China has the largest global shale gas reserves with independent higher estimates of 12.8–31.2 trillion cubic m of recoverable natural gas[566]. The Sichuan Basin is the most productive shale gas basin in China, with the Upper Ordovician Wufeng Formation–Lower Silurian Longmaxi Formation currently producing most of the shale gas in China. Four commercialized shale gas fields have been developed in the Sichuan Basin, including Fuling, Weiyuan, Changning and Zhaotong, with production rates growing rapidly. Between 2012 to 2017, shale gas production increased from 25 million cubic m to 9 BCM, a remarkable 360-fold increase[16, 104, 106, 434, 478, 567, 589–592].

While the potential of recoverable shale gas is high in the USA, China, and other countries, the production of shale gas from individual wells in a shale play is

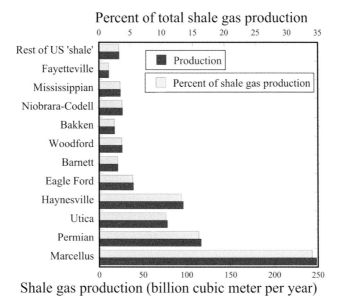

Figure 5.12 Shale gas production in the different shale plays in the USA in 2020. For location of the shale plays, see Figure 5.11. Data retrieved from US EIA dataset on natural gas production in the USA[574].

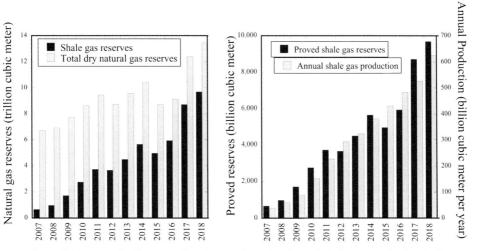

Figure 5.13 Technically proved and annual production of shale gas in the USA. In 2018, the annual shale gas production consisted of 6.5% (15-fold) of the total proved shale gas resources in the USA. The proved resources of shale gas consist of only ~18% of the total natural gas potential in the USA. Data from US EIA (2020)[588].

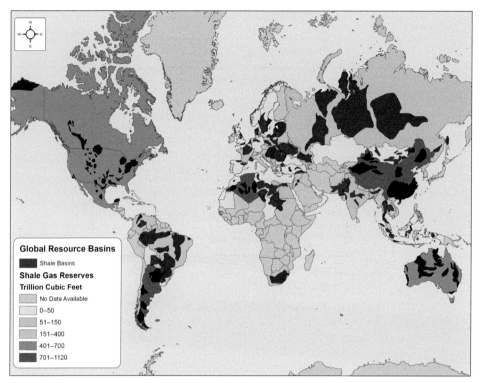

Figure 5.14 Global distribution of shale gas reserves sorted by basin potential (in trillion cubic feet). Data retrieved from US EIA (2013) report[363]. Map was made by A.J. Kondash (Duke University). (A black-and-white version of this figure will appear in some formats. For the colour version, refer to the plate section.)

limited, with a typical fast decline of natural gas production rates at about three to six months after the initial hydraulic fracturing. The decline of natural gas production over time defines the Estimated Ultimate Recovery (EUR) – the volume of economically natural gas expected from a well by the end of its producing life. The fast declining rate of natural gas (Figure 5.16) requires installation of multiple wells in order to generate economical viable production of shale gas. Since the beginning of shale gas development in the USA, it is estimated that about 100,000 wells have been installed to tap shale gas out of a total of 546,442 gas wells operated in 2018[593]. The high number of shale gas wells is one of the key factors for the environmental impact of shale gas development and possible effects on water resources, which are discussed in Section 5.5.4.

Water is the key component for extraction of natural gas from low-permeability shale rocks. The combination of high-pressure and frac chemicals that aim to increase the viscosity of the injected water (see more details on frac chemicals in Section 5.4) generates micro fractures in the rocks along the horizontal section of

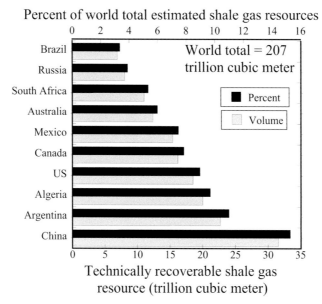

Figure 5.15 Distribution of the top 10 countries with the largest technically recoverable shale gas resources. Data from EIA (2013) report[363].

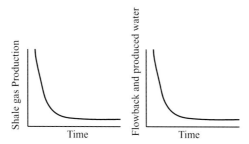

Figure 5.16 Decay of shale gas and flowback and produced water production over time after hydraulic fracturing. The rate of the decay of natural gas production determines the Estimated Ultimate Recovery (EUR) of a shale gas well during its lifetime.

the well and infiltration of the injected water into the shale pores (Figure 5.10). During the leak-off stage, the injected water enters the newly formed fractures and also the shale matrix. After the high-pressure fluid injection, the pressure in the wells reduces and natural gas that is retained into the organic matter is released and collected (defined as the flowback stage). In addition, the injected water blended with formation water within the shale formation is also released, generating flowback water (Figure 5.17). Yet, many studies have shown that the volume of the returned flowback water is significantly lower than that of the volume of the water used for hydraulic fracturing, indicating a large retention of the

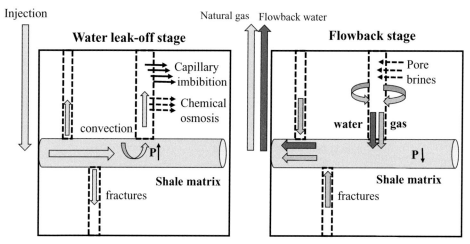

Figure 5.17 Illustration of the different stages of hydraulic fracturing process: (1) the water leak-off stage – the injected hydraulic fracturing water enters the fractures generated in the shale rocks and the shale matrix; (2) the flowback stage – the release of pressure causes natural gas and injected water to blend with pore water (formation water) from the shale matrix to flow to the wells and to the surface.

injected water. The large retention of the hydraulic fracturing water is evident in the USA[23, 24, 360, 594], Canada[595–598], and China[16, 599]. Laboratory experiments using the Horn River Basin shales from Alberta, Canada were conducted to evaluate the mechanism of fluid retention in the shale matrix. These studies suggest that the retention of the hydraulic fracturing water is due to a combination of factors, including (1) capillary-driven imbibition of the injected water through the pore networks in the shale; (2) osmotic effects induced from the large salt difference between the injected low-saline water and the high-saline formation water in pore spaces; and (3) the thickness of the electric double layer around clay mineral that controls the water imbibition into the shale matrix[595–598, 600–603]. In addition, other studies have shown that higher water imbibition (Figure 5.17) increases the moisture content in the shale, which results in reducing the methane adsorption capacity onto kerogen in the shale and promotes its release to the well[604, 605]. Consequently, the ability of water to enter into the shale matrix is the key to the reduction of methane adsorption capacity to become extractable as shale gas. A field experiment in Sichuan Basin has shown that hydraulic fracturing with saline water, rather than low-saline water, can cause a decrease in shale gas production and an increase in the volume of the flowback water from the shale gas wells, suggesting that the salinity of the injected water plays a major role on the imbibition capacity, and consequently also on the extraction capacity of shale gas[478]. In Section 5.5.3 we further evaluate the chemistry of flowback water and its interaction with the shale rocks.

5.3 Water Use for Natural Gas Production

The production of natural gas involves several stages (Figure 5.7) that include: (1) extraction of natural gas from conventional or unconventional wells; (2) separation from co-existing oil and water; (3) processing for removal of impurities (e.g., H_2S gas) and nonhydrocarbon species from the raw natural gas; (4) processing and separation of the hydrocarbons in which methane (C1) and ethane (C2) are used directly as "natural gas," propane (C3) and butane (C4) for Liquefied Petroleum Gases (LPG), and pentane (C5) and heavier hydrocarbon components for natural gas condensates; (5) transport of the different natural gas components to different storage capacities; and (6) delivery to users. The separated heavier hydrocarbons are commonly used for different industrial utilities, while methane and ethane are used exclusively for heating and electricity production. The major water utilization stages that are discussed in this chapter are the recovery stage, including drilling (for both conventional and unconventional wells), and hydraulic fracturing (for unconventional wells), processing and storage, and cooling gas plants for electricity production. For calculation of the intensity of the water use and wastewater production associated with natural gas production we use the energy capacity of natural gas following the formula outlined by Kondash and Vengosh (2015)[23] that includes the natural gas itself (one cubic m of gas equals 0.0286 GJ), residuals of Natural Gas Liquids (NGLs) (0.8 L of NGL for one cubic m of natural gas of 0.0205 GJ), and residuals of oil (10 BBL equals 0.01821 GJ). The water use intensity evaluation considers also the energies of NGLs and petroleum occurrence in natural gas wells. For natural gas occurrence in crude oil wells see discussion in Chapter 4.

5.3.1 The Recovery Stage: Water Use for Drilling and Hydraulic Fracturing

Installation of conventional natural gas wells requires a relatively small volume of water, estimated at 300–400 m³ per well for drilling (water intensity of 0.7 L/GJ) plus 27–37 m³ per well for cement[19, 606] (Table 5.1). In contrast, the hydraulic fracturing process requires a much larger water volume to inject into the well and generate fractures in the shale rocks. During the early stages (2011–2012) of hydraulic fracturing in the USA, the water use for hydraulic fracturing varied between 5,000 m³ per well (Permian Basin), 24,000 m³ per well (Marcellus Shale), and up to 33,000 m³ per well (Haynesville Shale) with an overall (i.e., life time) water intensity of 4–8 L/GJ[19, 23, 580, 594, 606]. Yet the intensification of the hydraulic fracturing process has increased the water use per well; in 2016 the water use ranged from 29,000 m³ per well (Marcellus Shale) to 42,000 m³ per well (Permian Basin) with a water intensity range of 5–20 L/GJ[24] (Figure 5.18; Table 5.1). The intensification of the water use for hydraulic fracturing processes was associated with the increasing length of the horizontal shale gas wells.

Table 5.1. *Summary of the water intensity values of water withdrawal and consumption for different sources and processing of natural gas. Water intensity values are reported in L/GJ unit (and m³/ MWh for electricity generation)*

Source	Water use per well (m³/well)	Water withdrawal (L/GJ-m³/ MWh)	Water Consumption (L/GJ-m³/ MWh)	Reference
Conventional natural gas extraction				
Drilling	300–400	0.7	0.7	Clark et al. (2013)[606]
Cement	27–37	0.06	0.06	Clark et al. (2013)[606]
Drilling		0.31	0.31	Spang et al. (2014)[48]
Drilling		1.0	1.0	Grubert and Sanders (2018)[29]
Produced water		13.9	19.3	Veil (2015)[375]
Produced water		9.3	9.3	Grubert and Sanders (2018)[29]
Unconventional shale gas extraction				
Hydraulic fracturing, USA (2011–2014)	4,000–33,000	4–8	4–8	Kondash and Vengosh (2015)[23]
Hydraulic fracturing, USA (2016)	29,000–42,000	5–20	5–20	Kondash et al. (2018)[24]
Hydraulic fracturing, Sichuan Basin, China	34,000–45,000	5–20	5–20	Zou et al. (2018)[16]
Flowback and produced water (first 12 months)		18.1–39.1	18.1–39.1	Kondash et al. (2018)[24]
Natural gas processing				
Dehydration		0.6	0.6	Grubert and Sanders (2018)[29]
CO_2 separation		3.1	2.1	Grubert and Sanders (2018)[29]
H_2S separation		0.073	0.067	Grubert and Sanders (2018)[29]
Hydrostatic pipeline tests		zero	0.1	Grubert and Sanders (2018)[29]
Storage in salt caverns		4.9	4.9	Grubert and Sanders (2018)[29]

Table 5.1. (*cont.*)

Source	Water use per well (m³/well)	Water withdrawal (L/GJ-m³/ MWh)	Water Consumption (L/GJ-m³/ MWh)	Reference
Electricity generation				
Combined cycle		23,780	11	Kondash et al.
Once through		(87.4)	(0.22)	(2019)[234]
Steam turbine		130,000	536	Kondash et al.
Once through		(468)	(1.9)	(2019)[234]
Combined cycle		275	223	Kondash et al.
Recirculating		(1)	(2.8)	(2019)[234]
Steam turbine		1,408	819	Kondash et al.
Recirculating		(5.1)	(2.9)	(2019)[234]
Overall natural gas		9,185	216	Kondash et al.
in the USA (2016)		(35.3)	(0.8)	(2019)[234]
Power Plant, USA,		780	41	Grubert and
2014		(2.8)	(0.15)	Sanders (2018)[29]

For example, in the Marcellus Shale the horizontal length increased from 1,200 m per well to 1,700 m per well. In some basins, the increase of the horizontal length was also associated with water intensity per the length of the horizontal well. For example, in the Permian Basin, the water intensity increased from 4 m³ per m horizontal length in 2011 to 30 m³ per m horizontal length in 2016. Similar water use volumes and intensities were reported from shale gas exploration in the Sichuan Basin, China, showing an increase in water use for hydraulic fracturing from 20,000 to 34,000 m³ per well in 2012–2016[16, 599, 607] to 45,000 m³ per well in 2018[478] (Table 5.1). The variations of the water use for hydraulic fracturing in different shale basins in the USA and China are presented in Figure 5.19.

One of the difficulties in evaluating the water intensity of shale gas exploration is evaluating the volume of natural gas that is generated from each well over time. While data for conventional natural gas have been established for a long time, the relatively new development of shale gas limits the ability to quantify long-term natural gas exploration. Furthermore, the production of natural gas from shale gas wells decreases rapidly after a few months of operation following hydraulic fracturing[360], and therefore, the prediction of life-long gas production through modeling the estimated ultimate recovery (EUR) is challenging. Consequently, the water intensity that is normalized to the volume of natural gas that was produced during the first 12 months after hydraulic fracturing is more accurate to evaluate the water intensity. Parallel to the increased shale gas production in the USA, the

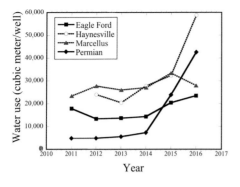
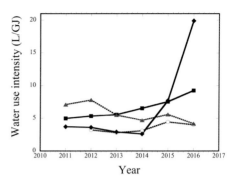
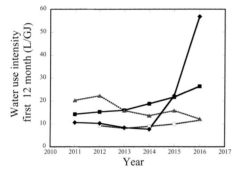

Figure 5.18 Changes of water use per well, the overall water use intensity, and the water use intensity for the first 12 months after hydraulic fracturing for major shale basins in the USA. Based on data reported in Kondash et al. (2018)[24].

water intensity (as normalized for the first 12 months of shale gas production) has also increased in some of the shale gas basins over time (Figure 5.18)[24].

Given the high water use of hydraulic fracturing relative to conventional natural gas exploration, much of the public debate on hydraulic fracturing has been related to the relatively large volume of water used for hydraulic fracturing[26]. This is due to the common utilization of high-quality potable water for hydraulic fracturing. As such, in water-scarce regions, large-scale shale gas development might compete with the agricultural and domestic sectors. Kondash et al. (2018)[24] have shown that many of the global shale basins are located in arid zones where water stress is high or extremely high, including important shale basins in the western USA, Northern Africa, South Africa, India and Pakistan, northwestern China, and Australia (Figure 5.14).

The actual water use for hydraulic fracturing, however, was found to be small in comparison to the overall industrial water used in the USA[23], and even on a regional scale as compared to the overall water availability in the Sichuan Basin in China[599]. Nonetheless, large-scale development of shale gas (and tight oil) in areas known to have severe water scarcity like the Permian Basin in western Texas could

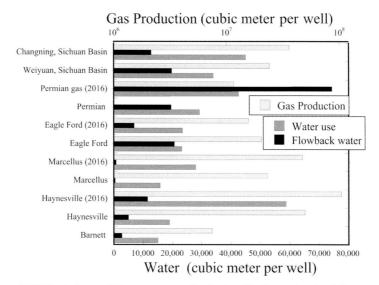

Figure 5.19 Variations of the water use for hydraulic fracturing in different shale basins in the USA and China as compared to the natural gas and flowback and produced water (FPW) production during the first 12 months of operation after hydraulic fracturing. Data for some shale plays in the USA reflect early stages (2011–2012) relative to later stages (2016) of shale gas development. Data retrieved and integrated from Kondash et al. (2017)[360], Kondash et al. (2018)[24], Zou et al. (2018)[16], and Liu et al. (2020)[478].

maximize the water utilization for hydraulic fracturing and further exacerbate the depletion of limited local water resources[24]. Using alternative water sources for hydraulic fracturing is therefore an important mitigation strategy for reducing use of potable water scarcity. Reusing flowback and produced water generated from shale gas wells for hydraulic fracturing of new wells has been conducted in the Marcellus Shale in the eastern USA and in the Changning gas field in Sichuan Basin in China [478]. The chemistry and the salinity of flowback and produced water co-extracted with shale gas is the key factor in recycling shale gas wastewater and is discussed in Section 5.5.3.

5.3.2 *Natural Gas Processing*

Following the recovery stage, natural gas is processed through different stages for separation (from oil and water), purification, transmission through compressor stations and high-pressure pipelines, storage (underground, liquefied, tanks), and supply for use (Figure 5.7)[606]. The storage of natural gas in the USA includes depleted natural gas or oil fields, shallow confined aquifers, and mostly salt caverns, where natural gas is injected into sealed salt domes or salt bedded formations[608]. All

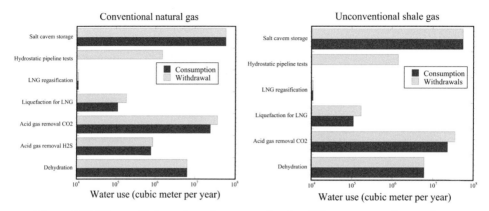

Figure 5.20 Water withdrawals and consumption for different stages of natural gas processing, transport, and storage for conventional natural gas (upper plot) and unconventional natural gas (lower plot).Data retrieved from Grubert and Sanders (2018)[29].

of these processing stages consume water. Grubert and Sanders (2018)[29] presented data for the volumes and intensities of the water withdrawal and consumption associated with the different stages of processing, transport, and storage of natural gas in the USA (Figure 5.20). The data generated in 2014 were from the processing and transport of natural gas derived from both conventional and unconventional wells. The data show that most of the water is consumed and that the majority of the water is used for removal of CO_2 (27%) and storage in salt caverns (65%), with an overall water consumption of 100 million cubic m for natural gas from conventional gas wells and 97 million cubic m for natural gas from unconventional gas wells. The higher water use for conventional gas is due to the water use for removal of H_2S (Figure 5.20), which is not as common in shale gas and thus involves less water use. The comparison between the water use for processing natural gas from conventional and unconventional wells was made for the year 2014 where the overall conventional natural gas extraction (1.88×10^{10} GJ) was almost equal to that of unconventional natural gas in the USA (1.78×10^{10} GJ) (data from Grubert and Sanders, 2018[29]). Overall, the water intensity of natural gas processing is very small, at several orders of magnitude lower than the recovery stage (Table 5.1).

5.3.3 Electricity Production from Natural Gas

The rise of shale gas and overall natural gas production (Figure 5.6) has resulted in increasing utilization of natural gas for the electricity sector in the USA. In 2019, out of 4.12 trillion kWh of electricity that was generated in the USA, 1,582 billion kWh (38.4%) was generated from natural gas, while electricity generated from coal combustion was only 966 billion kWh (23.5%). The increase in natural gas utilization has been accompanied by a decline in coal utilization over the last 20 years in the USA; in the early 2000s, coal contribution to the electricity sector was about 50% while natural gas was only 20%[609].

By far, water use for electricity production, mostly for cooling natural gas plants, has the highest water footprint in all of the life cycles of natural gas production (Table 5.1). Similar to coal, much of the water use for electricity production is for cooling gas plants, while the different types of power plant technology and the cooling systems (i.e., open vs. recycling) control the intensity of the water use (figure 5.21). The majority (81%) of natural gas production in the USA is derived from combined cycle plants, which consist of a combination of gas turbines and steam turbines built to spin generators and generate electricity. Gas combustion turbines that are not connected to the combined cycle plants make up only a small percentage (9%) and are typically used to provide instant balancing to the power grid in the USA[234]. In order to cool down the steam, which is used to spin steam turbines, plants use dry cooling (i.e., use air instead of water), recirculating cooling (collect water from a source to cool and condense the steam in either ponds or towers for reuse), and once-through cooling (withdraw and return of the water from and to a source after cooling). These modes of cooling operations result in different withdrawal and consumption proportions; while recirculating cooling withdraws only a small fraction (31%) of the overall water withdrawal for cooling plants, it consumes more water (90%) due to evaporation and water lost during recycling. In contrast, once-through cooling involves a much larger withdrawal (69%) but smaller fraction (10%) of the overall consumption[234].

The combination of the fuel technology (combined cycle vs. gas turbines) and cooling system (recirculating vs. once through) determines the water intensity of electricity production (Table 5.1). The combination of natural gas combined cycle plants (NGCC) with recirculating cooling systems has the lowest water withdrawal and consumption values among the other combinations (Figure 5.21; Table 5.1). Since the majority of natural gas plants in the USA use the combination of NGCCs with recirculating cooling systems, Kondash et al. (2019)[234] estimated water intensities of 9,185 L/GJ (35.3 m³/kWh) and 216 L/GJ (0.8 m³/kWh) for the withdrawal and consumption for overall electricity production in the USA for 2016. Grubert and Sanders (2018)[29] estimated overall lower water intensities of 780 L/GJ (2.8 m³/kWh) and 41 L/GJ (0.15 m³/kWh) of withdrawal and consumption, respectively (Table 5.1). The overall annual water withdrawal and consumption volumes for electricity generated from natural gas in 2014 for the USA were estimated at 18 BCM and 0.95 BCM, respectively[29].

Both water withdrawal and consumption for cooling natural gas plants are lower than that of the respective water use for cooling coal plants in the USA

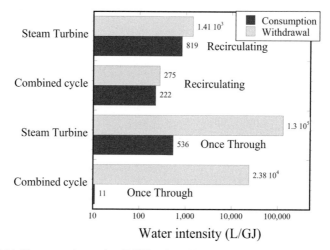

Figure 5.21 The water intensity (L/GJ) of cooling gas plants based on the combinations of the plant type and the cooling system. The combination of combined cycle plant with recirculating cooling system has the lowest water consumption and withdrawal footprint. Data retrieved from Kondash et al. (2019)[234].

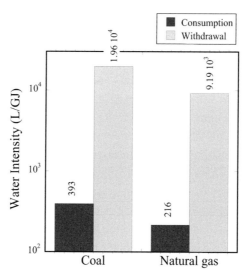

Figure 5.22 The intensities (L/GJ) of water withdrawal and consumption for electricity production from natural gas as compared to coal in the USA. Data from Kondash et al. (2019)[234].

(Figure 5.22). Therefore, the shift from coal as the predominant source in the US electricity sector towards increasing utilization of natural gas has reduced the water intensity for electricity production. In spite of the intensification of shale gas development and the relatively high volume of water used for hydraulic fracturing,

Kondash et al. (2019)[234] have shown that the overall annual water withdrawal (874 BCM) and consumption (175 BCM) for coal in 2016 was larger than that for of natural gas (45.5 BCM and 1.07 BCM). They suggested a reduction of 40 m^3 in water withdrawal and 1 m^3 for water consumption for every MWh of electricity that has been generated with natural gas instead of coal[234].

5.4 The Quality of Frac Water

One of the major public concerns related to hydraulic fracturing is the toxicity and human health risks associated with the man-made chemicals added to the frac water that is injected into shale gas wells as part of the hydraulic fracturing process. During the early stages of unconventional shale gas development, the content of the chemical additives known as "frac chemicals" used for hydraulic fracturing was not revealed by the oil and gas industry due to proprietary issues, which led to increasing public mistrust about whether the chemicals that are used for hydraulic fracturing could pose a risk to human health.

In most cases, the water used for hydraulic fracturing is local surface water and/ or groundwater, and thus the inorganic constituents would be in most cases low and insignificant. In some cases, however, the FPW water is recycled to be used as frac water and thus the salinity and inorganic chemicals would reflect the relative proportions between the recycled FPW and fresh water. For example, in the Marcellus Shale, hypersaline FPW is reused almost entirely and thus the water quality would be similar to that of the FPW (see Section 5.4.5). In Sichuan Basin, some gas fields used only a small fraction of FPW, which is added to the hydraulic fracturing water forming brackish water (chloride = 500 mg/L), while in other fields 100% FPW (chloride = 18,000 mg/L) is used for hydraulic fracturing[478]. Therefore, the initial water salinity of the water used for hydraulic fracturing can play an important role in the water quality of frac water.

There are several types of fluids commonly used for hydraulic fracturing. The first is the *slick water fluid system*, which is commonly used to frac low-permeability and high brittle rock formations with little elastic deformation. This type of fluid is commonly used in major shale gas plays, such as the Marcellus Shale, Barnett Shale, Eagle Ford, Hayesville, Utica/Point Pleasant, and many other low-permeability reservoirs. The slick water fluid is low-viscosity fluid, and the combination of high-volume, high-pressure, and high rate of fluid injection creates a complex fracture network in the shale rocks that increases the shale permeability and enables the release of natural gas. Both fresh water and saline water can be used to generate slick water fluid[587]. The second type of frac fluid is *cross-linked gel* or the *linear gel fluid system*, which is a heavy viscous fluid. The high viscosity fluid is used to carry and place the proppants into the fractures in the rock formation. Cross-linked gel fluids are typically applied in ductile geological formations with higher permeability, such as the oil windows of various shale plays like Eagle Ford and Bakken shales[587]. This method requires a large volume of proppant (sand) that is carried by the high viscosity fluid. In some shale plays a hybrid fluid system is used, in which slick water is used first to pump at a lower sand concentration, followed by cross-linked or linear gel to pump at a higher sand concentration to maximize near wellbore conductivity[587]. Overall, the principle of using frac fluids depends on the rock characteristic; for low-permeable and brittle shale rocks, slick water is preferable, whereas for rocks with higher permeability and less brittle characteristics, higher viscosity linear gel and cross-link gel are preferable for hydraulic fracturing.

Stringfellow et al. (2017)[610] have identified 81 chemical additives commonly used for hydraulic fracturing, in which 55 were organic, and 27 of these are considered readily or inherently biodegradable. Seventeen of the chemical additives have a high theoretical chemical oxygen demand and are used in concentrations that present potential treatment challenges[610]. Chen and Carter (2017)[611] conducted a survey of 5,071 shale gas wells from the Marcellus Shale and identified 517 chemical additives, of which 96 were inorganic compounds, 358 were organic species, and the remaining 63 were not identified. Based on Stringfellow et al.'s (2017)[610] classification, the major groups of chemical additives used for hydraulic fracturing are: (1) *gelling and foaming components* used to increase fracturing fluid viscosity, allowing for better proppant suspension and transport into developed fracture used for cross-linked gel or the linear gel fluid system; (2) *friction reducers*, which are chemicals such as polyacrylamide that are added to slick water fluid systems for reducing fluid surface tension and that facilitate removal of fracturing fluid from the formation; (3) *crosslinkers*, which are used to bind individual gel polymer molecules together to form larger molecules with higher viscosity and elasticity with better proppant transport compared with linear gels. Crosslinkers frequently used include boron, aluminum, titanium, and zirconium compounds, ammonium chloride, ethylene glycol, and potassium hydroxide; (4) *breakers* are used to reverse crosslinking and reduce viscosity of gelled fluids as well as to degrade friction-reducing polymers in slick water fluid to allow removal of residual polymers from newly created fractures and thus to enable gas and water recovery to the well. Breakers can be either organic or inorganic, such as calcium chloride, sodium chloride, and ammonium sulfate; (5) *pH adjusters* are used to adjust pH and improve effectiveness of certain chemical additives, particularly crosslinked polymer molecules. Common chemicals include acetic acid, fumaric acid, potassium hydroxide, sodium hydroxide, sodium carbonate, and potassium carbonate; (6) *biocides* such as quaternary ammonium compounds are used to remove bacteria that can trigger degradation of the chemical additives and contribute to corrosion of well tubing, casings, and equipment, in particular sulfate reducing and acid forming bacteria; (7) *corrosion inhibitors* are added to prevent corrosion by acids, salts, and corrosive gases; (8) *scale inhibitors* composed of phosphonic acid salts, sodium poly-carboxylate, and copolymers of acrylamide and sodium acrylate are used to prevent scaling and blocking flow in piping and tubing; (9) *iron control* chemicals composed of thioglycolic acid, citric acid, acetic acid, and sodium erythorbate are used to prevent iron oxide precipitation and scaling; (10) *clay stabilizers* such as choline chloride, tetramethyl ammonium chloride, potassium chloride, and sodium chloride are used to prevent the swelling of clays found in shale rocks; (11) *surfactants* composed of amphoteric, anionic, or non-ionic compounds are used to control for optimal viscosity of fracturing fluids, reduce surface tension between the shale rocks and the fluid, and assist fluid recovery after fracturing; and (12) *acids*, including HCl (hydrochloric) or HF (hydrofluoric) acids, are used to clean the perforations of any cement or debris[610].

Overall, while the majority of the chemical additives used for hydraulic fracturing are non-toxic, some chemicals are toxic and carcinogenic, such as formaldehyde, naphthalene, and acrylamide[611]. While spills of these chemicals could induce environmental and human health risks such as endocrine-disrupting activities[612], many of the toxic contaminants (e.g., ammonium, radium nuclides, organic matter) found in wastewater generated from shale gas operations are derived from the brines or formation waters that make up the bulk of flowback and produced water due to retention of the injected frac water into the shale rock matrix during hydraulic fracturing (see discussion on the mechanisms of hydraulic fracturing processes in Section 5.2.3).

5.5 Flowback and Produced Water

5.5.1 The Water Intensity of Flowback and Produced Waters

Natural gas extraction also involves production of wastewater known as "produced water" for conventional natural gas exploration and "flowback and produced water" from unconventional shale gas and tight-sand gas extraction. In conventional natural gas settings, the extraction of natural gas in permeable formations is accompanied by produced water originated from formation water entrapped in the geological formations. Similar to conventional oil wells, over the lifetime of a conventional natural gas well, the ratio of formation water to gas increases (Figure 4.10) until the relative proportion of water is too high to make the natural gas–water blend profitable. The relative proportion of the formation water in the blend depends on the geological condition and hydraulic permeability that allow the formation water to flow to the well-pumping zone. In addition to formation water, produced water from gas operations also includes condensed water with soluble hydrocarbons. Veil (2015)[375] reported a volume of 147 million cubic m generated from conventional natural gas wells in the USA, with a national weighted average water-to-gas ratio of 0.54 L/m^3 (97 bbl of water to million cubic feet of natural gas), which is equivalent to water intensity of 13.9 L/GJ. Similarly, Grubert and Sanders (2018)[29] estimated that in 2014 the annual produced water volume from conventional natural gas wells in the USA was 110 million cubic m with a water-to-gas ratio of 0.38 L/m^3 and water intensity of 9.4 L/GJ (Table 5.1).

While the volume of produced water in conventional gas wells typically increases with time, similar to the pattern shown for conventional crude oil, the volume of the flowback and produced water in shale gas and tight sand gas wells decreases over time, parallel to the decrease in natural gas production (Figure 4.10). The term "flowback and produced water" has been used differently among scholars. Some have used the term "flowback water" to reflect the return of the injected water to the surface after hydraulic fracturing, while "produced water" is used to describe the long-term flow of formation water from shale gas wells. Others have described the water that flows with shale gas at all stages of operation as combined flowback and produced water (FPW), a term that is also used in this book. The reason for combining the flowback water and produced water is that it is almost impossible to separate the two water sources that are mixed (at different proportions) in the fluids that are co-extracted with shale gas. The relative proportions of these two water sources change over time, and the proportion of the formation water component increases over time. We discuss these changes and effects on water salinity and quality in Section 5.5.2 of this chapter.

A common phenomenon of hydraulic fracturing is the retention (imbibition) of the injected water in the shale matrix following hydraulic fracturing, which triggers the release of methane from the shale. Therefore, only a fraction of the water that is injected as part of the hydraulic fracturing is returned to the surface. Kondash et al. (2017)[360] estimated that only 4–8% of the injected water is returned to the surface, while the majority of the FPW is composed of the formation water that is extracted from the shale rocks. The FPW production is high during the first three to six months after hydraulic fracturing, comprising up to 50% of the total volume of the FPW generated during the lifetime of a well. After that peak, the FPW production declines to follow the decline rate of natural gas in shale gas wells[360]. Consequently, the volume of the FPW following hydraulic fracturing is lower than that of the volume of the water use for hydraulic fracturing. For example, the FPW production of a 12,800 m^3 per well during the first 12 months after hydraulic fracturing is 28% of the volume of a 45,000 m^3 per well used for hydraulic fracturing in shale gas wells in the Changning gas field of the Sichuan Basin, China[478]. The relationship between the FPW volume generated during the first 12 months after hydraulic fracturing and the water use for hydraulic fracturing for different shale gas basins in the USA and China is illustrated in Figure 5.19. Overall, the proportion between the FPW volume generated during the first 12 months after hydraulic fracturing and the water volume used for hydraulic fracturing varies from 3% (Marcellus Shale) to 58% (Sichuan Basin), while a much larger volume of FPW is identified in the Permian Basin (Figure 5.19). Parallel to the increase of the water use for hydraulic fracturing and shale gas production per well, the intensification of shale gas production in the USA has also increased the volume and the water intensity of the FPW (Figure 5.23)[24]. Overall, the ratios of FPW-to-shale gas produced during the first 12 months of production are similar in many shale plays to the produced water-to-gas ratios observed in conventional natural gas, although some unconventional basins (e.g., the Permian Basin) show a much higher volume of FPW (Figure 5.19).

Extraction of both conventional and shale gas generates large volumes of produced water and FPW with different ratios of natural gas to produced water and FPW (Figure 5.24). Using a median water-to-gas ratio of 0.5 L/m^3, we reconstruct the production of produced water and FPW during the last 30 years in the USA (Figure 5.25). Starting in 2005 with the beginning of the shale revolution, we show exponential growth in natural gas production along with an accelerated production in mostly FPW, up to a total volume of both conventional and unconventional FPW of 526 million cubic m in 2018 (Figure 5.25). This estimated volume is much lower than the volume of combined conventional water and FPW that was

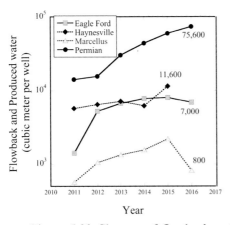
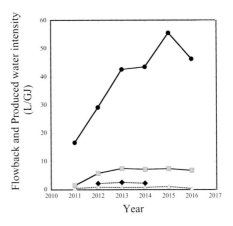

Figure 5.23 Changes of flowback and produced water (FPW) per well and the FPW intensity for the first 12 months after hydraulic fracturing from major shale gas basins in the USA. Based on data reported in Kondash et al. (2018)[24]

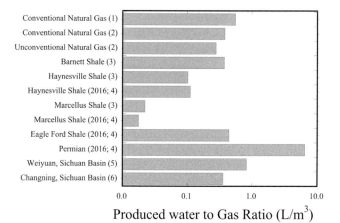

Produced water to Gas Ratio (L/m^3)

Figure 5.24 Variations of the produced water and FPW-to-natural gas ratios in conventional and shale gas basins in the USA. Note that the FPW-to-shale gas ratios were calculated for the first 12 months of production, where the volume of shale gas and FPW is at the highest production rates. Data were retrieved from Grubert and Sanders (2018)[29], Veil (2015)[375], Kondash et (2017)[360], Kondash et al. (2018)[24], Zou et al. (2018)[16], and Liu et al. (2020)[478]

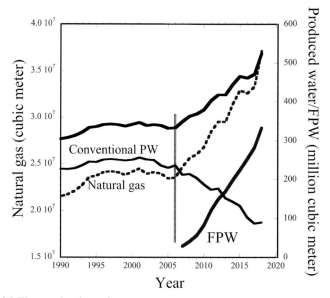

Figure 5.25 The production of natural gas (dashed line; cubic m on the left y-axis) and estimated conventional produced water and unconventional FPW (assuming water-to-gas ratio of 0.5 L/m^3; cubic m on the right y-axis) in the USA during the last 30 years. The beginning of the shale revolution is marked to show the exponential rise of natural gas, the consequent rise of FPW production, and the decline of conventional produced water in the USA. Data retrieved from US EIA[574]

generated from crude oil production in the USA during that year (2.32 BCM; Figure 4.14). While we show in Chapter 4 that the overall produced water generation from unconventional oil production did not change due to the much lower FPW-to-tight oil ratio (~1) relative to the produced water-to-conventional oil ratio (9.2), the rise of shale gas development with a similar produced water-to-gas ratio (~0.5 L/m^3) has caused a net increase in the total wastewater generated from natural gas in the USA (Figure 5.25).

The volume of FPW is expected to increase in the future with the continued increase of natural gas production. In the Marcellus Shale, it was predicted that the wastewater annual production would increase to 12 million cubic m by 2025[477]. Likewise, in the Permian, Eagle Ford, and Haynesville plays, the FPW production is expected to increase by 50-fold over the next decade[24]. The large volume of FPW requires adequate management and disposal. In the USA, the majority of FPW is disposed into deep Class II injection wells (pathway 1; see pathways marked in Figure 5.26), whereas in some basins (e.g., Marcellus Shale) FPW is reused for hydraulic fracturing (marked pathway 2 in Figure 5.26) and in the past (prior to regulations that restricted disposal of FPW) transported to centralized

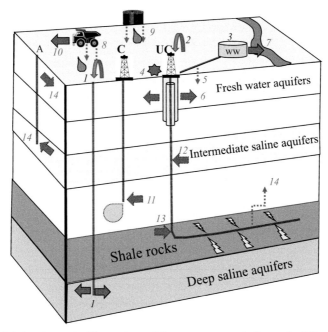

Figure 5.26 A schematic illustration of the transport and disposal of flowback and produced water from shale gas wells as well as possible contamination and leaking pathways (marked 1 to 14) of water and gases in the framework of shale gas exploration. (A black-and-white version of this figure will appear in some formats. For the colour version, refer to the plate section.)

waste treatment sites, where the treated water was discharged to nearby rivers[26] (pathway 3 in Figure 5.26). In the Marcellus Shale a large fraction of the FPW is reused for hydraulic fracturing; between 2011 and 2018 50–90% of the total FPW that was generated was reused (with and without treatment) for hydraulic fracturing in other frac sites[477]. Likewise, FPW from the Changning gas field in the Sichuan Basin in China is almost fully recycled for hydraulic fracturing[478]. And yet, further increases in shale gas exploration, and consequently, increasing production rates of FPW[24], will bring further pressure on the disposal infrastructure and could trigger seismicity in areas where large volumes of wastewater are disposed through deep-well injection[380, 383, 385, 387, 388, 613]. In addition, in areas where deep-well injection is not available, such as in the majority of the Marcellus Shale, the transport of FPW to disposal sites (e.g., eastern Ohio) induces additional cost for the management of the wastewater[477]. While reuse of FPW for hydraulic fracturing seems to reduce the fresh water footprint and resolve the cost and environmental risks of FPW, it has been shown that hydraulic fracturing with saline FPW reduces shale gas production by 20% per well as compared to hydraulic fracturing with low-saline water, and therefore there could be economical consequences for FPW recycling[478]. In addition, a temporary decline in constructing new wells for hydraulic fracturing (e.g., the slowdown of oil exploration in the USA due to the COVID pandemic) would restrict the ability of recycling FPW. This can be demonstrated in the Marcellus Shale, in which the combined reuse and recycling rates of FPW fell from around 87% in 2011 to only 55% in 2015, reflecting a slowdown in drilling activities[477].

5.5.2 *The Origin of Flowback and Produced Water*

The origin of produced water that is co-extracted with conventional natural gas as well as flowback and produced water that is co-produced with shale gas is related to the formation water that is entrapped in the geological formations. For conventional natural gas, the formation water can originate within the reservoir, or more commonly, migrate from other formations in the basin and accumulate in geological traps together, or independently, with the natural gas. In many cases, the formation water that occupies the gas fields originates from relicts of evaporated seawater modified by water–rock interactions and is blended with other water sources, including meteoric water and/or meteoric water that was associated with the dissolution of salts (halite and gypsum minerals) in the basin (e.g., Michigan Basin[578–580]). Much of the typical geochemical features that characterize oil-field water described in Chapter 4 (Section 4.3.2) apply also for produced water associated with conventional gas fields. In most cases, the origin of the formation water that is entrapped in the unconventional shale rocks is directly linked to the geological evolution of the formation water in the basin. Therefore, the salinity and water quality of the flowback and produced water co-extracted from shale formation commonly mimic the formation water from other formations in the basin (see also Chapter 4). The distribution of chloride and TDS concentrations in formation water from the major oil and gas basins in the USA are presented in Figures 4.14 and 4.15. Likewise, the salinity of FPW from shale gas basins varies by nearly an order of magnitude following the salinity of the formation water in the basin as shown in FPW from the Fayetteville (25,000 mg/L), Barnett (60,000 mg/L), Woodford (110,000–120,000 mg/L), Haynesville (110,000–120,000 mg/L), and Marcellus (up to 180,000 mg/L) shale formations[26]. Nonetheless, we show below that *in situ* interactions of the injected water during hydraulic fracturing and the formation water with the shale rocks result in distinctive geochemistry of the flowback and produced water from shale formation, which is different from conventional produced water.

The Appalachian Basin in the Eastern USA hosts the Marcellus Shale, one of the largest shale gas production formations in the USA (Figure 5.11). Investigations of the produced water co-extracted with natural gas and oil from the different geological formations in the Appalachian Basin have shown that the brines with salinity of up to 200,000 mg/L have the composition of seawater evaporated beyond the stage of halite precipitation, modified by water–rock interactions (e.g., calcium, barium, strontium, boron, and radium enrichments) and dilution by meteoric water. The source of the brines has been attributed to the highly saline brines from the Silurian Salina evaporite formation in the Medina Group (Figure 5.27) and associated also with halite deposits that were precipitated from the original evaporated seawater[412, 413, 466, 614–616]. The consistency in the geochemical characteristics (e.g., similar high Br–Cl ratio) of formation waters from the multiple formations overlying the Silurian rocks (Figure 5.27) suggests major upflow and/or lateral migration of the brines for thousands of meters from their original formation[614]. The salinity of the produced water decreases with stratigraphic age from the Silurian, Lower Devonian, Upper Devonian, and Middle and Lower Mississippian formations (Figure 5.27). The upflow of brines to the shallow formations that composed the regional aquifers in the Appalachian Basin results in salinization of the groundwater[410, 466]. The upflow of brines from deep sources in the Appalachian Basin is accompanied by natural gas flow with a thermogenic fingerprint[615, 617, 618] that in some areas can reach to the shallow formations and is associated with regional groundwater[466, 617–622]. The role of naturally saline water and natural gas on the quality of shallow groundwater is discussed in Section 5.6. In addition to the evidence for upflow of the Appalachian brines towards the shallow formations, Osborn et al. (2012)[616] have suggested that the high concentrations of radioactive iodine (high ratios of ^{129}I/I) in the Marcellus

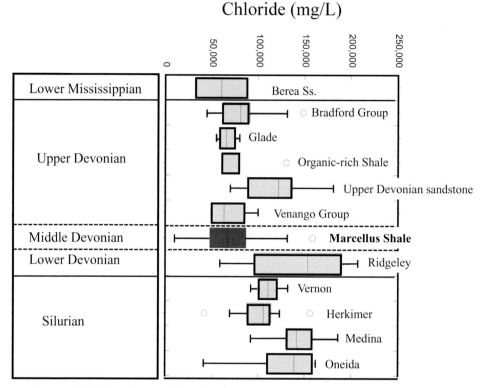

Figure 5.27 Variations of the chloride concentration of produced water from different oil and gas formations along the stratigraphic section of the Appalachian Basin. Data integrated from Osborn and McIntosh (2010)[615], Chapman et al. (2012)[422], and unpublished Duke University data

formation water reflect basin-scale fluid migration from the eastern section of the Appalachian Basin, where the original brine interacted with high-uranium source rocks that led to the enrichment of iodine-129 in the formation water from the Marcellus Shale[616]. Overall, the data from the Appalachian Basin suggest co-genic and common geological evolution of formation water in both the permeable (conventional) geological formations and impermeable Marcellus Shale. The occurrence of hypersaline brine within the shale formation has important implications for the salt gradient that promotes the imbibition process of the injected hydraulic fracturing water (Figure 5.16), as well as the chemistry and quality of the flowback and produced water generated from shale rocks.

Analyses of the chemistry of FPW from other shales also indicate that the geological history of formation water in a basin would control the chemistry of the saline pore water within the shale rock matrix. Ni et al. (2018)[434] have shown that the salinity and chemistry of the formation water entrapped in the Lower Silurian

Longmaxi shale in the Sichuan Basin is consistent with the composition of produced water from direct underlying formations (Sinian, Cambrian) with chloride content of up to 50,000 mg/L, and Br–Cl ratios that reflect 18-fold evaporated seawater. In contrast, produced water from overlying geological formations (Permian, Triassic) have higher chloride content of up to 140,000 mg/L and higher Br/Cl ratios indicating an origin from 30-fold evaporated seawater that was diluted with meteoric water[434]. Similar to the Marcellus Shale, the chemistry and salinity of the pore water in the Longmaxi Shale in the Sichuan Basin reflect the regional evolution of formation water in the basin.

While the source and evolution of formation water from oil and natural gas are very similar, formation water that is associated with natural gas resources can be distinguished by elevated concentrations of dissolved inorganic carbon (alkalinity), combined with enriched carbon-13 (δ^{13}C-DIC; see Section 5.2.2). The intrusion of low-saline meteoric water could induce microbial methanogenesis processes and the formation of ^{13}C-rich CO_2 that would dissociate to dissolved bicarbonate in the formation water. The association of dilution of the original saline formation water and elevated bicarbonate with high δ^{13}C-DIC has been detected in Michigan and Illinois basins[578–580].

The imbibition (retention) of the injected hydraulic fracturing water into the shale rock matrix triggers the release of methane that is adsorbed onto organic matter in the shale, in addition to the return of the injected water, blended with formation water that is also released from the shale matrix. Following hydraulic fracturing, the relative proportion of the formation water in the blend increases with time, and, consequently, the salinity of the FPW also increases over time. While the water use for hydraulic fracturing is typically low-saline water (surface water, groundwater), data from shale gas wells show exponential salinity growth of the FPW with time until it becomes similar to the salinity of the formation water entrapped in the shale matrix[405, 623–625] (Figure 5.28). Several key geochemical components in the geochemistry of the FPW, such as the stable oxygen isotopes and Br/Cl ratios, clearly indicating that the saline source is derived from residual formation water and is not the product of salt dissolution from the shale rocks[408, 623].

The chemistry of the FPW is controlled by three major factors. The first is the relative mixing proportions between the injected hydraulic fracturing and the saline formation water, resulting in typically lower salinity of the FPW during early stages of FPW formation after hydraulic fracturing. Second, the formation water that is entrapped within the shale rock has a different geochemistry relative to formation water from conventional geological reservoirs such as sandstone and limestone due to the water–rock interactions, in which the unique chemistry of the organic-rich shale would dominate the composition of the entrapped pore water. For example, shale rocks are enriched in uranium relative to common sandstone

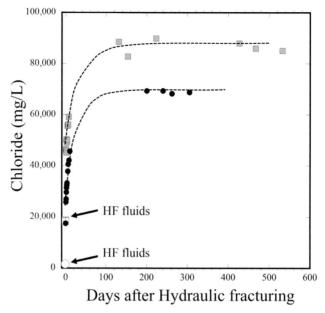

Figure 5.28 Time series of flowback and produced water from two shale gas wells in the Marcellus Shale fracked with fresh water (circles) and saline water (squares). The rapid rise in chloride contents in FP water towards constant values reflects the increasing fraction of the formation water in the mix with the injected frac water (marked by arrows). Data from Duke University (unpublished results)

and limestone rocks. Over geological time, the uranium, which is a radioactive element, decays into a series of decay-product nuclides that are part of the uranium-238 decay series (Figure 5.29). One of the products is ^{226}Ra nuclide, which can be mobilized into the formation water under high salinity and reducing conditions. Sandstone rocks from conventional reservoirs have typically higher thorium relative to uranium, and the decay of thorium-232 results in the formation of ^{228}Ra nuclide. Consequently, the enrichment of uranium in the organic-rich sale rocks generates formation water with high concentrations of ^{226}Ra and distinctively low ^{226}Ra–^{228}Ra activity ratios ($\ll 1$) relative to formation water from conventional basins with typically lower ^{226}Ra and higher ^{228}Ra–^{226}Ra activity ratios (>1)[408, 409, 448, 450] (Figure 5.30). This difference provides a way to distinguish between produced water originated from conventional oil and gas wells and FPW originated from unconventional shale gas wells. Another important diagnostic tracer for detecting the equilibrium between the host rocks and formation water is the strontium isotope (^{87}Sr/^{86}Sr) tracer that has been used to distinguish between the ^{87}Sr–^{86}Sr ratios of FPW from the Middle Devonian Marcellus Formation and conventional produced water from Upper Devonian formations (see the stratigraphy of the Appalachian Basin in Figure 5.27)[422, 626]. Likewise, the ^{87}Sr–^{86}Sr ratios of FPW from the Silurian Longmaxi Shale in the

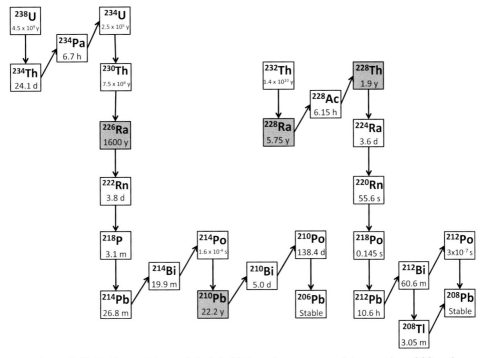

Figure 5.29 Radionuclides and their half-lives that are part of the uranium-238 and thorium-232 decay series.

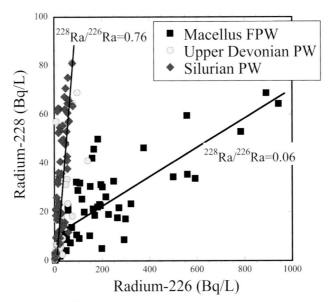

Figure 5.30 ^{228}Ra vs. ^{226}Ra activities in unconventional FPW from the Marcellus Shale (squares) and conventional produced water from the Upper Devonian (circles) and Silurian (diamond) formations in the Appalachian Basin (see Figure 5.28 for the stratigraphic formations). Data from Lauer et al. (2018)[448]

Sichuan Basin were different, with high ^{87}Sr–^{86}Sr ratios relative to conventional produced waters from the Permian and Triassic formations, reflecting the interaction of the formation water with the shale rocks[434]. Finally, FPW from the Marcellus Shale is also characterized by a distinguished barium isotope signature with high ^{138}Ba–^{134}Ba ratios as compared to that in conventional produced water from the Upper Devonian/Lower Mississippian rocks[627].

The third factor that affects the composition of FPW is the interaction of the injected low-saline water and shale rocks. The intrusion of low-saline water into the saline environment of the shale rock, a phenomenon known as "freshening," triggers inverse base-exchange reactions in which sodium is released while calcium is retained on clay minerals as well as desorption of boron into the FPW. The lower the salinity of the hydraulic fracturing (or "frac water"), the higher the boron desorption from the shale rocks that would result in higher boron concentrations in the FPW. While the bromide–chloride relationship is directly linear with a high correlation coefficient, reflecting mixing between two water sources, the boron-to-chloride ratios in the FPW are not linear, and show an enrichment relative to the expected mixing curve (Figure 5.31). This can also be seen in FPW from shale gas wells in the Sichuan Basin that were fracked with recycled saline FPW. Consequently, the FPW becomes enriched in boron with a distinctive lower boron isotope ratio (δ^{11}B) relative to the saline formation water with higher δ^{11}B and a lower B/Cl ratio (Figure 5.32). The association of

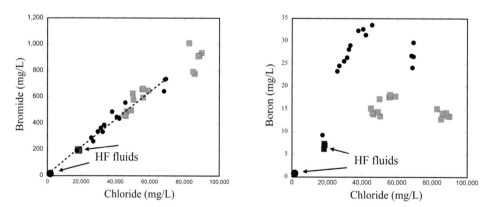

Figure 5.31 Variation of bromide and boron vs. chloride in flowback and produced water from two shale gas wells in the Marcellus Shale fracked with fresh water (circles) and saline water (squares). The high correlation ($R^2 = 0.95$, $p < 0.001$) between bromide and chloride reflects mixing relationships between the frac water (marked by arrows) and the saline formation water extracted from the shale matrix. In contrast, the enrichment of boron in the FPW reflects mobilization of boron from the shale rocks most likely due to the injection of low-saline water. Data from Duke University (unpublished results)

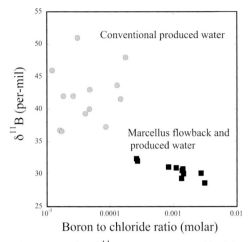

Figure 5.32 Boron isotope ratio (δ^{11}B) vs. boron-to-chloride ratio (molar unit) of unconventional FPW from the Marcellus Shale (squares) and conventional produced water from the Upper Devonian (circles) formations in the Appalachian Basin (see Figure 5.28 for the stratigraphic formations). Data from Warner et al. (2014)[405]

boron-enrichment with high boron-to-chloride ratios and low δ^{11}B has been detected in relatively early FPW in the Marcellus Shale[405] and Longmaxi Shale in the Sichuan Basin[434], and is different from the boron isotope ratios and boron-to-chloride ratios measured in conventional produced water and/or FPW generated during later stages of FPW production (i.e., ~100% contribution of the formation water).

Overall, the boron, radium, and strontium isotopes are powerful geochemical tracers that are capable of detecting the occurrence of FPW originating from unconventional shale gas wells in the environment, as opposed to the impact of conventional produced water[26, 405, 434, 628]. The ability to use geochemical fingerprints for tracing the origin of wastewater derived from conventional and unconventional oil and gas wells is important for evaluating the impact of shale gas development on water resources[26, 629,87, 88, 630], which is discussed in Section 5.5.4.

5.5.3 *The Quality of Flowback and Produced Water*

While most public concerns have focused on the chemical additives used during the hydraulic fracturing process, the large retention of the injected frac water and the contribution of the formation water within the shale matrix to the FPW generates wastewater with different types of contaminants, mainly derived from naturally occurring inorganic and organic contaminants in the shale formations. Consequently, the difference between oil and gas wastewater derived from conventional gas operations relative to that of unconventional oil and gas operations is small and only low concentrations of chemical additives have been reported in FPW. Nonetheless, in some parts of the USA, unconventional oil and gas wastewater is regulated differently. For example, in Pennsylvania, conventional oil and gas wastewater can still be discharged to the environment after some treatment (that has shown not to be very effective, see below), whereas unconventional oil and gas wastewater is commonly regulated to be disposed only to deep-well injection and/or reuse for hydraulic fracturing.

Similar to OPW, produced water and FPW from conventional and unconventional gas wells are composed of (1) organic constituents; (2) salts and halides; (3) metals and metalloids; (4) nutrients; (5) naturally occurring radioactive material (NORM); and (6) gases (Figure 4.13). In comparison to produced water associated with crude oil, produced water from gas wells has higher contents of low molecular-weight aromatic hydrocarbons such as benzene, toluene, ethylbenzene, and xylene (BTEX)[372, 375, 416]. About half of the organic matter in FPW from the Sichuan Basin in China is composed of C6–C21 straight-chain alkanes and C7–C13 naphthenes[631]. The occurrence of the highly volatile and toxic hydrocarbons makes gas wastewater more toxic than oil produced waters, in some cases by a factor of ten[632]. While conventional gas wastewater could include chemicals used for well stimulation such as mineral acids, dense brines, and additives, unconventional FPW from shale gas wells could contain tracers of frac chemical additives[623]. In contrast, the inorganic chemistry and thus quality of produced water from natural gas wells reflects the chemistry and quality of the formation water in the basin, and thus is similar to that of oil produced water. Likewise, the quality of FPW from shale gas wells reflects the chemistry of the formation water in the basin, in addition to the occurrence of contaminants derived directly from the leaching of shale rocks. The occurrence of many inorganic contaminants in both conventional produced water and unconventional FPW is directly correlated with that of chloride, and thus the rise of the salinity of the FPW during the early stages of gas production (Figure 5.29) is associated with higher abundances of other inorganic contaminants in the FPW such as barium, strontium, ammonium, bromide, iodide, ammonium, and radium (^{226}Ra and ^{228}Ra) nuclides, known as NORMs (Table 5.2, Figure 5.33)[26, 289, 404, 408, 411, 413, 444, 624, 633, 634]. For example, inorganic contaminants such as bromide, iodide, ammonium, and ^{226}Ra from the Marcellus FPW and conventional produced waters from different geological formations in the Appalachian Basin show direct correlations with chloride concentrations (Figure 5.33). Therefore, the higher the salinity of the FPW or produced water, the more likely it is to have higher concentrations of these contaminants. While the ratios of some elements like bromide to chloride are similar to conventional and unconventional FPW, ammonium and ^{226}Ra-to-chloride ratios are higher in the

Table 5.2. *Median values of major and trace elements as well as NORMs from FPW and produced water from the major unconventional and conventional gas basins in the USA as compared to the US EPA ecological (CCC), secondary drinking water, and drinking water (Maximum Contaminant Level) standards**

Component	Unconventional			Conventional				CCC/ secondary	MCL
	Appalachian	Michigan	Gulf Coast	Appalachian	Michigan	Fort Worth	Gulf Coast		
Major elements (mg/L)									
Total Dissolved Salts (TDS)	99,000	163,965	139,591	153,000	271,730	173,698	63,700	500	N/A
Ammonium	66	N/A	N/A	180	96	N/A	162	1.7	N/A
Boron	12.4	5.5	86	36	31	6.2	52	N/A	N/A
Barium	1000	55	34	239	11	46	33	N/A	2
Calcium	8,065	23,848	5,305	17,500	30,300	14,461	1,800	N/A	N/A
Chloride	64,000	64.800	25,590	111,000	172,079	105,958	44,274	250	250
Iodide	27.2	2	11.6	10	12	16.6	18	N/A	N/A
Potassium	257	389	390	1040	1680	386	214	N/A	N/A
Lithium	65.4	12.1	9	45	29	5.5	6.9	N/A	N/A
Magnesium	785	2230	35	2,350	5,884	2,250	331	N/A	N/A
Sodium	N/A	388	N/A	42,500	50,330	47,200	25,021	N/A	N/A
Sulfate	23.9	25.8	30	145	277	92	46	250	N/A
NORMs (Bq/L)									
Combined radium-226 and radium-228	52	N/A	N/A	36	N/A	N/A	41	N/A	0.185

* Data from the US Geological Survey archive of the chemistry of produced water[362]

177

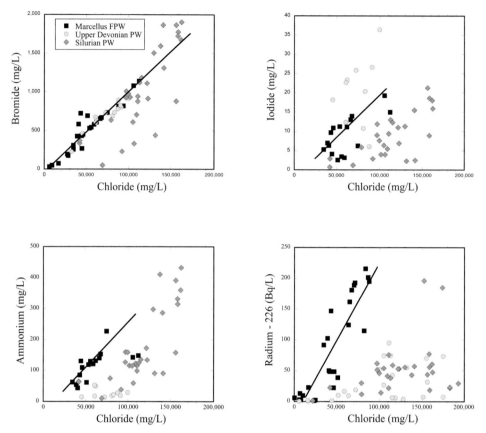

Figure 5.33 Variations of bromide, iodide, ammonium, and radium-226 vs. chloride concentrations in unconventional FPW from the Marcellus Shale (squares) and conventional produced water from the Upper Devonian (circles) and Silurian (diamond) formations in the Appalachian Basin (see Figure 5.28 for the stratigraphic formations). Data for bromide, iodide, and ammonium from Harkness et al. (2015)[411] and radium from Lauer et al. (2018)[448]

Marcellus FPW (Figure 5.34), reflecting the specific contribution of the shale rocks to the chemistry of the formation water entrapped in the shale matrix. Table 5.2 presents the major contaminants in produced water and FPW from major conventional and unconventional gas plays in the USA.

One of the distinctive water quality parameters in wastewater derived from unconventional FPW is the high levels of radium concentrations (activities, since we report the level of radioactive decay in becquerels per liter – Bq/L unit, which is decay per second per liter[289, 404, 406, 408, 409, 434, 444, 448, 450, 633]). Several studies have shown that the activity of radium in formation water directly correlates with the salinity and that the adsorption of radium onto clay minerals decreases with increasing salinity[445, 635, 636]. Since the organic-rich shale rocks contain high

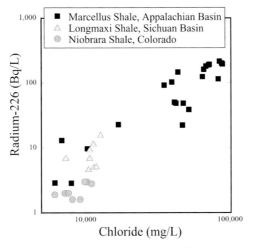

Figure 5.34 Radium-226 vs. chloride concentrations in unconventional FPW from the Marcellus Shale (squares), the Sichuan Basin in China (triangle)[434], and the Denver-Julesburg (DJ) Basin in Colorado (circles)[623].

concentrations of uranium[637–640], over geological time, ^{238}U decays to ^{226}Ra, and the ^{226}Ra activities in the shale rocks become equal to that of uranium. In contrast, sandstone and other rocks from conventional oil and gas reservoirs have typically high concentrations of both thorium and uranium[641]. The decay of ^{232}Th to ^{228}Ra generates relatively high ^{228}Ra–^{226}Ra activity ratios in such rocks that are very different from the distinctively low ^{228}Ra–^{226}Ra activity ratios in shale rocks. This distinctive relationship is reflected also in the formation waters from shale rocks relative to sandstone rocks, as demonstrated in the very low ^{228}Ra–^{226}Ra activity ratios in the Marcellus FPW (0.06) relative to conventional formation waters (0.76) from the Appalachian Basin[448, 450] (Figure 5.31). An exceptional case is the FPW from the Utica Shale from the Appalachian Basin with a ^{228}Ra–^{226}Ra activity ratio ~1, which is significantly different from that of FPW from the Marcellus Shale, reflecting a higher abundance of Th and higher Th/U in the Utica Shale rocks[642]. As shown in this chapter, the distinctive ^{228}Ra–^{226}Ra activity ratios in the oil and gas wastewater can be used to delineate the source of contamination and to reconstruct the timing of a spill[450]. Overall, the occurrence of radium in FPW is directly associated with the salinity of the FPW, and radium activities range from high values (up to 220 Bq/L) in the hypersaline Marcellus Shale, down to 16 Bq/L in the much less saline-saturated FPW from the Sichuan Basin (chloride = 12,000 mg/L)[434], and even lower (3 Bq/L) in FPW from the Denver-Julesburg Basin, Colorado[623] (Figure 5.34). The salinity variation in formation water from both oil and gas basins across the USA is presented in Figure 4.15.

5.5.4 The Impact of Flowback and Produced Water on Water Resources: The Impaired Water Intensity

5.5.4.1 Impact on Surface Water

Produced water and FPW can affect the environment through direct discharge to waterways as part of the wastewater management scheme, or indirectly via leaks and spills (e.g., overflow or breaching of surface pits, insufficient pit lining, onsite spills, transportation). Figure 5.26 illustrates possible pathways, including spills at the well sites (marked pathway 4 in Figure 5.26), leaking of pipes (pathway 5), leaking and discharge from treatment centers to both shallow groundwater and surface water (pathways 6 and 7), leaking during wastewater transportation (pathway 8), leaks from storage ponds and pits (pathway 9), and spreading of oil and gas wastewater on roads. The discharge and leaking of the wastewater to the environment can affect the quality of water and soil in the impacted areas, which would result in an additional water footprint, defined in this book as *impaired water*. Given the high concentrations of salts, metals, nutrients, and NORMs in produced water and FPW (see Section 5.5.3; Table 5.2), the release of produced water and FPW to the environment could have multiple effects. First, it causes direct salinization and contamination of impacted water resources to levels that would make the water unusable for domestic use and/or induce ecological risks. Second, the release of FPW or produced water containing high concentrations of halides (chloride, bromide, and iodide[324, 411, 431, 643]) to river water, combined with chlorination treatment of the downstream river water (a common practice for using river water for drinking water), could trigger the formation of highly toxic disinfection byproducts in the downstream treated drinking water[112–114]. Third, the release of FPW or produced water containing high barium and NORM concentrations (Table 5.2) to the environment and dilution with low-saline water would cause secondary precipitation of radium-rich barite mineral and/or adsorption of radium to riverbed sediments and/or soil, and, consequently, accumulation of radioactivity on sediments and soils in the impacted sites[404, 452, 455]. Elevated levels of radium have been detected in riverbed sediments in disposal sites in rivers and streams in Pennsylvania, where mostly conventional oil and gas waste-water was discharged to the river[448, 450]. Likewise, a higher than background level of radium was detected in soil impacted by FPW spill[410]. As discussed in Section 4.3.4, the quality of oilfield water and (as illustrated in Figure 4.20) the quality of produced water and FPW (Table 5.2) require dilution of 300- to 700-fold in order to restore the quality of the impacted water to the baseline levels that are acceptable for ecological and drinking water standards.

The largest impact of produced water and FPW on the environment is through direct discharge to waterways either without or with some treatment. In the USA, areas where deep injection is not available due to the geological structure (e.g., Pennsylvania, New York) or restricted because of proximity to fault systems, wastewater is then recycled and reused for hydraulic fracturing, or treated in publicly owned treatment works (POTWs), municipal wastewater treatment plants (WWTP), or centralized waste treatment (CWT) sites and then discharged to surface water under the US National Pollutant Discharge Elimination System (NPDES) permits. The US Environmental Protection Agency[644] has identified 426 facilities across the USA, of which 198 facilities accept oil and gas extraction wastes[644]. Before the rise of shale gas development in Pennsylvania, the wastewater that was treated and discharged to surface water was only from conventional oil and gas wells. The rise of shale gas development was associated with the transport of the FPW from shale gas wells to CWT sites and the release to local streams and rivers. In 2013, the total wastewater in Pennsylvania was 1.79 million cubic m, which included 0.42 million cubic m of FPW from unconventional shale gas (23% of total wastewater). From the total waste volume, 1.2 million cubic m was recycled for hydraulic fracturing (67%) and 0.58 million cubic m (33%) was transported to CWT facilities and released to surface water[411]. Findings of the environmental impacts resulted from the discharge of wastewater derived from unconventional shale gas development to streams in Pennsylvania[404, 645] (see below) have elicited public concern, and consequently, since 2014 the Pennsylvania Department of Environmental Protection restricted the transport of unconventional FPW to CWT sites and discharge to surface water. Given that shale gas production in 2013 was 8.7×10^{10} cubic m[593], the discharge of 0.58 million cubic m FPW from shale gas development implies a water intensity of 0.17 L/GJ. Table 5.2 shows that wastewater from the Marcellus Shale contains contaminants in concentrations that are higher by 200- to 300-fold as compared to ecological and drinking water standards. Assuming a dilution factor of 300-fold, the impaired water intensity of natural gas wastewater is estimated as 50.8 L/GJ (Table 5.3).

The practice of transporting conventional produced water to CWT sites in Pennsylvania, however, has not stopped, and the CWT facilities have continued to receive, treat, and dispose of conventional produced water to local streams[448]. In 2015, the volume of conventional produced water discharged from three of the major CWT sites in Pennsylvania (Franklin, Hart, and Josephine) was 150,000 m³. Analysis of US Environmental Protection Agency data[644] reveals that the major CWT treatment facilities in Pennsylvania were designed for a permitted treatment capacity of 2.4 million cubic m of oil and gas wastewater. In only one site (Eureka Resources) the facility has the treatment entailed the elimination of the salts by desalinization, while in the other sites the treatment includes chemical precipitation, filtration, and activated carbon filter that removes organic constituents but only a fraction of the inorganic constituents in conventional oil

Table 5.3. *Estimated impaired water intensity associated with controlled discharge and spills of wastewater from conventional and unconventional natural gas wastewater*

Mechanism	Wastewater volume	Impaired water intensity (L/GJ)	Case study/source
Conventional produced water discharge to rivers through CWT	0.58×10^6 m^3	972	Pennsylvania[644]
Unconventional FPW discharge to rivers through CWT	0.67×10^6 m^3	50	Pennsylvania[411]
Conventional produced water spread on roads	0.06×10^6 m^3	70	Pennsylvania[454]
Leaks in Pennsylvania	200 m^3 per wells	139	Pennsylvania and West Virginia[650]
Spills in Colorado	Total fluids 23,500 m^3	3.9	Colorado[531, 649]
Spills in New Mexico	Total fluids 56,100 m^3	11.3	New Mexico[531, 649]
Spills in Pennsylvania	Total fluids 31,300 m^3	2.7	Pennsylvania[531, 649]

and gas wastewater[644]. The most common treatment is the addition of Na_2SO_4 to promote the precipitation of metals, including barium and radium, before the treated wastewater is discharged to local surface waters[448].

Research on the impact of the oil and gas wastewater discharged from the CWT treatment facilities in Pennsylvania has shown that the treatment does not effectively remove the salts and only partially reduces the radium content of the treated effluents, causing a major impact on the quality of surface water and riverbed sediments at the disposal sites[404, 411, 448, 452, 455]. The discharge of hypersaline effluents (e.g., chloride concentrations up to 100,000 mg/L in the Josephine site) with high concentrations of bromide (1,164 mg/L), iodine (27 mg/l), and ammonium (68 mg/L) has affected the surface water quality with high concentrations as far as 2 km downstream from the discharge point (depending on the volume of the natural flow of the river)[404, 411]. An even larger impact on river quality was observed during CWT operations; chloride and bromide concentrations in the Allegheny River in Pennsylvania increased by 8 mg/L and 0.075 mg/L at a distance of 12 km downstream of the CWT discharge point, and 5 mg/L and 0.025 mg/L as far as 50 km downstream[646]. In addition to the direct impact on the ecosystem from the discharge of highly saline water with high concentrations of ammonium, the increase in bromide and iodide concentrations in the downstream river further increases the probability of formation of disinfection byproducts, in particular the highly toxic brominated and iodinated trihalomethanes in down-stream river water that is chlorinated and used for drinking water[411, 431]. Such impacts were identified in downstream sites in the Allegheny River in Pennsylvania, where the discharge of the treated wastewater effluents caused an increase in total trihalomethanes and shifts towards higher proportions of the toxic brominated trihalomethanes analogs were observed[646].

In addition to the impact of water quality, the discharge of produced water has caused the accumulation of NORMs in the streambed sediments at the disposal sites. Lauer et al. (2018)[448] have shown that in spite of the significant removal of radium from the treated effluents (70–90%), elevated activities of radium-228 and radium-226 were detected in stream sediments in the vicinity of the outfall (total radium-228 and radium-226 = 90−25,000 Bq/kg) relative to upstream sediments (20−80 Bq/kg; up to ~1,250-fold enrichment). This suggests radium accumulation in the river sediments due to long-term discharge of treated oil and gas wastewater effluents, even with relatively low radium concentrations[448]. The accumulation of radium on the impacted sediments from the disposal sites resulted in high radium concentrations, far exceeding the upper limit for disposal of waste solids containing NORMs in the USA (185–1850 Bq/kg). Solid wastes with radium concentrations above the 1,850 Bq/kg threshold limit should be transferred to a licensed radioactive waste disposal facility that has strict requirements related to site location[448]. In sum, the disposal of conventional oil and gas wastewater can generate a radioactive "hot spot" in the stream or river sediments that that can affect the environment over a long period. Evidence for bioaccumulation of metals associated with the Marcellus wastewater shown in Sr/Ca and strontium isotope ratios was measured in fresh water mussel shells collected downstream of the facility that corresponded to the time period of the largest Marcellus wastewater disposal to the river[647].

We estimate the impaired water intensity of treatment and discharge produced water from conventional gas wells in Pennsylvania by using the relative ratio between natural gas and produced water (Figure 5.25), and the relative proportions between conventional oil and natural gas in Pennsylvania (ratio of 1.4) using EIA data[593]. We estimate that out of a total of 2.4 million cubic m of oil and gas wastewater that was discharged to rivers and streams in Pennsylvania, 686,000 m^3 was directly from conventional natural gas production. Given the volume of conventional natural gas in Pennsylvania (5.4 BCM[593]), this implies a discharge of water-to-gas ratio of 0.13 L/m^3 and water intensity of 3.3 L/GJ. Given the TDS in conventional natural gas in Pennsylvania is about 150,000 mg/L (Table 5.2), which is 300-fold higher than the common ecological threshold of 500 mg/L, we suggest an impaired water intensity of 972 L/GJ (Table 5.3). This is a conservative estimate since the long-term accumulation of radium and other metals in sediments could further deteriorate water quality over time. It is interesting to note that although the volume of conventional wastewater (0.67 million cubic m) was not much larger than the volume of discharged shale gas wastewater in 2013 (0.58 million cubic m), it has a much larger water intensity (972 L/GJ vs. 50 L/GJ) given the relatively lower (by one order of magnitude) natural gas production from conventional natural gas wells relative to the shale gas production rates in Pennsylvania (i.e., the volume of wastewater is the same but the volume of shale gas is much higher per well, thus a lower water intensity of unconventional natural gas).

In addition to transport to treatment centers and discharge into surface water, wastewater from conventional oil and gas operations has also been utilized for deicing, dust suppression, and land spreading on roads as well as for road maintenance across the USA, but particularly in Pennsylvania, New York, and West Virginia[454]. As shown in Section 5.5.3, the spreading of produced water that contains high levels of salts, organic and inorganic chemicals, and NORMs can directly contaminate water and soil in impacted areas. Between 2008 and 2016, the volume of oil and gas wastewater spread on roads in Pennsylvania was estimated between 20,000 m^3 and 60,000 m^3, which is equivalent to 6% of the total conventional wastewater generated in Pennsylvania[454]. Since the relative proportions between wastewater from oil wells relative to natural gas wells are not known, we use the same empirical ratio of conventional oil to conventional natural gas of 1.4. Therefore, we estimate that out of the total of 60,000 m^3 produced water annually spread on roads, ~25,000 m^3 was directly from conventional natural gas production. Given that the volume of conventional natural gas in Pennsylvania was about (5.4 × 10^9 cubic m[593]), this implies a spread water-to-gas ratio of 0.005 L/m^3 and water intensity of 0.12 L/GJ. Yet the median TDS concentration of the spread water in Pennsylvania was 293,000 mg/L[454], which is 586-fold higher than the common ecological threshold of 500 mg/L, implying an impaired water intensity of 70 L/GJ (Table 5.3).

Spills and leaks are additional pathways for wastewater from conventional and shale gas wells that negatively affect the environment (Figure 5.26). The magnitude of spills associated with unconventional shale gas development is directly associated with the density of well installation. Vengosh et al. (2014)[26] have shown that the occurrence and frequency of spills and leaks coincide with the density of the Marcellus Shale gas drilling in northeastern and western Pennsylvania. They showed that the number of reported violations increased in areas closer to higher (>0.5 well per km^2) shale gas drilling density, and the frequency of violations per shale gas well doubles in areas of higher drilling density[26]. Between 2008 and 2012, 161 of the ~1,000 complaints regarding shale gas wells in Pennsylvania received by the state described contamination that implicated oil or gas activity, of which 14% was attributed to brine salt leaking[648].

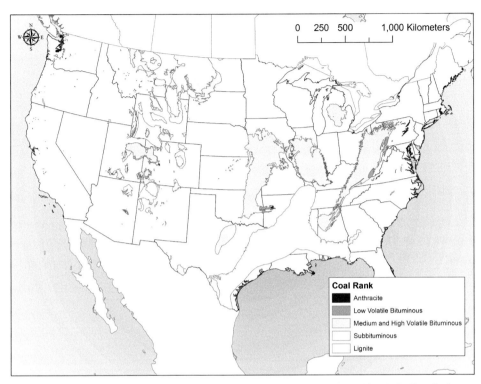

Figure 3.4 Map of major coal basins in the USA. Map generated by A.J. Kondash (Duke University) based on US Geological Survey and US Energy Information Administration (EIA)[130].

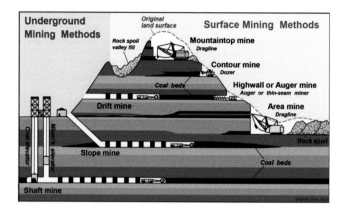

Figure 3.7 Illustration of surface and subsurface coal mining and mountaintop mining in the Appalachian Basin. Data are from US EPA website: www.epa.gov/sc-mining/basic-information-about-surface-coal-mining-appalachia.

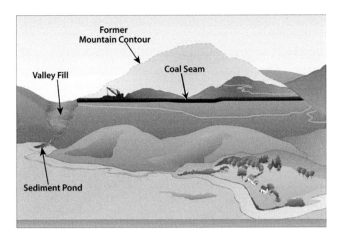

Figure 3.7 (*cont.*)

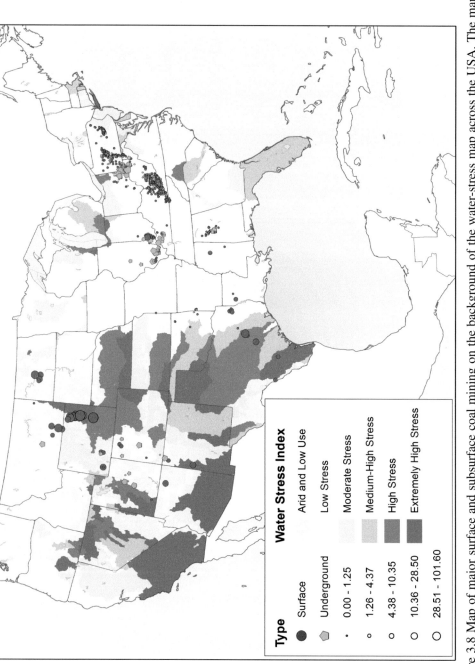

Figure 3.8 Map of major surface and subsurface coal mining on the background of the water-stress map across the USA. The map shows that the larger surface mining with higher water footprint occurs in highly stressed areas in the western USA. Map was created by A.J. Kondash (Duke University) based on EIA-7A, Coal Production and Preparation Report[172] and Aqueduct Water Stress Projections[173] .

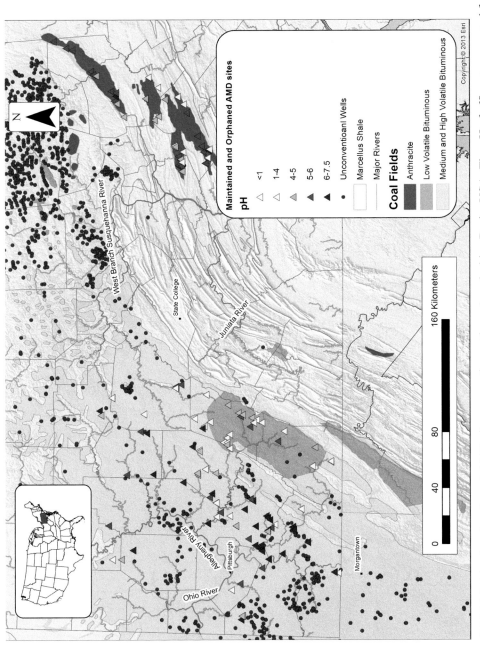

Figure 3.11 Maintained and orphaned AMD sites across the northern Appalachian Basin, USA. The pH of effluents generated from AMD (marked by triangles) are sorted by their values. The different coal types are presented. Map was prepared by A.J. Kondash (Duke University) based on ESRI Topographic Basemap[230] and EIA Layer Information for interactive state maps[130].

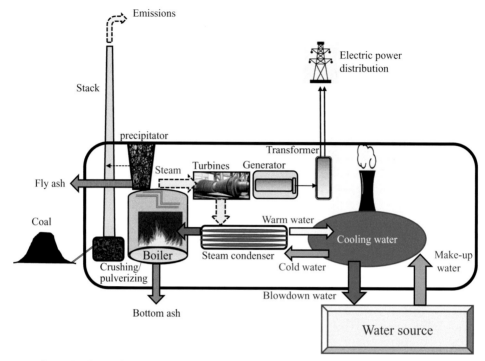

Figure 3.13 A schematic illustration of a closed-loop (recirculating) cooling tower system for thermoelectric coal-fire plants. Illustration follows the example of Plant Scherer in Georgia, USA, one of the largest coal-fire plants in the USA with a capacity of 520 MW[239].

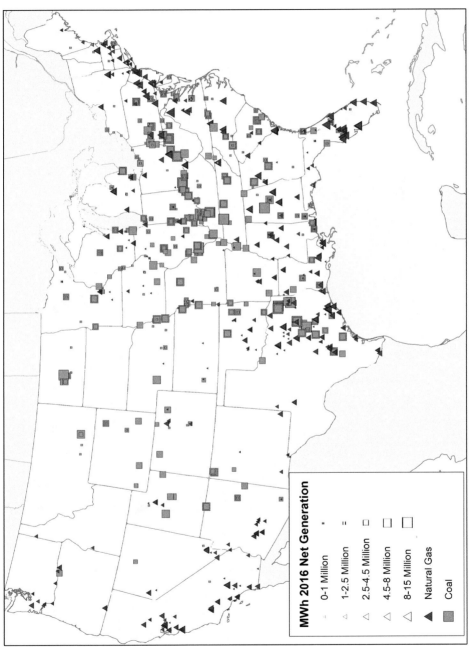

Figure 3.14 Distribution of major coal plants sorted by their capacity in the USA. Map was created by A.J. Kondash (Duke University) and is based on Kondash et al. (2019)[234] and US EIA.

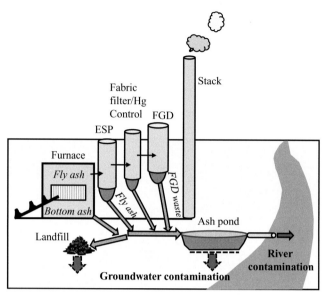

Figure 3.15 Schematic illustration of different types of CCRs generated in different stages following coal combustion in a coal-fire plant and their disposal in landfills and coal ash ponds. Modified from US EPA[254].

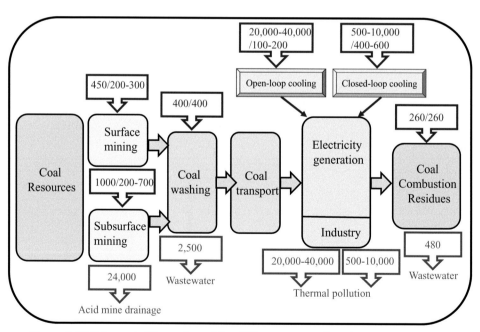

Figure 3.19 Schematic illustration of the different stages of coal generation and the water intensity values (L/GJ) for water withdrawal/consumption (in blue) and water impaired (red) at the different stages of coal production, including coal mining; coal washing and processing; coal transport, power generation, and processing of coal combustion residuals (CCRs).

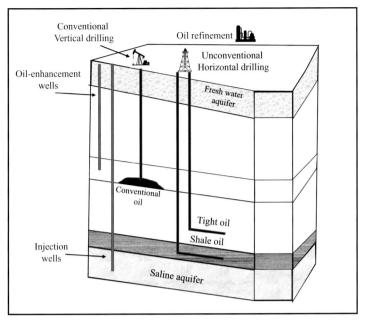

Figure 4.4 Illustration of the geological structure of conventional oil and unconventional tight and shale oil exploration through vertical and horizontal wells.

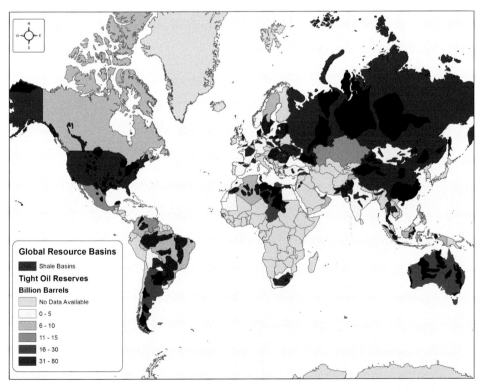

Figure 4.8 Global tight oil resources and shale basins sorted by potential production. Map was prepared by A.J. Kondash (Duke University) based on US EIA report (2013)[363] and EIA oil and gas maps website[364].

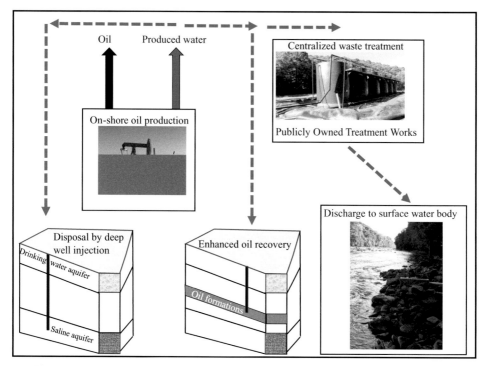

Figure 4.12 A schematic illustration of the transport and fate of oil produced water generated on-shore in the USA.

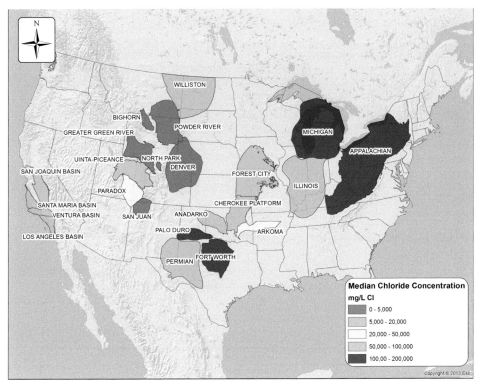

Figure 4.16 Major oil and gas basins in the USA sorted by the chloride concentrations in the formation waters. Data for map from the US Geological Survey archive of the chemistry of produced water[362]. Map prepared by A.J. Kondash (Duke University) based on US EIA oil and gas maps website[364].

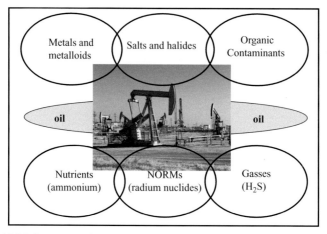

Figure 4.18 Major types of contaminants in oil produced water.

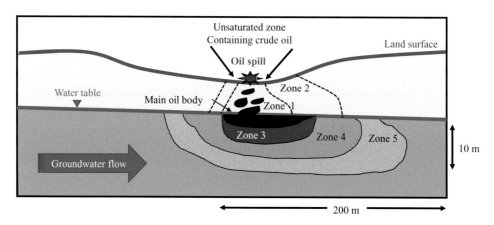

Figure 4.26 A cross-section of the crude oil spill site near Bemidji, Minnesota. Areas in the unsaturated and saturated zones were defined by changes in the groundwater quality as monitored in 1997 by the US Geological Survey[533–536]. The figure is based on figure 4 in Delin et al. (1998)[533].

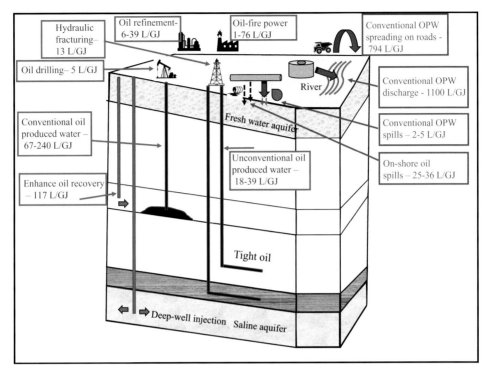

Figure 4.27 Schematic illustration of the different stages of oil exploration and processing as well as the associated water intensity values (L/GJ) for withdrawal/consumption (in blue) and water impaired (red).

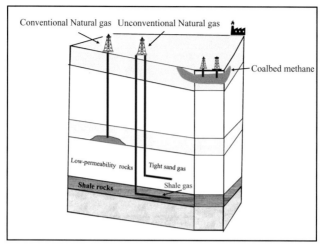

Figure 5.3 Different sources of natural gas exploration. Figure based on US EIA[15].

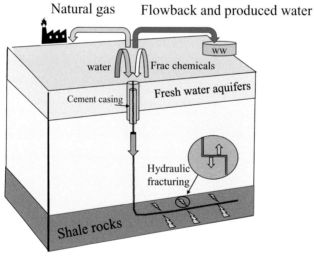

Figure 5.10 Schematic illustration of the configuration of shale gas wells and the process of hydraulic fracturing. The injection of high-pressure water and frac chemicals into separated sections along the horizontal sections of the well generates micro-fracturing in the low-permeable shale rocks. Imbibition of the injected water releases entrapped natural gas and water within the shale matrix and upon reduction of the well pressure the natural gas and water are tapped to the surface.

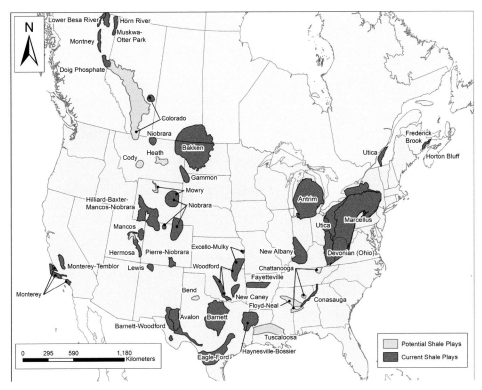

Figure 5.11 Distribution of shale gas plays across the USA. Map was generated by
A.J. Kondash (Duke University) based on US EIA oil and gas maps website[364].

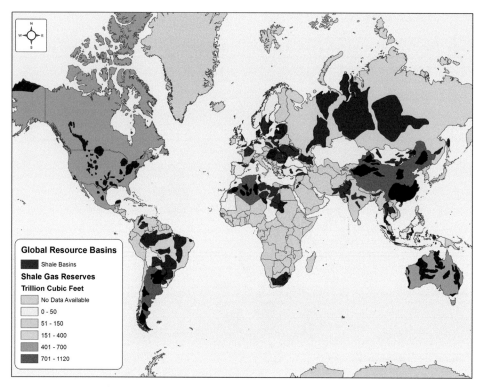

Figure 5.14 Global distribution of shale gas reserves sorted by basin potential (in trillion cubic feet). Data retrieved from US EIA (2013) report[363]. Map was made by A.J. Kondash (Duke University).

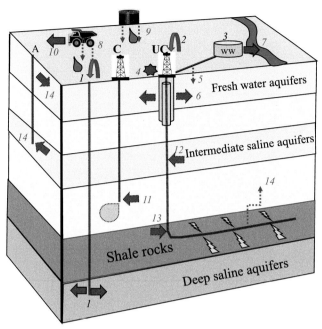

Figure 5.26 A schematic illustration of the transport and disposal of flowback and produced water from shale gas wells as well as possible contamination and leaking pathways of water and gases in the framework of shale gas exploration.

Figure 5.35 Well water on fire from a well in Pennsylvania with oversaturation of methane.

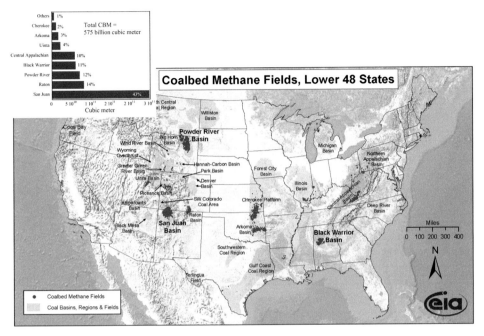

Figure 5.37 Map of the major coalbed methane basins and proved reserves in the USA. Data from the National Research Council Report (2010)[706].

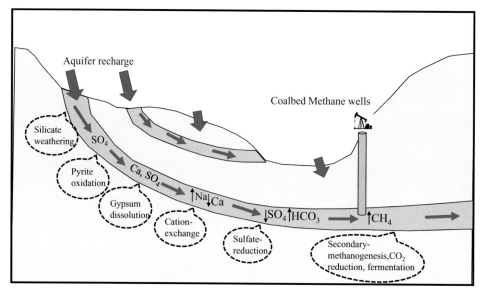

Figure 5.41 Schematic illustration of possible flowpath of groundwater in coal seam aquifers and sequence of geochemical processes along the flowpath that control the groundwater chemistry associated with CBM extraction. Modified from National Research Council Report[706] and Dahm et al. (2014)[726].

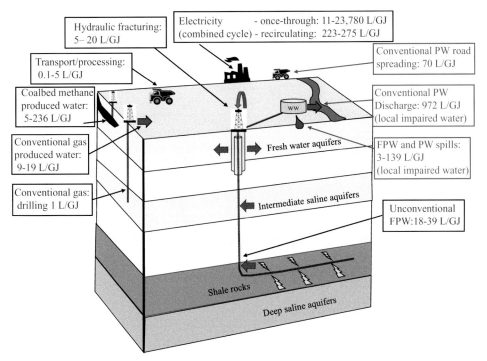

Figure 5.44 Schematic illustration of the different stages of natural gas exploration and processing as well as the associated water intensity values (L/GJ) for withdrawal/consumption (in blue) and water impaired (red).

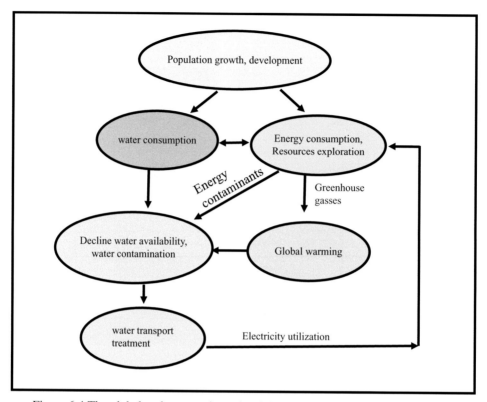

Figure 6.4 The global anthropogenic cycle of the relationships between population growth and societal development on intensification of the water and energy use, which induce water decline and degradation combined with greenhouse gas emissions that further exacerbate water availability and quality in arid and semi-arid regions. The increase of water management and treatment due to the reduction in clean water availability requires more energy that triggers more water use and contamination as well as further global warming.

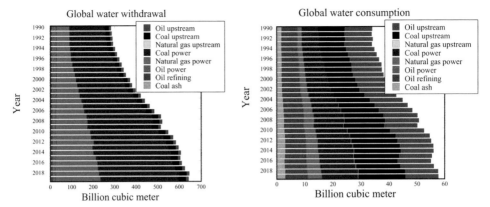

Figure 6.10 Estimates of water withdrawal and consumption for fossil fuels during the last few decades.

Figure 6.20 Flaring of unconventional oil well in the Bakken Basin of North Dakota.

Unconventional oil and gas development between 2005 and 2014 in New Mexico, Colorado, North Dakota, and Pennsylvania was associated with a high number of spills, particularly during the first three years of well operation, when shale gas wells were drilled, completed, and had their largest shale gas production. Analysis of the spill types has shown that the leaking from tanks, flowlines, and transportation was the major spill reported pathway category. The spill water was composed of FPW (wastewater), hydraulic fracturing fluids, drilling waste, and other types of fluids associated with shale gas development. About 2–16% of the wells had a spill each year, with median spill volumes ranging from 0.5 m^3 in Pennsylvania to 4.9 m^3 in New Mexico[531, 649].

Analysis of the probability of water contamination associated with the Marcellus Shale has revealed five pathways of water contamination: transportation spills, well casing leaks, leaks through fractured rock, drilling site discharge, and wastewater disposal. It was found that the potential contamination risk with hydraulic fracturing wastewater disposal was several orders of magnitude larger than the other pathways[650]. For a "best-case scenario," it was found that an individual shale gas well would release at least 200 m^3 of wastewater[650]. Based on the US EIA, ~20,000 shale gas wells have been installed since the beginning of shale gas exploration from the Marcellus Shale in Pennsylvania and West Virginia that generated a combined shale gas production of 215 BCM in 2018[593]. Assuming that each shale gas well can generate 200 m^3 of wastewater[650], we estimate that the accumulated wastewater that was generated from ~20,000 shale gas wells was about 4 million cubic m. Dividing this volume of wastewater to the volume of shale gas production (215 BCM) yields water intensity of 0.5 L/GJ. Table 5.2 shows that wastewater from the Marcellus Shale contains contaminants in concentrations that are higher by 200–300-fold as compared to ecological and drinking water standards. Using a dilution factor of 300 yields a water intensity of 139 L/GJ (Table 5.3). In contrast, analysis of the accumulated volume of fluids from spills reported for New Mexico, Colorado, and Pennsylvania[531, 649] as normalized to the volume of the natural gas produced in these states between 2008 and 2018, multiplied by a dilution factor of 300 yields a lower water intensity of 2.7–11.3 L/GJ (Table 5.3). Overall, while spills from conventional produced water and FPW can have severe impacts on the quality of the impacted water resources and associated ecology, the overall water intensity of wastewater spills on a state scale is relatively low, reflecting the local rather than state-wide impact of wastewater spills from natural gas exploration.

5.5.4.2 *Impact on Groundwater*

One of the major public concerns related to the development of unconventional shale gas is the potential leaking of wastewater from the fractures created by the hydraulic fracturing process (pathway 13 in Figure 5.26) or from leaking pipes of the shale gas wells (pathway 5), causing contamination of the overlying groundwater resources. Jackson et al. (2015)[651] investigated the fracturing depths and water use for ~44,000 wells reported between 2010 and 2013 in the USA and showed that 2,000 wells (~5%) were shallower than 1.6 km, while 350 wells (~1%) were shallower than 0.9 km, highlighting the risks of fluids migration through propagation of fractures generated from hydraulic fracturing to overlying active aquifers[651]. In contrast to stray gas contamination of groundwater that has been detected in multiple drinking water wells near shale gas development (see Section 5.6), only one case has provided evidence for groundwater contamination directly from fluids derived from hydraulic fracturing: Pavillion Field, Wyoming[652]. Conventional and unconventional (tight gas) hydrocarbon production in the Pavillion Field is primarily from natural gas from relatively shallow formations. Following complaints about the quality of local groundwater, in 2008 the US EPA began a study on the quality of the local groundwater that included installation of two monitoring wells at depth intervals corresponding to the groundwater use in nearby domestic drinking water wells. The objective of these monitoring wells was to evaluate the potential upward solute transport of compounds associated with hydraulic fracturing and well stimulation. DiGiulio and Jackson (2016)[652] indicated that high concentrations of inorganic (potassium and chloride) and organic compounds used for hydraulic fracturing were found in the monitoring wells, suggesting upflow of stimulation fluids and contamination of the overlying aquifer. They emphasized that well stimulation in the Pavillion Field occurred at a relatively shallow depth (<500 m), which is close to the depth of some of the deepest domestic groundwater wells used in the area[652]. As far as we are aware, this was the only case study with a clear indication of subsurface migration and the actual impact of frac water on nearby groundwater quality.

The more common occurrence of groundwater contamination from conventional gas and shale gas development is through leaks from surface activities and unlined ponds and storage reservoirs (pathways 5, 8, and 9 in Figure 5.26). Evidence for leaking of fracturing fluid chemicals from the surface storage was found in groundwater in northwestern Pennsylvania[653]. Analysis of the groundwater revealed the occurrence of volatile organic compounds, including gasoline range and diesel range compounds with the presence of bis(2-ethylhexyl) phthalate, which is a disclosed hydraulic fracturing additive. Integration of the organic data with inorganic geochemical data indicated transport of organic chemicals to groundwater via an apparent accidental release of fracturing fluid chemicals on the surface rather than subsurface flow of frac fluids from the underlying shale formation[653]. Likewise, organic chemicals used for hydraulic fracturing and flowback water

(2-n-Butoxyethanol) were found in foam generated in drinking water wells near shale gas wells in Pennsylvania, indicating local groundwater contamination from frac chemicals, but not directly from flowback water due to the lack of changes in groundwater inorganic chemistry and salinity[654].

Another mechanism for groundwater contamination associated with oil and gas development is through unplugged abandoned oil and gas wells[655] (pathway 14 in Figure 5.26). Given that there are over 4.3 million oil and gas wells and more than 15 million water wells drilled in North America, unplugged or improperly plugged oil and gas wells are a potential threat and pose a pollution pathway to overlying aquifers from the legacy of oil or gas well developments[656]. Groundwater contamination from abandoned oil and gas wells has been demonstrated in a case study in the Appalachian region of western Pennsylvania, where saline and acid coal mine drainage contaminated a shallow aquifer through artesian discharge from abandoned oil and gas wells[657]. Likewise, artesian flow of saline and sulfate-rich water to the surface through unplugged oil and gas wells drilled in the late 1800s and early 1900s before the advent of government regulations continues to affect surface water and potentially threaten potable groundwater aquifers, as shown in southwestern Ontario[658]. Evidence for direct contamination of shallow groundwater through direct leaking of abandoned conventional oil and gas wells has also been reported in other locations in the Appalachian Basin in Pennsylvania[463, 464] and Garfield County in Colorado[465].

The ability to detect the sources of groundwater contamination is not simple in some areas where fresh water aquifers are affected by upflow of naturally occurring saline groundwater. A critical question common to debate is the hydraulic connectivity between the shale gas formations and the overlying shallow drinking water aquifers[466]. Warner et al. (2012)[466] investigated the geochemistry of groundwater from northeastern Pennsylvania and found mixing relationships between shallow groundwater and a deep formation brine that caused groundwater salinization in some areas. Based on the geochemical fingerprint in the salinized groundwater, they suggested possible migration of the Mid-Devonian Marcellus brine to the shallow aquifers through naturally occurring pathways and not due to contamination of wastewater from shale gas operation. Yet the presence of deep-source formation water suggests conductive pathways and specific geostructural and/or hydrodynamic regimes in northeastern Pennsylvania that are at increased risk for the uprising of saline water and contamination of shallow drinking water resources[466]. Similarly, Harkness et al. (2017)[410] investigated the geochemical variations of groundwater before, during, and after hydraulic fracturing and in areas of shale gas development in northwestern West Virginia. Using a wide set of geochemical tracers including inorganic geochemistry, stable isotopes of strontium, boron, lithium, and carbon, they found that the saline groundwater originated via naturally occurring processes, likely from the migration of deep brines to the shallow aquifers. These observations were consistent with the lack of changes in water quality observed in drinking water wells following the installation of nearby shale gas wells[466]. A similar investigation of shallow groundwater from a Triassic karst aquifer in areas of shale gas development in the Sichuan Basin in China used geochemical tracers to detect possible contamination of the FPW on the shallow groundwater. The study did not detect evidence for the impact of FPW on the groundwater salinity and chemistry, and no statistical difference of the overall groundwater quality was observed between groundwater from active shale gas extraction areas relative to nonactive areas, excluding the possibility of contamination from shale gas wastewater[628]. As mentioned in Section 5.5.2, the ability to delineate between the impact of naturally occurring saline water and direct contamination of produced water or FPW from shale gas wastewater depends on the sensitivity of the geochemical tracers (i.e., a distinctive isotope difference between the sources) for the different contamination sources, as well as the degree of resistance for chemical degradation processes during groundwater flow in the aquifer. Therefore, conservative constituents with minimum interactions and exchange with the aquifer rocks (e.g., Br/Cl ratio) would better preserve and mimic the composition of the contamination sources,

Figure 5.35 Well water on fire from a well in Pennsylvania with oversaturation of methane. (A black-and-white version of this figure will appear in some formats. For the colour version, refer to the plate section.)

whereas organic constituents may not be preserved along the groundwater flow, and thus would not be ideal proxies for detecting the sources of groundwater contamination.

5.6 Stray Gas Contamination of Groundwater Resources Induced from Hydraulic Fracturing

Similar to the effects of oil spills, there is visual evidence of natural gas presence in water, often in the form of bubbles, or in some cases, the water can be flammable due to oversaturation of methane in the water (Figure 5.35), which has become one the symbols of the environmental risks of hydraulic fracturing. The images of flames emerging from a kitchen faucet shown in the iconic movie *Gasland* have captured the imagination of millions across the globe, helping to spur public opposition to hydraulic fracturing. Environmental activism against hydraulic fracturing has contributed to it being banned in some European countries as well as in some US states like New York. Whereas this book has sought to illustrate the widespread environmental impacts that are derived from conventional oil and coal exploration and combustion, particularly on water availability and quality, public attention has largely galvanized around unconventional shale gas development in recent years. In particular, stray (fugitive) gas contamination of drinking water wells near shale gas drilling sites has become a focal point of investigation of the negative impact of unconventional shale gas development on water resources.

Although dissolved methane in drinking water is not currently classified as a health hazard for ingestion, it is an asphyxiant in enclosed spaces and degassing can trigger explosions and fire hazard[659, 660]. Yet the occurrence of methane in drinking water wells poses a potential flammability or explosion hazard to homes with private domestic wells. The saturation level of methane in near-surface groundwater is about ~28 mg/L (~40 cc/L), and the US Department of the Interior recommends monitoring if water contains more than 10 mg/L (~14 cc/L) of methane, and immediate action if concentrations exceed a level of 28 mg/L. In Pennsylvania, the safety threshold is lower (7 mg/L), from which household utilization of methane-rich groundwater is not recommended. Following early development of shale gas exploration in Pennsylvania, residents living near fracking sites began to recognize that their well water contained methane, and in some cases was oversaturated with respect to methane, which caused their water to become flammable (Figure 5.35). The bitter debate that followed concerned whether the occurrence of methane in drinking water wells near shale gas drilling sites is derived from the leaking of the shale gas wells, uprising from the fractures created in the underlying shale formation, or whether it was derived from naturally occurring processes that were not associated with shale gas development. Naturally occurring methane had been detected in shallow groundwater from many aquifer systems across the Appalachian Basin. In many cases, the stable-isotope fingerprints of the methane showed high (positive) δ^{13}C-CH$_4$ composition, suggesting that the high methane in shallow aquifers from this region is predominantly thermogenic in origin[615–617] and reflects a regional upflow of thermogenic methane to the shallow aquifers (see Section 5.2.2 on the chemistry of natural gas).

The first study that presented the risks of stray gas contamination of shallow groundwater in active gas-extraction areas was Osborn et al. (2011)[660]; this study showed high methane concentrations in drinking water wells that were located less than 1 km from shale gas wells in northeastern Pennsylvania, with concentrations that increased with proximity to the nearest gas well. The study also found that the δ^{13}C-CH$_4$ values of dissolved methane and ratios of methane-to-higher-chain hydrocarbons in shallow groundwater near shale gas were similar to the thermogenic Marcellus methane source, while the lower-concentration wells away from shale gas had carbon isotopic signatures reflecting more biogenic or mixed biogenic/thermogenic methane sources. No evidence was found for contamination of drinking water samples with deep saline brines or FP water[660]. Evidence for stray gas contamination was further reinforced by Jackson et al. (2013)[620] who analyzed 141 drinking water wells across the Appalachian Basin of northeastern Pennsylvania and showed statistically higher concentrations of methane, ethane, and propane and the C_1–[C_{2+} C_3] ratios (i.e., indicators of thermogenic gas; see Figure 5.9), as well as higher ratio of the noble gas ^{4}He to methane for wells located <1 km from shale gas wells. The proximity to the nearest shale gas well was found to be a more significant factor for controlling methane concentrations as opposed to geological and hydrogeological factors. These studies have emphasized the distinction between naturally occurring background hydrocarbon gases with thermogenic composition and groundwater impacted by stray gas contamination in wells located near shale gas drilling sites[620, 660]. Additional evidence for stray gas contamination was documented from a study of groundwater samples collected in the Denver-Julesburg Basin of northeastern Colorado[661]. In this study, 593 out of 924 samples contained methane, and 42 water wells from 32 separate cases near the Wattenberg field had high methane with a thermogenic or mixed microbial-thermogenic geochemical signature, reflecting stray gas contamination from underlying oil and gas producing formations[661]. The data from 32 cases were consistent with complaints on water quality concerns, reports filed to the state agency, occurrence of bubbles in water, explosions, and thermogenic gas detected during baseline sampling. The impacted groundwater wells were all located within 1 km of the known oil and gas wellbore failure, consistent with observations in the Marcellus Shale[660].

In contrast, several other studies have disputed these observations and argued that the higher methane concentrations in shallow groundwater in the Appalachian Basin were naturally occurring, and could be explained by topographic and hydrogeological factors associated with groundwater discharge zones[629, 662, 663]. Likewise, analysis of natural gas from eight monitoring wells installed in areas of shale gas development did not detect stray gas contamination, but rather naturally occurring source derived from a deep source underlying the aquifer that is associated with the upflow of saline groundwater from a deeper source[664]. The occurrence of deep saline water with high concentrations of thermogenic methane that upflows to shallow aquifers overlying shale gas plays has been shown in several studies across the Appalachian Basin overlying the Marcellus Shale[410, 617, 621], as well as other shale plays such as the Haynesville Shale[665], Eagle Ford Shale (more biogenic methane)[666], and Barnett Shale in Texas, with a strong thermogenic signal[667]. Likewise, investigation of groundwater overlying the Fayetteville Shale in north-central Arkansas showed no evidence of methane contamination with a mostly biogenic isotopic fingerprint[668]. Many studies in the USA, Canada, Poland, the UK, South Africa, and China have emphasized the importance of baseline studies of the regional groundwater and establishing reliable datasets that would provide better assessment of changes in the water and gas geochemistry[410, 621, 669–687]. The natural flux of methane into shallow aquifers and the possible secondary process (e.g., anaerobic oxidation of dissolved methane[688]) that can modify the original chemistry and isotopic composition of natural gas (i.e., further enriched in δ^{13}C-CH$_4$[688]) make the distinction and ability to delineate stray gas contamination more challenging[622].

In order to overcome these obstacles and delineate stray gas contamination against a backdrop of naturally occurring salt- and gas-rich groundwater, Darrah et al. (2014)[618] have integrated noble gas geochemistry with traditional hydrocarbons abundance and isotopic compositions. Noble gases are tracers that are not affected by secondary reactions and their occurrence and isotope ratios in groundwater reflect their sources (e.g., young meteoric water, deep and old "crustal" water) and solubility in water. Stray gas contamination was delineated in cases where the methane concentrations in the impacted groundwater was much higher and deviated from the general correlation pattern between the salinity and methane that characterizes the background groundwater, combined with high δ^{13}C-CH$_4$ and C$_1$–[C$_{2+}$ C$_3$] ratios and high concentrations of helium reflecting the contribution of deep and old water sources. In contrast, background groundwater had higher proportions of atmospheric gases like neon and argon, reflecting recent recharge of meteoric water. The study also documented changes in water quality over time, reflecting the spatial spreading of the contaminated groundwater plume. The study investigated drinking water wells overlying the Marcellus and Barnett shales in Pennsylvania and Texas, and identified eight discrete clusters of fugitive gas contamination, of which four clusters reflect gas leakage from intermediate-depth formations through failures of annulus cement (pathway 12 in Figure 5.26), three clusters with geochemistry identical to shale production gases that seem to implicate faulty production casings (pathway 13), and one to an underground gas well failure. Importantly, due to the expected differential solubility of noble gas between free gas and dissolved gas in water, the study was able to rule out gas contamination by upward migration from depth triggered by horizontal drilling or hydraulic fracturing (pathway 14)[618]. Likewise, noble gas geochemistry has provided clear evidence for stray gas contamination in drinking water wells near the Marcellus Shale gas drilling sites in Lycoming County, Pennsylvania, reflecting upward migration of gas from the Marcellus Formation in a free-gas phase[689]. The abundance of noble gas was also used in the study of shallow groundwater overlying the Barnett Shale in Texas, indicating the arrival of gas from geological formation underlying the major aquifers, and yet ruling out a link to hydraulic fracturing due to a lack of correlations of noble gas ratios with distance to the nearest gas production wells[690, 691]. In another research area in the Marcellus Shale in West Virginia, a detailed study of groundwater geochemistry before, during, and after the installation of shale gas wells did not reveal any changes in methane occurrence, hydrocarbons ratios, or noble gas geochemistry[410].

One of the major arguments that was made against the concept of stray gas contamination cases was the inconsistency between the gas chemistry of the impacted shallow groundwater and the composition of the shale gas, which is typically much more thermogenic than the dissolved gas associated with shallow groundwater. Yet Muehlenbachs (2013)[692] showed direct evidence for methane leakage through surface casing vent flow (i.e., methane gas collected at the well surface, not dissolved in groundwater) that originated from geological formation located at intermediate zones overlying the shale rocks in newly completed and hydraulically stimulated horizontal shale gas wells in the Montney and Horn River areas of northeastern British Columbia, Canada. Consistently, stray gas contamination associated with conventional oil wells was also detected in Alberta, in which methane sourced from intermediate formations leaked into shallow aquifers and not from underlying production formations such as the Lower Cretaceous Mannville Group[693]. Methane leaking from the annulus of conventional oil and gas wells was also demonstrated in the Appalachian Basin[463, 464]. Overall, stray gas contamination of shallow groundwater can result from either natural gas leaking up through the well annulus, typically from shallower (intermediate) formations, or through poorly constructed or failing well casings from the well penetrated into the target formations[694]. Leaking of natural gas through well annulus or failing casing has been demonstrated in many cases[19, 348, 648, 695, 696].

For example, Ingraffea et al. (2014)[697] investigated 41,000 conventional and unconventional oil and gas wells that were drilled between 2000 and 2012 in Pennsylvania and showed a common phenomenon of structural integrity of casing and cement in oil and gas wells. The study found higher risk (1.6-fold) of compromised structural integrity in an unconventional gas well relative to a conventional well drilled within the same time period, and that unconventional gas wells in northeastern Pennsylvania, where previous studies indicated occurrence of stray gas contamination[619, 620, 660] are at a 2.7-fold higher risk for leaking relative to the conventional wells in the same area[697].

In addition to the direct occurrence of methane in drinking water wells, the presence of fugitive methane could change the oxidation-reduction conditions in the aquifer and trigger different related processes such as sulfate reduction and reductive dissolution of oxides in the aquifer that would mobilize redox-sensitive elements such as manganese, iron, and arsenic from the aquifer matrix and affect the groundwater quality. Following a catastrophic underground blowout during the drilling of a gas well in the Netherlands, large volumes of natural gas were released from the reservoir to the surface and contaminated the overlying groundwater[688]. Investigation of the groundwater chemistry revealed that elevated levels of thermogenic methane coincided with high dissolved iron and manganese concentrations at the fringe of the methane plume. It was suggested that microorganisms catalyzed anaerobic oxidation of methane coupled with reduction of ferric oxides in the aquifer solids to produce soluble Fe^{2+}. Following the move of the methane plume, the iron concentrations decreased due to the depletion of solid-phase ferric oxide minerals and the reverse in the oxidation state[688]. A similar phenomenon was observed in domestic water wells impacted by methane leaking from Marcellus Shale gas wells in Pennsylvania; iron concentrations increased following the increase of methane in the water and later decreased, together with a decrease in sulfate concentrations[689]. These temporal trends in iron and sulfate concentrations were attributed to iron release due to reducing conditions followed by source depletion and sulfate reduction processes induced by the newly elevated concentrations of methane[689]. The concentrations of redox-active species of sulfate and iron in groundwater were used as an indirect proxy for stray gas contamination across drinking water wells in Pennsylvania, where only a small percentage (0.08%) of the wells had shown this geochemical feature[698]. Evidence for dissimilatory bacterial sulfate reduction of fugitive methane near conventional oil wells in Alberta was associated with the combined H_2S formation, ^{13}C-depleted bicarbonate, and low sulfate concentrations relative to the regional groundwater[699]. Likewise, elevated arsenic, selenium, strontium, and TDS concentrations exceeding the US EPA Drinking Water MCL in some private water wells located within 3 km of active natural gas wells in aquifers overlying the Barnett Shale formation of North Texas were suggested to indicate mobilization of naturally occurring contaminants from the aquifer rocks due to changes in the aquifer redox condition triggered by fugitive methane contamination[700].

Overall, in spite of clear evidence for stray gas contamination of groundwater near shale gas sites in the Marcellus[618, 689], Barnett[618], and Denver-Julesburg (northeastern Colorado)[661] shales in the USA, as well as the Montney and Horn River shales[692, 693] in Canada, the extent and magnitude of migration of natural gas from leaking shale gas wells still remains contested[701]. It is clear that the utilization of traditional geochemical tools of hydrocarbon abundance and stable isotope ratios of carbon and hydrogen (described in Section 5.2.2) are important, but not always sufficient to detect contamination and to distinguish naturally occurring thermogenic methane derived from a regional flow of underlying formations from fugitive methane leaking from shale gas wells. Therefore, additional geochemical tools such as noble gas geochemistry are needed[619, 622]. The noble gas methodology has detected stray gas contamination in a subset of wells from limited sites in northeastern Pennsylvania[618, 619], Sugar Run in Lycoming County in northern Pennsylvania[689], and Parker County in Texas[618], but ruled out stray gas contamination in West Virginia[410]. Furthermore, the differential solubility of noble gas when dissolved in water provides the ability to detect the mechanism of stray gas contamination and to indicate that most of the cases reflect leaking wells through faulty well construction and not from the uprising of natural gas directly from the shale formations. Despite the estimated one million unconventional oil and gas wells that were installed in the USA[701], the research on the scope and magnitude of fugitive natural gas contamination in groundwater resources is restricted to only a few cases without the ability to fully access its impact. Also it is important to recognize that one-time or "snap-shot" evaluation of the groundwater quality may not be sufficient to detect contamination, as it shown that temporal changes in the groundwater quality can be important to evaluate the migration of methane in the aquifer and associated changes in the redox state of the aquifer system[688, 689]. A common argument against the concept of stray gas contamination is the need for a baseline study to fully confirm the evidence of contamination, which is often unavailable. While baseline evaluation is important, water contamination can be detected by comparing the quality of the impacted groundwater to the regional groundwater located away from the contamination sites, albeit under the same hydrogeology settings (e.g., aquifer lithology).

5.7 Coalbed Methane

5.7.1 Coalbed Methane Production

Coalbed Methane (CBM) is natural gas extracted from coal deposits through the pumping of water-saturated coalbeds at relatively shallow depths[702] (Figure 5.3). Since 2005, global resources of CBM were estimated at 143 trillion cubic m,

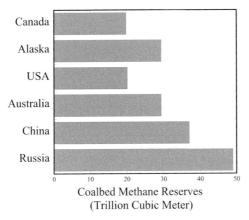

Figure 5.36 Distribution of global coalbed methane reserves. Data from Al-Jubori et al. (2009)[702]. Note the separation of the continental USA and Alaska

whereas only 1 trillion cubic m had been recovered[703]. The largest global CBM reserves are located in Russia, China, Australia, Alaska, Canada, and the USA[702, 703] (Figure 5.36). Yet in 2015, the annual global production was only a small fraction of this potential, mostly from the USA, Canada, Australia, and China[704]. In spite of having the largest global coal production, China's CBM production lags behind the USA and Australia, and in 2015 CBM production in China was 5 BCMs compared to 33.6 BCMs in the USA and over 30 BCMs in Australia[705]. Therefore, much of the world's coalbed methane recovery potential remains untapped, as only 0.6% of global CBM resources are recovered[704]. In the USA, the total proven CBM reserve is estimated at 575 BCMs (Figure 5.37). Coalbed methane production in the USA began in the early 2000s and reached a peak in 2008–2009 with an annual production of 57 BCMs, which was about 8% of the total natural gas production in the USA (Figure 5.38)[574]. Yet the annual CBM production declined to 27 BCM in 2018, which was only 2.6% of the total annual natural gas production in the USA[574]. Given its large potential, CBM was considered during the early 2000s as the major alternative substitute for the declining conventional natural gas[702]. Yet since the rise of shale gas in the mid-2000s (Figure 5.6), CBM production has declined. The majority of CBM production (80%) is generated in the western part of the USA, where San Juan (43%), Powder River (12%), and Raton (14.2%) are the major producing basins[706].

Water plays a critical role in coalbed methane production. While the unconventional method of hydrological fracturing involves *injection* of water to extract shale gas retained in organic matter within shale rocks (see Section 5.2.3), CBM production involves the *extraction* of groundwater through the pumping of water from saturated coalbed formations[702]. The pumping of coalbed formations removes water that is adsorbed to the coal, reduces pressure, and enhances methane desorption from the coal to the well (Figure 5.39). Experimental work has

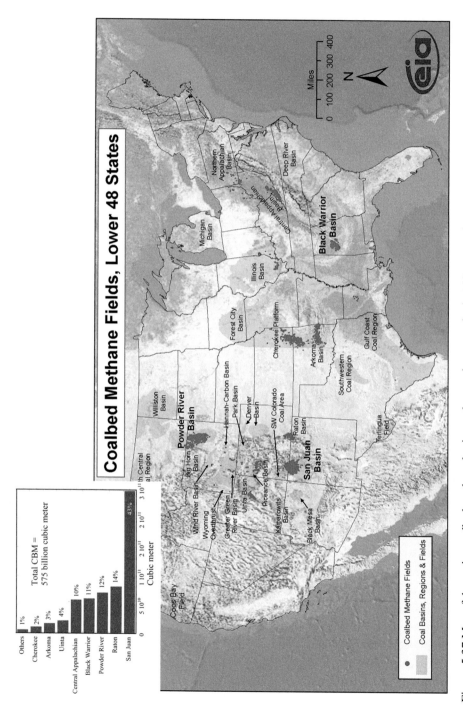

Figure 5.37 Map of the major coalbed methane basins and proved reserves in the USA. Data from the National Research Council Report (2010)[706] . (A black-and-white version of this figure will appear in some formats. For the colour version, refer to the plate section.)

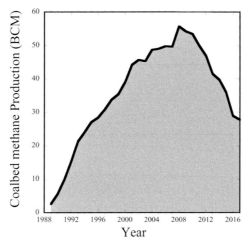

Figure 5.38 The rise and fall of coalbed methane production (cubic m) and percentage of total natural gas production in the USA. Data from US EIA on natural gas production in the USA[574]

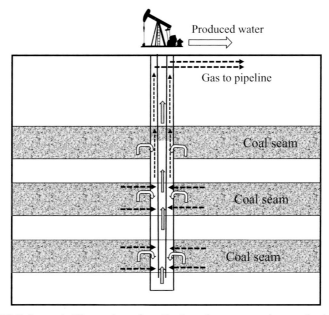

Figure 5.39 Schematic illustration of coalbed methane extraction method through a vertical well across multiple coal seams. Hydraulic fracturing through the coal seams allows groundwater to flow from the coal seams through the annulus[708] and is pumped out through the tubing well to generate CBM produced water. Methane (dashed arrow) is released from the coal seams and flows into the annulus between the casing and the tubing and rises to the surface where it is piped to a compressor station. Modified from Al-Jubori et al. (2009)[702]

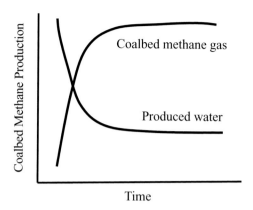

Figure 5.40 The inverse relationship between CBM and CBM water production in a typical CBM well. The pumping of groundwater and decline of the groundwater table reduce pressure in the coal seam aquifers that trigger the release of CBM attached to the coals.

shown that both methane and water adsorptions decrease with temperature, while having different responses to pressure; while water adsorption continuously decreases with pressure, methane adsorption first increases, and then decreases with pressure. Since water competes with methane for adsorption sites, its removal affects the methane adsorption capacity of coal and facilitates the methane release from the coal seams[707]. More advanced technology has utilized hydraulic fracturing in vertical wells to enhance the flow of groundwater from the coal seams into the CBM wells[702, 708] (Figure 5.39). Hydraulic fracturing has been used to enhance the flow of methane gas in the San Juan, Raton, Piceance, and Uinta basins, whereas in the Powder River Basin it was used only infrequently due to the high natural permeability of the shallow, methane-bearing coal seams[706]. Given the high volume and quality of produced water that is co-extracted with coalbed methane, the disposal and handling of the produced water is one of the major management challenges associated with CBM production[704, 706]. The water production depends on the aquifer permeability, and large variations in water production per well have been observed between basins. The water production rate from CBM wells is typically high during early stages and declines over time with the rise of CBM production (Figure 5.40). In addition, the permeability of the coalbeds is a critical factor in CBM extraction[709]. While higher-rank coals would have higher potential CBM extraction, high-rank coal seams have also typically low permeability, which reduces the ability to extract water. This has been identified as one of the limiting factors in the ability of China to develop CBM production[705].

Overall, the boom and bust of CBM production in the USA is striking; only a fraction of the 24,000 wells drilled during the peak production period of the late

2000s in the Powder River Basin in Wyoming are currently (2020) producing natural gas[108], with an estimated 3,000 wells orphaned, leaving behind water contamination and a broader environmental legacy[710]. A decade later, this lesson may be even more relevant for the shale gas boom in the USA that has faced a massive reduction in oil and gas production following the COVID-19 pandemic and rebounded following the recovery of the US economy in 2021.

5.7.2 Coalbed Methane Produced Water

As part of the CBM extraction, a large volume of groundwater is co-generated with the methane and defined as CBM produced water. In the western USA alone, the annual CBM produced water up until 2006 was about 200 million cubic m, from which 77% was produced in the San Juan Basin in New Mexico and 13% in the Powder River Basin in Wyoming and Montana[706]. The CBM produced water is commonly discharged into disposal ponds, stream channels, or reinjected into shallow groundwater aquifers. The quality and salinity of the groundwater from the coal formations are major limiting factors for the ability to dispose the CBM produced water and/or reuse it for other applications such as irrigation[706]. Furthermore, the proximity of shallow aquifers that supply drinking water to the underlying coal aquifers, and the possible direct hydrological connections, could facilitate the flow of saline groundwater to the shallow aquifers and contamination of important drinking water resources, particularly in the water-scare regions of the western USA. The key for the salinity level of the CBM produced water is the salinity of the original groundwater in the coalbed aquifers superimposed with secondary geochemical processes that could modify the original groundwater chemical composition[661, 711–715].

The salinity of CBM produced water varies significantly between and within the basins. For example, the salinity of the formation water in the Warrior Coal Basin in Alabama varies from low salinity (TDS < 3,000 mg/L) in the recharge southeastern zone to high saline water (up to 31,000 mg/L) in the center and northwestern part of the basin[716–718]. Likewise, in the Powder River Basin, Wyoming, the salinity of the CBM co-produced water increases with distance from the recharge zone, with a TDS increase from 500 mg/L to 4,000 mg/L at 60 km downstream from the recharge zone[719–722]. Similarly, in the Fruitland Formation of the San Juan Basin, the salinity of CMB water in the recharge area is low (TDS –1,000–5,000), while the salinity of groundwater in the confined aquifer is much higher (TDS~ 23,000)[706, 723].

In spite of the original thermogenic origin of natural gas generated with coalbed seams, microbial secondary methanogenesis affects the carbon isotope composition of methane extracted from coal seams with typical δ^{13}C-CH$_4$ less than –50‰,

Table 5.4. *Chemical composition of CBM produced water wells penetrating coal beds in the Fort Union Formation, Powder River Basin, Wyoming and Montana**

Element	Median	Minimum	Maximum
TDS	980	230	2,900
Calcium	30	1.8	78
Magnesium	13	0.6	46
Sodium	250	81	1,160
Bicarbonate	1,110	270	3,310
Sulfate	0.15	<0.3	530
Chloride	9.5	1.8	130

* Data from Rice et al. (2008)[721]

relatively lower δ^2H-CH$_4$, and positive δ^{13}C-DIC in the residual DIC. In most cases, microbial coal bed gases have carbon and hydrogen isotope fractionations close to those expected for CO$_2$ reduction[721, 724]. Likewise, the chemical composition of the CBM produced water is almost exclusively sodium-bicarbonate (Na–HCO$_3$) with a very low dissolved sulfate content (Table 5.4). This universal chemical composition indicates a sequence of geochemical processes, including (1) sulfate reduction that generates dissolved bicarbonate in the groundwater; (2) oversaturation and secondary precipitation of calcite and dolomite minerals, in which cases secondary mineral precipitation reduces the calcium and magnesium contents; and (3) reverse base-exchange reaction that increases sodium in the residual groundwater[711, 715, 719, 721, 722, 725]. In contrast to produced water from oil and gas, CBM produced water is commonly characterized by Na–Cl ratios >1, implying a net enrichment of Na$^+$ relative to Cl contents. In addition to sulfate reduction, bicarbonate is also produced through methane-fermentation processes that occur in deeper seams[715], which causes an enrichment of ^{13}C in the residual dissolved inorganic carbon (δ^{13}C-DIC> 20‰) that is distinctive from regional groundwater[721, 724]. The sequence of secondary water–rock interactions induced from sulfate reduction and methane-fermentation processes results in major modification of the chemistry of the regional groundwater that flows through the coal seams as illustrated in Figure 5.41. Overall the chemistry of CBM produced water is composed of three major components: (1) solutes originated from recharge water and water–rock interactions during the early stages of groundwater recharge to the aquifer: (2) mobilization of soluble solutes from the aquifer rocks and geochemical processes at different stages of groundwater flow in the aquifer controlled by the oxidation-reducing conditions of the coal seam aquifers (Figure 5.41); and (3) the original coal depositional environment, in particular the occurrence of saline water in marine settings[726]. CBM produced waters from coal formations associated with marine basins have typically higher salinity,

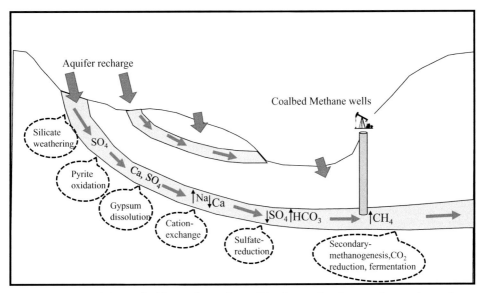

Figure 5.41 Schematic illustration of possible flowpath of groundwater in coal seam aquifers and sequence of geochemical processes along the flowpath that control the groundwater chemistry associated with CBM extraction. Modified from National Research Council Report[706] and Dahm et al. (2014)[726]. (A black-and-white version of this figure will appear in some formats. For the colour version, refer to the plate section.)

chloride, boron, and trace metal concentrations relative to CBM produced waters from continental depositional environments. For example, CBM produced waters from the deep Lower Cretaceous Mannville Formation in in Western Canada are highly saline (TDS of 31,000–89,000 mg/L) with Na-Cl composition and high concentrations of boron (6–31 mg/L) and toxic trace elements such as arsenic (50–160 μg/L), selenium (440–1,250 μg/L), and lead (15–92 μg/L). In contact, CBM produced waters from the overlying Upper Cretaceous Horseshoe Canyon/ Belly River Group have much lower salinity (TDS of 470–14,700 mg/L) with a typical Na–HCO$_3$ composition and much lower concentrations of boron (0.05–1.3 mg/L) and trace elements (e.g., arsenic 1–13 μg/L)[727]. Likewise, CBM produced waters from the Powder River basin in the western USA have relatively low salinity and overall low trace element concentrations[722], whereas CBM produced water from the Piceance, Uinta, Raton, and San Juan basins contain higher concentrations of chloride and overall higher TDS[706] (Table 5.5). The Powder River Basin contains primarily sodium bicarbonate-type formation water and low TDS, whereas the Piceance, Uinta, Raton, and San Juan basins contain produced water composed of mixed sodium bicarbonate and sodium-chloride types and generally high TDS. Overall, the preservation of the original marine formation waters that were trapped within the coal seam pores controls the salinity and

Table 5.5. *Water quality of CBM produced water from coal basins around the word*[727, 732, 733]

Coalfield/basin	Country	TDS	Sodium	Bicarbonate	Chloride
Powder River	USA	250–2,800	12–1,200	236–3,000	BDL–280
San Juan	USA	150–39,300	36–7,400	117–13,900	BDL–20,100
Bowen	Australia	2,650–8,090	1,100–3,060	300–2,860	747–4,530
Mannville Formation	Canada	31,100–89,000	11,510–32,000	100–1,650	17,000–52,000
Horseshoe Canyon Formation	Canada	470–14,700	100–4,400	200–8,900	40–1,700
Jincheng	China	690–3,400	120–950	350–2,280	11–150
Jharia	India	550–2,280	330–2,270	1,600–3,020	60–345

Table 5.6. *Coalbed methane natural gas and produced water production in major CBM basins in the USA**

Basin	Water production (m^3/y)	Gas production (m^3/y)	Water-to-gas ratio	Water intensity (L/GJ)	N wells	TDS range (mg/L)	Dilution factor	Impaired water intensity (L/GJ)
Powder River	1.14×10^8	1.23×10^{10}	9.26×10^{-3}	236.2	18,000	250–800	3	709
San Juan	7.31×10^6	3.43×10^{10}	2.1×10^{-4}	5.4	7,500	10,000–100,000	100	544
Raton	2.1×10^7	4.17×10^9	5×10^{-3}	127.4	3,400	900–30,000	30	3,822
Piceance	4.77×10^4	7.08×10^6	6.7×10^{-3}	171.8	110	6,400–42,700	40	6,872
Uinta	4.93×10^6	2.09×10^9	2.4×10^{-3}	60.1	1,255	10,000	10	601

* The produced water-to-natural gas ratios are used to evaluate the water intensity (L/GJ) of CBM produced water. Due to the large variations in the salinity (TDS) of the CBM produced water, the dilution factor for restoration to TDS = 500 mg/L levels varies and thus the impaired water intensity is induced by the discharge of CBM produced water to the environment.

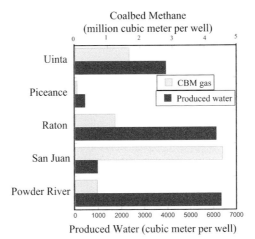

Figure 5.42 Variations of natural gas and CBM produced water per well in the different major CBM basins in the USA. Data from the 2010 National Research Council Report[706]

quality of the CBM produced water[726]. The salinity and quality of CBM produced water from different basins in the USA, Canada, Australia, China, and India are summarized in Table 5.2, reflecting the large range in salinity, yet also the common geochemical characteristics of elevated sodium and bicarbonate concentrations.

Data from the 2010 National Research Council report on CBM management[706] provide information on the volume of natural gas and produced water in the different major CBM basins in the western USA (Table 5.6). Converting these values to natural gas and produced water production per well, shows large variation between the CBM basins (Figure 5.42), with the San Juan Basin characterized by low produced water (975 m^3 per well) and high natural gas (4.6 million cubic m) production, whereas the Powder River and Radon basins have higher produced water (6342 m^3 per well) and lower natural gas (0.7 million cubic m) production (Figure 5.42). These different water-to-gas ratios imply different water intensity values, ranging from 5.4 L/GJ in the San Juan Basin to 236 L/GJ in the Powder River Basin (Table 5.6). In 2014, the total volume of CBM produced water in the USA was 110 million cubic m[29], which corresponds to an annual CBM gas production of 37 BCM[574]. This implies an integrated water intensity value of 74 L/GJ. This is consistent with the value reported in Grubert and Sanders (2018)[29] for CBM produced water of 70 L/GJ and reflects the balance of the different water intensity values from different CBM basins. Using the CBM produced water-to-CBM natural gas ratio in 2014, we extrapolate the data to other years (Figure 5.43), and show that in 2008–2009, during the peak time of CBM production in the USA, the volume of CBM produced water reached 170 million cubic m per year.

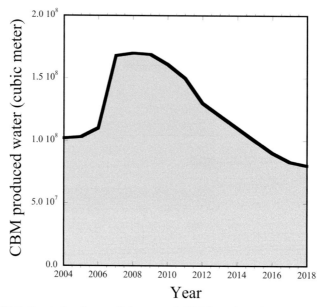

Figure 5.43 Estimated volume of the annual CBM produced water volume generated in the USA. Data calculated based on the ratio between volume of CBM
produced water[29] and total CBM natural gas generated in the USA[574] in 2014 and
extrapolation to other years using EIA data on CBM natural gas between 2004
and 2018.

As shown below, the management of the CBM produced water varies according
to the CBM water quality. In addition to the large variations in the ratios of
produced water to natural gas, the salinity of the produced water varies among the
CBM basins, between brackish water in Powder River (TDS up to 3,000 mg/L) to
highly saline water in the San Juan Basin (up to 100,000 mg/L; Table 5.5).
Therefore, the dilution factor required to reduce the salinity to an acceptable level
of TDS = 500 would be very different (a factor of 3–100) for the different basins,
and consequently, also the impaired water intensity values, with a range of
540–6,872 L/GJ (Table 5.6). This is evident in the San Juan Basin; despite the high
salinity of CBM produced water in this basin, the impaired water intensity value is
relatively low (544 L/GJ) due to the relatively low volume of the saline produced
water generated in the basin and high gas production (Figure 5.42).

  The management of CBM produced water and its potential impact on the
environment depends on multiple factors such as the volume, quality, and supply
stability of the CBM produced water, proximity of CBM wells to areas of disposal
and/or potential beneficial use, the compatibility between the quality of the CBM
produced water and receiving water bodies, soil properties of irrigable land,
storage and disposal capacity, and regulations and permission to reuse CBM

Table 5.7. *Distribution (in percentage) of the major disposal practices of CBM produced water in the western USA, expressed as percentage of the total volume of CBM wastewater**

Basin	Reinjection to subsurface	Surface impoundments	Discharge to streams	Irrigation	Industrial dust control
San Juan	99.9				
Uinta	97	3			
Powder River (Wyoming)	3	64	20	13	
Powder River (Montana)			61–65	26–30	5
Raton (Colorado)	28	2	70		
Raton (New Mexico)	100				
Piceance (Colorado)	100				

* Data from the 2010 National Research Council Report[706].

produced water for beneficial use[706]. These factors play a major role in the management and reuse of CBM produced water in the different basins in the USA; while the low saline CBM produced water from the Powder River Basin is mostly reused or discharged to surface water, the saline CBM produced water from the San Juan Basin with high gas capacity and low volume of produced water is mostly disposed through deep-injection wells (Table 5.7). In Wyoming, 64% of the CBM produced water is disposed to surface-lined impoundments designed to allow water to evaporate, 20% is directly disposed to waterways (with some treatment), 13% is used for irrigation, and 3% injected into the subsurface. In contrast, in Montana, 65% of the CBM produced water is directly discharged to waterways, 26% used for irrigation, 5% disposed to surface-lined impoundments, and 4% to dust control (i.e., road spreading)[706].

One of the major issues associated with the production of CBM produced water is the withdrawal of the water resource, specifically in arid areas, like in the western USA, where natural water resources are limited and some of the groundwater resources are fossil (i.e., originated from recharge that took place hundreds to thousands of years ago) and thus non-renewable[706]. As part of CBM extraction, the decline of groundwater pressure in shallow aquifers changes the hydrological balance and increases the likelihood of the upflow of typically saline groundwater and further salinization of the shallow aquifers. The impact on shallow aquifers depends on the depth of the coal-bearing aquifers and the hydraulic connectivity between the coalbeds and the shallow aquifers. While the impact on shallow groundwater resources is unlikely in the San Juan, Raton, Uinta, and Piceance basins due to the large depth and hydrological separation between the deep coal aquifers and overlying formations, a possible impact is expected in the

Powder River Basin where the coal-bearing formations are shallow and hydrologically connected to the overlying shallow aquifers.

An additional mechanism with a potential environmental impact is the leaking of CBM produced water from surface impoundments and contamination of shallow groundwater resources. In 2007, permits for more than 4,000 impoundments had been issued within Wyoming alone[728]. In addition to the direct contamination from CBM produced water, the infiltration of large volumes of water through the unsaturated zone in an arid zone like the western USA could induce leaching and dissolution of naturally occurring salts from the unsaturated zone and further salinization of the shallow aquifer. This has been demonstrated in the Powder River Basin in Wyoming, where the saline shallow groundwater had high concentrations of sulfate, sodium, and magnesium. The infiltration of CBM water with high sodium into the subsurface triggered gypsum dissolution. Following gypsum dissolution, the dissolved calcium is exchanged for sodium and magnesium on exchangeable sites in clay minerals, which in turn trigger further gypsum dissolution[728]. This "positive feedback" chain of reactions has generated contaminated groundwater with TDS of up to 100,000 mg/L, which is 43 times the salinity of the original CBM produced water (2,300 mg/L) in the overlying impoundment. This case study is a clear demonstration that our estimate for the impaired water intensity of CBM produced water is very conservative; given the quality of CBM produced water from the Powder River Basin we estimate an impaired water intensity of 709 L/GJ, assuming a threefold dilution factor. However, the contaminated groundwater underlying the CBM impoundment with a salinity of 100,000 mg/L[728] implies a much higher water intensity of 47,200 L/GJ (i.e., water intensity of 236 L/GJ × 200 dilution factor). The direct discharge of CBM produced water to surface water can potentially affect stream ecology and the overall water quality in impacted watersheds[729]. In addition to the high salts, high concentrations of bicarbonate in CBM produced water can also induce toxicity in fish and affect the overall ecological system[730, 731].

Finally, beneficial use, such as irrigation and water use for livestock with CBM produced water, can also be a solution for managing CBM produced water. At the TDS range between 1,000 and 7,000 mg/L, water can be used for livestock, although consumption of water having a TDS greater than 5,000 mg/L is often associated with intestinal distress[706]. Irrigation of CBM water can affect the soil salinity and properties, as demonstrated in a study in Wyoming, where irrigation of untreated CBM resulted in high soil salinity and sodicity (i.e., high sodium content that reduces the soil permeability upon retention of sodium onto exchangeable adsorption sites in the soil). However, using soil amendments and treatment of the CBM produced water resulted in lower sodium concentrations in the soil[712].

5.8 Integration of Water Use and Produced Water Generated from Natural Gas

Figure 5.44 presents the different pathways and associated water intensity values of conventional and unconventional natural gas exploration and associated wastewater production during the different life cycles of drilling, hydraulic fracturing for shale gas, transport and processing, electricity generation, wastewater disposal, and spills (see values and data sources in Table 5.1). Data collected for the USA[29] show that in 2014, the total water withdrawal for all stages of conventional and unconventional natural gas exploration was 18.4 BCM, which included water withdrawals for cooling gas plants (97.7%), drilling (0.1%), hydraulic fracturing (1.1%), transport (0.6%), and processing (0.5%). The total water consumption in the USA in 2014 was 1.35 BCM and included water consumption for cooling gas plants (70.2%), drilling (1.7%), hydraulic fracturing (15.5%), transport (8.1%), and processing (4.4%) (Figures 5.45 and 5.46). In 2014, conventional natural gas production (457 BCM) was 51.3% of the total annual

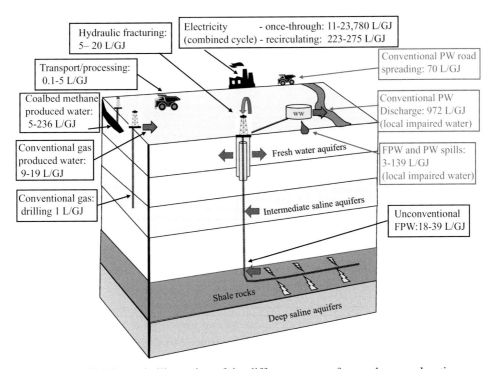

Figure 5.44 Schematic illustration of the different stages of natural gas exploration and processing as well as the associated water intensity values (L/GJ) for withdrawal/consumption (in blue) and water impaired (red). (A black-and-white version of this figure will appear in some formats. For the colour version, refer to the plate section.)

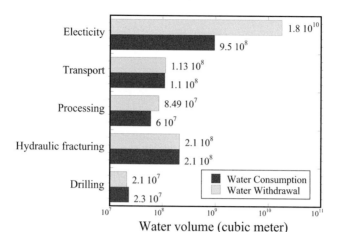

Figure 5.45 The volume of water consumption and withdrawal at different cycles of natural gas production in the USA. Data from Grubert and Sanders (2018)[29]

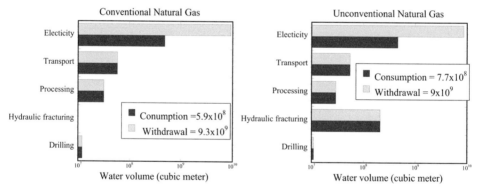

Figure 5.46 The volume of water consumption and withdrawal at different cycles of conventional (left panel) and unconventional (right panel) natural gas production in the USA. Data from Grubert and Sanders (2018)[29]

natural gas production, whereas unconventional shale gas (396 BCM) accounted for 44% of the total annual natural gas (889 BCM) production in the USA[593]. Since the exploration of unconventional shale gas was only ~7% lower than that for conventional natural gas in 2014, a comparison of the water use for conventional and unconventional natural gas shows higher water consumption volumes for unconventional shale gas due to hydraulic fracturing (210 million cubic m), which was 15.5% of the total water consumption (Figure 5.46). These values reflect relatively early stages of shale gas exploration; the intensification of hydraulic fracturing between 2011 and 2014 in the USA has increased the water footprint of hydraulic fracturing. Therefore, the future development of shale gas exploration is expected to further increase the volume of water consumption for

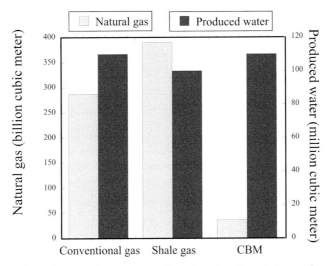

Figure 5.47 The volumes of conventional produced water, shale gas flowback and produced water (FPW), and coalbed methane produced water as compared to natural gas production in the USA in 2014. Data retrieved from Grubert and Sanders (2018)[29] and EIA[593]

unconventional gas exploration[24]. Nonetheless, the relative contribution of water consumption for cooling gas plants is higher (70%) and effectively controls the overall water intensity of natural gas production for the electricity sector.

Similarly, we evaluate the relationships between the volumes of conventional produced water, unconventional FPW, and coalbed methane produced water to the volumes of natural gas for the 2014 US data (Figure 5.47)[29]. In 2014 the total volume of produced water from natural gas exploration in the USA was 320 million cubic m, from which 110 million cubic m (34%) originated from conventional gas wells, 100 million cubic m (31%) from shale gas wells, and 110 million cubic m (34%) from coalbed methane wells (data from Grubert and Sanders, 2018[29]). The data show that the ratios of conventional produced water and FPW-to-natural gas from conventional and unconventional wells were similar (~2.5×10^{-4}), while the ratio between produced water and coalbed methane was one order of magnitude higher (~3×10^{-3}), reflecting the higher water intensity of CBM produced water (Figure 5.47).

5.9 Policy and Regulations

Unconventional gas exploration with the expansion of hydraulic fracturing and horizontal drilling has shown significant activity in the design of new policies and regulations for natural gas in the twenty-first century. The International Energy

Agency (IEA) noted in 2011 (p. 7[564]) that the "use of hydraulic fracturing in unconventional gas production has raised serious environmental concerns and tested existing regulatory regimes"[564]. This has been the case for the USA, where, since the 1980s, authority for energy regulation has rested with individual states[366, 734].

Most notably, when it came to regulating shale gas development, the US 2005 Energy Policy Act[735] exempted hydraulic fracturing from federal water use regulations. Informally known as the Halliburton loophole, this meant that the fluids used in hydraulic fracturing were not regulated under the Safe Drinking Water Act (SDWA), and as such this allowed for hazardous fluids from the hydraulic fracturing process to be injected into the subsurface, as they would be exempted from the Underground Injection Control (UIC) provisions of the SDWA (see Chapter 4 for discussion of the UIC Program)[736].

Given such exemptions, much uncertainty about the potential environmental and social impacts has shrouded unconventional gas production, resulting in high levels of public concern, especially concerning hydraulic fracturing activities near highly populated areas[737]. Owing to the absence of an overarching regulatory framework for unconventional gas production, as exploration and production began to take off in the 2000s, a wide variation of regulations have ensued, especially within the USA. Furthermore, some states in the USA and provinces in Canada opted to place a moratorium on shale gas exploration in response to the high levels of public scrutiny and uncertainty about socioeconomic and environmental impacts; likewise, the French parliament in May 2010 voted to ban hydraulic fracturing (IEA 2011, 65[564]). Table 5.8 provides examples of how some countries/states have sought to restrict unconventional gas exploration.

In other instances, the exploration of unconventional gas and its impacts on water has provided an opportunity to move forward with new forms of regulations. That said, much of the regulatory landscape has been uneven and heterogeneous across different countries and unconventional gas fields.[736, 738] The USA is probably the most peculiar case for regulating unconventional natural gas because, similar to the oil and gas industry more broadly, mineral rights can be privately owned, such that private companies can lease subsurface mineral rights from landowners[739]; in most other countries, mineral resources are owned and controlled by the state[91]. Regulatory experts have sought to lay out the complex regulatory framework for natural gas exploration and production in the USA given the variation that exists across states, taking into account site selection and preparation (e.g., set back requirements), drilling the well (e.g., casing), hydraulic fracturing (e.g., water withdrawal limits and frac chemicals disclosure), wastewater storage and disposal, excess gas disposal (i.e., venting and flaring), production (i.e., severance taxes), and plugging and abandonment of wells[738].

Table 5.8. *Countries/states with restrictions on shale gas exploration*

USA (New York)	In December 2010, the governor of New York imposed a moratorium on high-volume hydraulic fracturing. Then in 2014, Governor Cuomo banned the process of fracking, and in 2020 the New York state legislature passed legislation in its FY 2021 Executive Budget that would make the ban permanent.
USA (Maryland)	In 2017, Governor Hogan signed a bill to ban hydraulic fracturing after the legislature had voted to ban the practice after a moratorium from 2011was set to expire.
France	In May 2011, France's Lower House of Parliament voted to ban hydraulic fracturing. The ban was upheld by France's constitutional court in 2013.
Canada (Quebec)	Extended a prior moratorium on hydraulic fracturing in 2014.
Canada (New Brunswick)	Announced a moratorium on hydraulic fracturing in 2014 and then extended the moratorium indefinitely in 2018.
Canada (Nova Scotia)	Announced a moratorium on hydraulic fracturing in 2014.
Bulgaria	In 2012, Bulgaria's parliament banned the exploration of shale gas reserves and the use of hydraulic fracturing.
Germany	In 2016, Germany approved a law that bans the use of hydraulic fracturing, but does not outlaw the conventional drilling for oil and gas, which is left at the discretion of the states.
Scotland	In 2019, the government extended its 2015 moratorium on unconventional oil and gas exploration, including the use of fracking.
South Africa	In 2011, South Africa's cabinet placed a moratorium on oil and gas exploration licenses, restricting hydraulic fracturing within the Karoo region, but then lifted the ban in 2012.

Moreover, because hydraulic fracturing of unconventional gas has, at times, preceded the development of a new regulatory framework (unlike many European countries that have taken a more precautionary approach and sought to commission studies to understand the environmental impacts first), this has led to a patchwork of regulations to address public concerns that have ensued[366]. This was most evident with regulations for the disclosure of frac chemicals used in the hydraulic fracturing process; whereas some states required some level of public disclosure, a 2012 Natural Resources Defense Council study found that more than half of US states with hydraulic fracturing activity did not have any disclosure requirements[740]. Disclosure of the chemicals used in the hydraulic fracturing process before hydraulic fracturing begins is critical because this information is necessary for communities residing near production sites to have their water sources tested and for delineating a baseline of the water quality so as to ascertain whether the fracturing fluid or produced water has resulted in any contamination[740].

Thus, in the absence of federal regulation, citizens and local governments have pushed for more disclosure about the chemicals used in hydraulic fracturing, resulting in different disclosure practices across states[741]. In many instances, private regulation has filled this void through the voluntary online chemical disclosure registry[741]. For example, FracFocus (https://fracfocus.org/) was established by the Ground Water Protection Council and the Interstate Oil and Gas Compact Commission. Yet, frequently industry has argued for trade secret exemptions so as not to disclose the full list of the frac chemicals.

In contrast, the development of shale gas in China is carried out at the highest political level in which development targets are set in the Five-Year plans, of which the first shale gas targets were set in the twelfth Five-Year Plan (2011–2015); its National Shale Gas Development Plan was released in 2012[599, 742]. Yet too in China, specific policies for regulating shale gas have been slow. Thus, as in the USA, many of China's oil and gas state-owned companies are also responsible for developing and enforcing their own environmental standards, which may not be disclosed to the public[743].

Regulating the disposal of produced water from the hydraulic fracturing process has increasingly been the focus of US states given the large volumes involved. Here too, unconventional oil and gas wastewater is regulated differently across the USA. For example, in Pennsylvania, conventional oil and gas wastewater can still be discharged into the environment after some treatment, or used to spread on roads for deicing or dust suppression, whereas unconventional oil and gas wastewater must be disposed to deep-well injection and/or reuse for hydraulic fracturing. The disposal of produced water will be an issue that will take on additional significance for states that are water-scarce. In particular, China has focused on the water impacts and opportunities for treatment of wastewater generated through hydraulic fracturing as well as reuse wastewater for hydraulic fracturing[366, 478].

Chapter 5 Take-Home Messages

- Global natural gas production has constantly increased during the last 50 years. Since 2005, the development of unconventional shale gas and hydraulic fracturing has increased shale gas production in the USA, which became the largest natural gas producing country.
- Shale basins with potential shale gas are distributed in all continents, particularly in China, with the largest potential shale gas reserves. Unlike oil, natural gas development requires installation of adequate infrastructure (e.g., pipe network) for delivery.

- The water use for hydraulic fracturing and shale gas extraction is 20-fold higher than conventional natural gas exploration. The location of many of the world's shale basins in water-scarce areas raises concerns for water availability and competition with other sectors if potable water is exclusively used for hydraulic fracturing. Recycling of flowback water for hydraulic fracturing can reduce the potable water use, but also the natural gas production.
- The water use for hydraulic fracturing is directly associated with the length of the lateral component of horizontal wells; extending the horizontal wells has caused an increase in water use in the USA. Yet, the majority of water withdrawal and consumption for natural gas is for cooling gas plants. The type of cooling systems determine the magnitude of water use; closed-system recirculating cooling systems withdraw much less but consume more water than open-system cooling systems. The higher the water withdrawal the larger the volume that is discharged from gas plants to the hydrosphere, thus casing thermal pollution.
- During hydraulic fracturing the injected water with frac chemicals is retained into the shale matrix, which enhances the release of methane to the fractures and the wells. Consequently, only a small fraction of the flowback water that is returned to the surface originates from the injected hydraulic fracturing water, while formation water that is extracted from the shale rocks is the predominant water source of flowback and produced water.
- Some of the man-made chemical additives used for hydraulic fracturing (e.g., water viscosity) are toxic, which has raised public concerns about the environmental and water quality risks. However, since the majority of the frac water is retained in the shale matrix during hydraulic fracturing, most of the contaminants in shale gas wastewater are inorganic and derive from the contribution of formation water.
- Since the formation water in shale and tight sand rocks originated from residual evaporated seawater that interacted with the host formation rocks, the flowback and produced water from shale gas wells are typically highly saline with high concentrations of salts, metals, nutrients, and radioactive elements. The release of shale gas wastewater into the environments poses ecological and human health risks, and therefore the water impaired intensity of even small volumes of spilled wastewater is relatively high.
- The geochemical fingerprint of flowback and produced water from shale gas wells is distinguished in some cases from conventional produced water, as well as other sources of contamination, and therefore has been utilized to identify and monitor the impact of shale gas wastewater in cases it was released into the environment.
- The volume of wastewater generated from shale gas wells is similar to the volume of wastewater from conventional gas wells.
- The rise of unconventional shale gas development has been associated with high well-density, which resulted in high rates of wastewater spills and leaks into the environment.
- In spite of the public perception, the chemistry and quality of wastewater from unconventional shale gas are similar to those of wastewater from conventional natural gas. Yet many of the regulations in the USA only apply to unconventional

wastewater and not conventional wastewater, which in some states is discharged into rivers or spread on roads for dust suppression and deicing.

- Traditional geochemical and isotope techniques are sensitive to detecting the sources of natural gas in groundwater, in particular to distinguish between naturally occurring biogenic and thermogenic natural gas derived from shale gas well leaks. However, thermogenic natural gas can also originate from naturally occurring sources, and more advanced geochemical methods, such as the utilization of noble gas and their isotopes, are needed to adequately delineate the source of natural gas in groundwater. Evidence for actual stray gas contamination in drinking water wells near shale gas sites has been observed in several cases in Pennsylvania and Texas.

- Coalbed methane is another unconventional method used to extract natural gas that has a large global potential, particularly in Russia, China, and Australia. The USA is the largest producer of coalbed methane, particularly in the western basins in New Mexico, Utah, and Montana. Yet, the rise of shale gas and the decline of coalbed methane extraction in the USA have left thousands of abandoned wells that pose environmental and methane emission risks.

- The future expected retiring of millions of conventional and unconventional wells across the USA pose major environmental and atmospheric contamination risks and thus an environmental legacy for decades to come.

- Policies for regulating unconventional gas are fragmented across countries in which some states have sanctioned unconventional gas exploration and production and others have sought to restrict its exploration and production. In the USA, much of the regulatory policy has been left to individual states or to the private sector to self-regulate, as has been the case with frac chemical disclosure policies.

6

Integration

The Role of Energy in the Anthropogenic Global Water Cycles

6.1 Introduction to the Anthropogenic Global Water Cycle

The twentieth century saw the world population grow from about 1.6 billion people to more than 6 billion; this trend of rising population growth has continued into the twenty-first century, and as of 2021, there were approximately 7.88 billion people on planet Earth[744, 745]. With a growing population comes increasing demand for water: demand for global water use has risen from 500 BCM in the early 1900s to about 4,000 BCM in 2010 (Figure 6.1). For many countries – whether it is in the Middle East, in South and Southeast Asia, or in the Americas, securing access to fresh water resources is often seen as integral to a country's national security[746]. This is particularly the case when states depend on transboundary rivers, such as the Euphrates, Jordan, Nile, Indus, and Mekong.

Worldwide, the agricultural sector has been the predominant consumer (70%) of water[747–749]. While not the primary focus of our discussion on the water–energy nexus, water and energy use for agriculture cannot be overlooked owing to its contribution to meeting national food security. In some countries such as India, Pakistan, Iran, and Mexico, irrigation mostly from groundwater resources exceeds ~90% of the total water demand and sustains much of the food production and livelihood of millions of people[750]. Indeed, by the mid-1980s, owing to the introduction of electric pumps for tube wells in India, which allowed individual farmers to access the groundwater for agriculture, India became the largest groundwater user in the world[751].

The increase of global water extraction has not only depleted the quantity of water resources, but in many areas has also caused irreversible water quality degradation such as seawater intrusion in coastal aquifers, large-scale salinization of groundwater, and anthropogenic contamination of surface water and shallow aquifers. In most of the arid and semi-arid regions, the accelerated extraction of

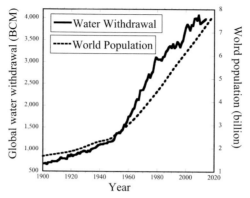

Figure 6.1 Global water withdrawal and world population variations since the early nineteenth century. Data retrieved from Ritchie and Roser (2018)[772].

limited water resources has exacerbated the already severe water scarcity conditions[747–749]. Groundwater pumping beyond the natural replenishment has caused a major draw-down of the groundwater table and intensification of groundwater contamination from both man-made and naturally occurring (geogenic) sources. For example, the millions of wells in the Indo-Gangetic Basin of Northern India and Pakistan extract over 200 BCM per year[85], which results in a massive decline of groundwater level as recognized by NASA's Gravity Recovery and Climate Experiment (GRACE) satellite measurements[752]. The groundwater table decline is associated with large-scale water quality degradation, including nitrate, fluoride, and uranium contamination[85, 87, 753]. Likewise in the Middle East, over-exploitation of groundwater resources beyond the natural replenishment has caused large-scale salinization of the groundwater, rendering the water unusable for the domestic and agriculture sectors[88].

Combined with over-exploitation of water resources, the rise in global temperature and particularly sea surface temperature during the twentieth century has triggered major changes in atmospheric circulation patterns, resulting in changes in precipitation patterns and water availability. Rising emissions of anthropogenic greenhouse gases (see Section 6.4) over the twentieth century have been associated with an increase in global aridity with longer and intensified periods of drought in Africa, southern Europe, East and South Asia, eastern Australia, and many other parts of the northern mid–high latitude, including the southwestern USA. Droughts are commonly associated with anomalous tropical sea surface temperatures, with La Niña-like surface temperature anomalies leading to drought periods in south western North America, and El-Niño-like surface temperature periods causing drought in eastern China. In Africa, the southward shift of the warmest sea surface temperature in the Atlantic and warming in the

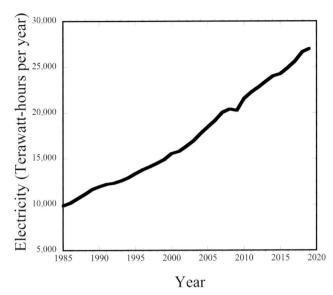

Figure 6.2 The increase of global electricity over time. Data from BP global dataset[11].

Indian Ocean are associated with droughts such as the Sahel droughts[754]. Likewise, the warming of the Indian Ocean is associated with the reduction of precipitation over northern India[84]. The Intergovernmental Panel on Climate Change (IPCC) projects that water availability in arid and semi-arid regions will further decline with increased global warming[89], therefore exacerbating the current water crisis in many of the arid and semi-arid regions under future global warming scenarios[754, 755, 756].

Together with increasing water utilization, population growth and development have increased global energy production; since the mid-1980s, global electricity use has increased at a rate of 509 terawatt-hours per year, up to 27,000 terawatt-hours in 2019 (Figure 6.2). In 2019, fossil fuel sources contributed 63% to global electricity production, in which coal, natural gas, and oil, respectively, contributed 36%, 23%, and 3% of global electricity production (Figure 6.3; data from BP global dataset[11]). As of August 2020, only 10% of worldwide electric utilities rely upon renewable energy sources, particularly in Europe (30% of global renewable energy used for power generation in 2019), China (26%), and the USA (17%), whereas the majority of the utilities remain heavily invested in fossil fuels despite international efforts to reduce greenhouse gas emissions[760]. The rising energy consumption for household, agricultural, and industrial use has also led to an increase in water use for energy production. Therefore, the use of water for extraction and processing energy resources, in particular fossil fuels, has been an

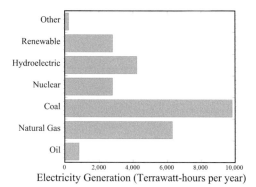

Figure 6.3 Distribution of major energy sources for global electricity. Data from
BP global dataset[11]

important component of global water utilization. In Section 6.2 we quantify the
global volume of water withdrawal and consumption for fossil fuel extraction. This
book highlights the impacts of the different cycles of energy production on the
water quality and the impaired water intensity associated with wastewater and
waste solids generated by fossil fuel production. The volume of wastewater
generated from fossil fuels and evaluation of possible water quality impacts on a
global water scale is evaluated in Section 6.3.

In addition, this chapter lays out the ways in which understanding the energy–
water nexus is vital for addressing the climate crisis. Simply put, the increase of
fossil fuel extraction has increased greenhouse gas fugitive atmospheric emissions,
in particular carbon dioxide and methane, which play a critical role in global
warming. Multiple climate models predict that precipitation over arid and semi-
arid regions will be significantly reduced with global warming[747–749, 761–765],
which is supported by evidence of a reduction in precipitation and intensification
of drought periods in areas already suffering from water shortages[766]. Therefore,
Section 6.4 presents the impact of fossil fuels on the accumulation of greenhouse
gases in the atmosphere. Finally, the decline of available clean water and
degradation of the water quality have led to larger investments in transporting
water from water-rich areas to water-poor areas along with water treatment, both of
which are energy-dependent[767]. The energy use for groundwater pumping, surface
water transport (e.g., the California Aqueduct), and water treatment (e.g.,
wastewater treatment, desalination) is the other component of the energy–water
nexus. Section 6.5 describes the magnitude of electricity required for each of the
water management and treatment processes.

The worldwide degradation of the quality of water resources has created a
demand for intensive treatment of water to make it usable[768, 769]. In particular, the
volume of water generated from the 15,906 operational desalination plants around

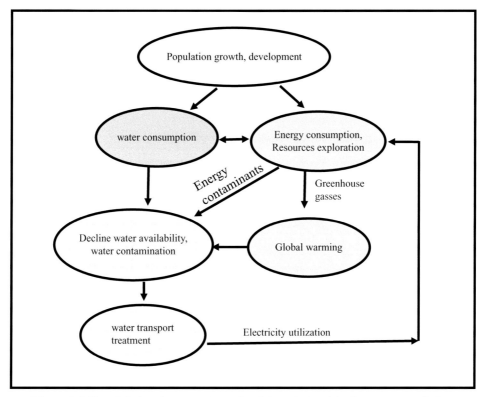

Figure 6.4 The global anthropogenic cycle of the relationships between population growth and societal development on intensification of the water and energy use, which induce water decline and degradation combined with greenhouse gas emissions that further exacerbate water availability and quality in arid and semi-arid regions. The increase of water management and treatment due to the reduction in clean water availability requires more energy that triggers more water use and contamination as well as further global warming. (A black-and-white version of this figure will appear in some formats. For the colour version, refer to the plate section.)

the world has exponentially increased to 35–38 BCM per year[770, 771]. In turn, the more energy that is utilized for the water sector, the more water is then also utilized for energy production (e.g., cooling thermoelectric plants); this nexus thus leads to more greenhouse gas emissions, especially if the energy source is primarily derived from fossil fuels. Consequently, the increase of electricity utilization for the water sector triggers more water use and associated contamination from fossil fuel exploration, along with contributing to global warming, which further exacerbates the water crisis. Overall, this chapter describes this negative-feedback loop (Figure 6.4) associated with the anthropogenic water–energy cycles in which the current paradigm of economic development is heavily vested in fossil fuels and intensive water consumption. As long as countries continue to rely heavily on water-intensive energy production and energy-intensive water use, we will continue to see diminishing water resources and quality, along with a sharp growth in greenhouse gas emissions, and as the climate continues to become warmer, water resources globally will be impacted, further leading to an increase in energy utilization for water quantity and quality remediation.

6.2 Global Water Use for Fossil Fuels

The International Energy Agency (IEA)[45] estimated that about 400 BCM of water was globally withdrawn for energy production in 2014, in which the majority (350 BCM; 88%) was for cooling power plants for electricity generation. It was estimated that 48 BCM of water was annually consumed for energy production, in which the majority (30 BCM; 64%) was consumed for primary energy production[45]. Spang et al. (2014)[48] estimated a global water consumption of about 52 BCM, from which 26.7 BCM (51%) was for extraction of fossil fuels, whereas IEA estimated 31 BCM (60%) out of the total water consumption was for fossil fuel extraction[45]. The IEA estimated global water withdrawal for fossil fuels was approximately 251 BCM, 63% of the global water withdrawal for energy production[45]. Based on the literature of the water intensity for water withdrawal and consumption presented in this book for coal (Chapter 3), crude oil (Chapter 4), and natural gas (Chapter 5), we provide here a new evaluation of the magnitude of global water use for fossil fuels. We divide our discussion into three stages of the life cycle for fossil fuel extraction: (1) recovery and extraction (e.g., mining, drilling); (2) power generation, and processing (e.g., oil refining); and (3) waste disposal (e.g., coal ash ponds). While coal production is a continuous process in which coal mining is annually balanced by consumption, natural gas and crude oil production can be generated from wells that were drilled in previous years, and thus their current production does not necessarily reflect the water use needed in the initial recovery stage (e.g., drilling, hydraulic fracturing). Nonetheless, we use the annual energy production of different fossil fuel sources in an attempt to estimate global water use, assuming that the production and consumption rates are in a steady state. We use conservative water intensity values that reflect long-term energy production.

6.2.1 Global Water Use for the Extraction and Recovery Stage of Fossil Fuels

In order to evaluate the water withdrawal and consumption of fossil fuel production in the recovery stage, we use the relative contribution of the different production of fossil fuels and the associated water intensity. Since the water intensity values are expressed in water volume to gigajoules (L/GJ), we use the

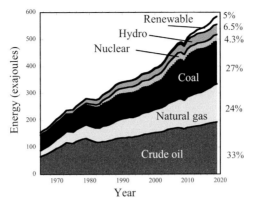

Figure 6.5 Major energy consumption over time expressed in exajoules ($\times 10^{12}$ joules). During the last decade, the overall energy consumption increased by 13%, from which crude oil, natural gas, and coal consumption increased by 10%, 20%, and 4%, respectively. Nuclear consumption decreased by 4%, and hydro and renewable consumption increased by 14% and 67%, respectively. Data were calculated from BP global production dataset[10]

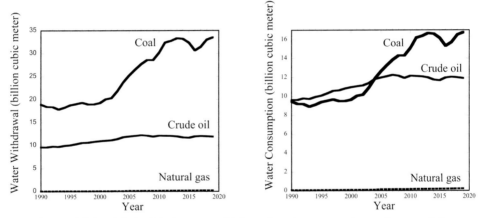

Figure 6.6 Modeled global water withdrawal and consumption volumes of the upstream (recovery) stage of fossil fuels.

energy unit (exajoules) for fossil fuel production based on the BP dataset[10] (Figure 6.5). Given the range of water intensity values for coal mining and processing (Table 3.1 in Chapter 3), we selected median values of 200 L/GJ and 100 L/GJ for water withdrawal and consumption, respectively, which is consistent with values reported for coal mining water intensity in China. We used the BP global coal production dataset[10] to estimate the global water withdrawal and consumption variations of the coal recovery stage during the last three decades (Figure 6.6). Based on global coal production in 2019 (8,129 million metric tons; 167.6 exajouls[10]), we estimate global water withdrawal and consumption of

33.5 and 16.8 BCMs, respectively (Table 6.1). For evaluating the global water use for natural gas, we used the water intensity value of 1 L/GJ, which characterizes the water intensity of the recovery stage of conventional natural gas[29]. Given the relative higher water intensity of hydraulic fracturing (5–20 L/GJ over the lifetime of a well)[24], we calculated the water intensity associated with the relative proportion of the US shale gas within the global natural gas budget (only after the beginning of shale gas exploration in the USA in 2005). We assume that the water intensity of water consumption is equal to water withdrawal since water that is used for well drilling and hydraulic fracturing is not returned to the surface. We used the BP global natural gas production dataset[10] to estimate the global water consumption variations of the natural gas recovery stage during the last three decades (Figure 6.6). The International Energy Agency (IEA 2018) estimates that offshore natural gas production[773] increased from about 600 BCM in 2000 (25% of global natural gas production of 2,400 BCM in 2000[10]) to 1,000 BCM in 2016 (28% of total global natural gas production of 3,559 BCM[10]). Therefore, we assume that global on-shore natural gas production comprises 72–75% of total global natural gas production. We estimate that the global on-shore natural gas production of 2,872 BCM (from which US shale gas contributes 15.3%) implies a global water consumption (and withdrawal) of 0.22 BCM (Table 6.1). It is interesting to note that in spite of only 15% contribution of US shale gas to global on-shore natural gas production, the water used in 2019 in the USA for hydraulic fracturing (0.33 BCM) amounted to 58% of the global water use for shale gas exploration (Table 6.1), which reflects the relative high water intensity of hydraulic fracturing relative to conventional gas extraction.

For crude oil extraction, we used water intensity values of 100 L/GJ that reflect a median value of the water consumption for conventional crude oil (13–350 L/GJ; Table 4.1 in Chapter 4), which includes both drilling and enhanced oil recovery. For unconventional tight oil we used the water intensity of 13 L/GJ that characterizes the long-term water intensity of hydraulic fracturing (i.e., taking into account the decay of tight oil production after the first few months of high production rates)[24]. We calculated the water intensity associated with the relative proportion of US tight oil (18% of global crude oil production in 2019) and added this fraction to the global budget (only after the beginning of tight oil exploration in the USA in 2010). We assume that the water intensity of water consumption is equal to water withdrawal since water that is used for well drilling, enhanced oil recovery, and hydraulic fracturing is not returned to the surface. We assume that global on-shore crude oil production comprises only 71% of total global oil production[774], and used the BP global crude oil production dataset[10] to estimate the global on-shore water withdrawal and consumption variations for crude oil recovery during the last three decades (Figure 6.6). In 2019, global on-shore crude

Table 6.1. *Data on fossil fuel production in 2019, water intensity values, and estimated volumes of water withdrawal and consumption for 2019*[†]

Source	Production (tons*/ m³** per year)	Production (exajoules per year)	Water withdrawal intensity (L/GJ)	Water withdrawal (BCM per year)	Water consumption intensity (L/GJ)	Water consumption (BCM per year)
Coal	$8.129 \times 10^{6*}$	167.6	200	33.5	100	16.8
On-shore natural gas (conventional)^	$2.872 \times 10^{9**}$	103.4	1	0.14^	1	0.14^
Shale gas (USA)	$460 \times 10^{9**}$	16.6	5	0.08	5	0.08
On-shore crude oil (conventional)^	$3.3 \times 10^{9**}$	116.3	100	11.6^	100	11.6
Tight oil (USA)	$0.59 \times 10^{9**}$	20.7	13	0.27	13	0.27
Total upstream water use				**45.6**		**28.9**

[†] Fossil fuel data from BP global dataset[11]. Symbols: * refers to metric tons, ** to cubic m, and ^ to global conventional natural gas exploration, excluding the US shale gas and tight oil production in 2019.

223

oil production was 3.92 BCM per year (116.3 exajoules per year[10]), which implies
a global water consumption (and withdrawal) of 11.6 BCM per year (Table 6.1).
The US tight oil contributed 0.59 BCM per year (20.7 exajoules) in 2019, which
implies a water withdrawal and consumption of 0.27 BCM. Combined, the water
withdrawal and consumption for the recovery stage of all fossil fuels in 2019 were
45.6 and 28.9 BCM, respectively. For water withdrawal, coal, crude oil, and
natural gas consisted of 73%, 26%, and 0.5%, respectively, while for water
consumption, coal, crude oil, and natural gas consisted of 58%, 41%, and 0.8%,
respectively (Table 6.1).

6.2.2 Global Water Use for Power Generation, Oil Refining, and Coal Ash

As shown in this book, generating electricity requires a large volume of water for
cooling thermoelectric plants. In order to evaluate the global water withdrawal and
consumption for power generation, we used the relative contribution of the
different fossil fuels to the electricity sector and the water intensity of using water
for cooling thermoelectric plants. While coal (58% of the total electricity generated
from fossil fuels, 36% of the total global electricity in 2019[10]) and natural gas
(37% and 23%, respectively) contributions have increased over the last few
decades, the crude oil (5% and 3%, respectively) contribution to the global
electricity sector has reduced (Figure 6.7). We used water intensity values that
represent the most updated literature and representative cases. Since 50% of
the world electricity from coal is generated in China[11], we used the China
integrated data of water withdrawal of 37 m^3/MWh and water consumption of
1.3 m^3/MWh[235] (Table 3.2) to represent global water use intensities for coal.

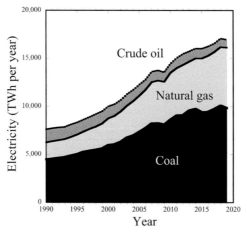

Figure 6.7 Variations in global electricity production derived from fossil fuel
sources. Data from BP global dataset[10]

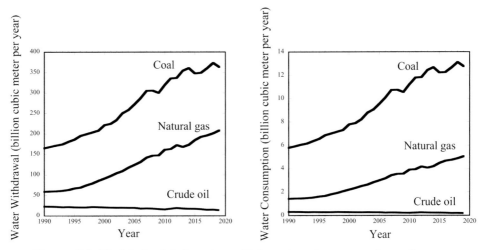

Figure 6.8 Modeled global water withdrawal and consumption of power generation from fossil fuels.

For natural gas, the USA generates 27% of the world electricity from natural gas, and thus we used the US example with water withdrawal of 33 m³/MWh and water consumption of 0.8 m³/MWh for 2016 (Table 5.1)[234]. For electricity generated from oil, we also use available data from the US dataset with water withdrawal of 16.9 m³/MWh and water consumption of 0.23 m³/MWh reported in Gruber and Sanders (2018)[29]. The integration of the global electricity generation data reported in the BP global coal production dataset[10] over the last three decades and the intensity values reported above enabled us to reconstruct the water withdrawal and consumption of the last three decades (Figure 6.8). The overall global water withdrawal and consumption is estimated for 2019 as 585.9 and 17.9 BCM, respectively, in which coal, natural gas, and oil consist of 62%, 36%, and 2% of water withdrawal and 71%, 28%, and 1% of water consumption.

As shown in the chapters on coal and natural gas, changing the cooling technology from open- to closed-loop systems would significantly reduce the water withdrawal but nonetheless would increase the water consumption of these plants. Using alternative technologies such as dry cooling would reduce both water withdrawal and consumption, but would require additional coal and natural gas production to compensate for lower efficiency of these systems, and therefore would increase greenhouse gas emissions. Given that coal production and combustion constitute the largest global water withdrawal and consumption (Figure 6.8), the transition from coal to natural gas and utilization of modern thermoelectric plants with closed water-cooling loops would significantly reduce the water footprint of electricity production. This was demonstrated in the USA with the rise of shale gas and decline of coal as the major source for the electricity

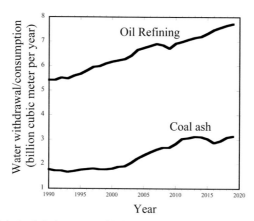

Figure 6.9 Modeled global water withdrawal (and consumption) for oil refining and coal ash washing.

sector[234]. In addition, several studies have shown that the water intensity for solar and wind energy is considerably lower than that of fossil fuels[29, 30]. A global transition from fossil fuels to solar and wind energy would significantly reduce the water footprint for generating electricity, and would also eliminate greenhouse gas emissions, and thus would reduce the global warming impact on the water resources. The inevitable negative loop cycles of the energy–water nexus illustrated in Figure 6.4 could therefore be alleviated through the transition from fossil fuels to renewable energy sources.

In addition to the water used for the recovery stage and power generation of fossil fuels, we evaluate the global water withdrawal and consumption of oil refining and coal ash handling. For oil refining, we estimate a water intensity of 40 L/GJ (Table 4.1 in Chapter 4). We assume that the water intensity of water consumption is equal to water withdrawal since water that is used for oil refining is not returned to the surface. We used the BP global crude oil consumption dataset[10] to estimate the global water use for oil refining during the last three decades (Figure 6.9). In 2019, we estimate global water consumption (and withdrawal) for oil refining of 7.7 BCM. For the water use for coal ash handling, we use the China example of water intensity of 125 L/GJ (based on the volume of 980 million cubic m reported for water use for coal ash handling in China in 2005[119]). We assume that the water use evaluation applies for only coal ash, which consists of 15% of the volume of the combusted coal. We therefore use 15% of the global consumed coal reported in the BP global dataset[10] to estimate the global water use for coal ash washing during the last three decades (Figure 6.9). In 2019, the global water use is estimated as 3.1 BCM. Since these vast volumes of wastewater are directly associated with oil refining and CCR disposal, substitution of fossil fuels with renewable energy sources would eliminate the associated wastewater and thus the environmental and human health risks in addition to the reduction of the water use.

6.2.3 Integration of Global Water Use for Fossil Fuels

Based on the estimates presented in this book, the combined global water withdrawal for the different stages of fossil fuel extraction and production increased from 283 BCM in the early 1990s to 642 BCM in 2019. During this period, the combined global water consumption increased from 33.9 BCM to 57.8 BCM (Figure 6.10). In 2019, coal was the major component of global water withdrawal (62%), followed by natural gas (33%) and oil (5%). Likewise, coal was the major component of water consumption (57%), followed by crude oil (34%) and natural gas (9%) (Figure 6.11). Most of the water withdrawal is associated with power generation, while the upstream components are much larger for global water consumption (Figure 6.11). The distribution of water withdrawal for all stages of energy extraction and production suggest the predomination of coal (430 BCM in 2019) relative to natural gas (209 BCM) and crude oil (33 BCM). Most of the water consumption is for coal (50 BCM in 2019), crude oil (20 BCM), and natural gas (5 BCM) (Figure 6.12). The magnitude of water use for fossil fuels calculated in this book is higher than previous estimates reported in earlier studies such as the 2011 International Energy Agency (IEA) report[45] and Spang et al. (2014)[48]. The higher water footprint of fossil fuel utilization evaluated in this book could reflect more updated water intensity values reported in the literature as well as intensification of the fossil fuels and electricity production utilization during the last decade. While the differences in global energy production between crude oil (193 exajoules in 2019), coal (168 exajoules), and natural gas (144 exajoules; data from BP global dataset[10]) are relatively small, our estimates of global water withdrawal and consumption show large differences in the water footprint among the different fossil fuels, with a distinctively higher water footprint for coal. We demonstrate these differences in calculating the integrative water intensity values, which represent the estimated global water withdrawal and consumption for all stages of energy extraction and production normalized to the global energy use. Figure 6.13 shows the relative high water-intensity values for water withdrawal for coal (2,588 L/GJ) and natural gas (1,453 L/GJ), reflecting the role of water withdrawal for cooling power plants that mostly use coal and natural gas. In contrast, the water intensities of water consumption for coal (295 L/GJ calculated for global water use normalized to energy use in 2019) and crude oil (103 L/GJ) are higher than natural gas (37 L/GJ), reflecting the higher water use for coal mining and conventional crude oil extraction and refining as compared to the water consumption of natural gas.

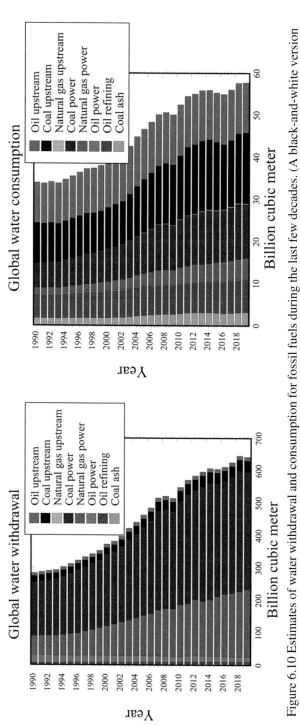

Figure 6.10 Estimates of water withdrawal and consumption for fossil fuels during the last few decades. (A black-and-white version of this figure will appear in some formats. For the colour version, refer to the plate section.)

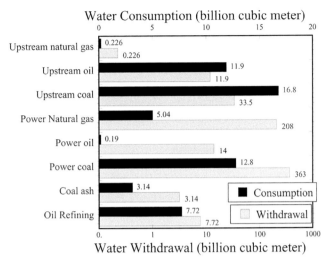

Figure 6.11 Distribution of the estimated volumes of global water withdrawal and consumption (BCM) of fossil fuels during the different stages of extraction (upstream), power generation, and post-water use for coal ash and oil refining in 2019.

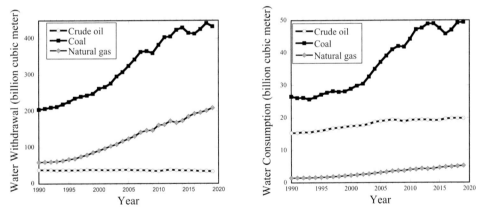

Figure 6.12 Summary of the estimated volumes of total global water withdrawal and consumption (BCM) sorted by fossil fuel types during the last three decades.

6.3 The Global Impaired Water Intensity

In evaluating the volume of global impaired water, we identify two types of water quality deterioration: (1) chronic water contamination induced by common exploration and operation of the energy source (e.g., formation of acid mine drainage in coal mines, release of oil and gas wastewater to the environment); and (2) accidental water contamination resulting from spills of oil, coal ash, and oil and gas wastewater. Given the estimated global water use volumes for the extraction

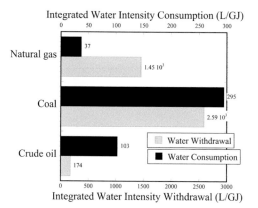

Figure 6.13 Water intensity values (L/GJ) for global water withdrawal and consumption integrated for all stages of fossil fuel extraction and production evaluated in this book.

and processing of fossil fuels presented in Section 6.2, we evaluate the relative addition of impaired water on a global scale.

6.3.1 Global Wastewater of Fossil Fuels

The first critical component is the volume of global wastewater generated from fossil fuel exploration. In Chapters 3–5 we evaluated the intensity of wastewater that is generated from coal mining, coal ash ponds, crude oil, and natural gas (flowback and oil produced water). For natural gas, we used the ratio between produced water and natural gas in the USA that applies for both conventional and unconventional shale gas (gas-to-water ratio of 2.5×10^{-4})[29], and thus quantified the global volume of produced water generated from global natural gas production. For oil, we use a conservative oil produced water-to-oil ratio of 5 relative to higher ratios found for the on-shore oil wells in the USA[29, 373, 375, 416]. For coal, we evaluate two types of wastewater: coal mining and CCR. In China, the volume of coal mining wastewater was estimated as 4 m^3 per ton of coal producing[119]. We assume 50% of the wastewater is reuse (i.e., 2 m^3 wastewater per ton coal) and we combine it with global coal mining data[10] to reconstruct global coal mine wastewater. For the global CCR wastewater estimate, we use the EPA report[301] for the volume of CCR annually generated in the USA (2.5 BCM) with a water intensity of 96 L/GJ, combined with global coal mining data[10], assuming that only 40% of the global CCR is stored in coal ash impoundments (see Chapter 3 for discussion of CCR reuse). The integrated wastewater volumes are presented in Figure 6.14. In 2019, we estimate that the global wastewater from all of the major components of fossil fuels was 43 BCM, in which coal and crude oil wastewaters consist of 52% and 46%, respectively (Figure 6.14). Since the operation and

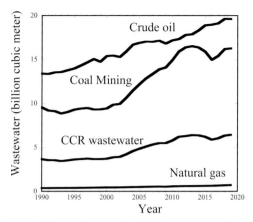

Figure 6.14 Estimates of the volumes of global wastewater that have been generated from fossil fuels over the last three decades. For evaluation of the volume of
crude oil wastewater, we used an OPW-to-oil ratio of 5 and the volume of global
on-shore crude oil production[10]. For evaluation of the volume of natural gas
wastewater, we used a produced water-to-gas ratio of 2.5×10^{-4} found in conventional natural gas in the USA[10]. For coal mining wastewater we used the coal
mining wastewater-to-coal ratio of 2 m^3/metric ton reported for China[119] and
global coal production[10]. For CCR wastewater we used data from USEPA on
the volume of CCR wastewater generated in the USA and assume that only 40% of
global CCR is discharged to coal ash impoundments.

management of wastewater varies between and within countries, it is difficult to
evaluate how much of the global wastewater is released into the environment. For
example, most of the wastewater from coal mining in China is released into the
environment[119], and half of the wastewater from coal ash ponds in the USA are
discharged to nearby surface water and groundwater[280, 281]. In contrast, only 3%
(96.2 million cubic m out of a total of 3.17 BCM) of oil produced water in the
USA is discharged to surface water in the USA[373, 375, 416], whereas unconventional
oil and gas wastewater is almost fully returned to the subsurface through recycling
for hydraulic fracturing or disposal through deep-well injection. Using the USA as
a possible model for global scaling, the volume of on-shore global oil produced
water would be over half a BCM of highly saline water that would be discharged to
the environment. As shown in this book, each of the fossil fuel wastewater types
contains high levels of contaminants that can cause severe water contamination
upon their release into the environment, and thus the impaired water intensity of
wastewater adds an additional water footprint to fossil fuel operations.

Owing to the large variation in fossil fuel production across countries, the
regulation of the high volume of wastewater that is generated from fossil fuel
operations (Figure 6.14) has remained fragmented across countries. Whereas
32 countries in Europe came together in 1979 to sign the Convention on Long-

Range Transboundary Air Pollution (LRTAP[775]) to control air pollution, particularly sulfur emissions, such concerted international efforts have not ensued for managing water contamination from fossil fuel production. In sum, the need remains for scientists and policymakers to work together to assess the global impact of fossil fuel production and consumption on water resources and devise the necessary regulations to prevent contamination of the environment. In the next section, we discuss in greater detail what is meant by impaired global water quality.

6.3.2 Global Water Quality Impaired by Fossil Fuel Operations

The ability to determine the volume of impaired water from fossil fuel activities on a global scale is challenging since many of the cases evaluated in this book are site-specific or apply to different countries with different policies on wastewater management. For example, in the USA a small fraction of wastewater from conventional oil and gas operation is allowed to be released into the environment through transport to treatment centers and discharge to rivers, as well as spreading on roads for deicing and dust suppression. However, this practice may not be acceptable in other countries. Similarly, effluents from CCR impoundments in the USA are allowed to be discharged to waterways in the USA under the Resource Conservation Recovery Act, whereas in Europe, CCR may even be defined as hazardous waste according to the Waste Framework Directive, depending on the content of heavy metals[776]. Therefore, the impaired water intensity (Table 6.2) determined for this book applies for specific case studies but cannot be extrapolated to a global scale. In contrast, the worldwide formation acid mine drainage (AMD) at both active and abandoned coal mines is inevitable and generates a chronic water quality impact. The magnitude of the AMD impact, however, can vary between counties and climate zones and depends on multiple factors, such as rainfall, coal

Table 6.2. *Summary of the water impaired intensity (L/GJ) for the production and processing of fossil fuels*

Activity	Impaired water intensity (L/GJ)
Coal mining	2,500
Coal mining legacy	24,000
Coal combustion	Thermal pollution
Coal ash disposal	500
Oil wastewater discharge	794–1,106
Oil spills	25–49
Natural gas wastewater released to the environment	70–900
Natural gas wastewater spills	4–139

geology and sulfur content, hydrology, location of exposed populations. AMD is widespread in the USA, Australia, Brazil, Canada, Chile, China, Romania, South Africa, and India with thousands of mines generating AMD affecting the environment[120, 777, 778]. As shown in the book, large-scale coal mining in the Appalachian Basin in the eastern USA has generated large volumes of AMD with a very high water-impaired intensity of 24,000 L/GJ (Table 6.2). Yet in arid areas of northeastern China, for example, the volume of AMD and thus impaired-water intensity is lower. Similarly, we estimate a total volume of 585 BCM of water withdrawal for cooling thermoelectric plants sourced from fossil fuels (Figure 6.8), which means that all of this water volume returns to the environment (i.e., the definition of water withdrawal). This vast amount of discharged water causes thermal pollution and degradation of the water quality (e.g., decrease of oxygen content) for thousands of worldwide watersheds located near thermoelectric plants[232].

Oil and wastewater spills are, unfortunately, an inevitable component of fossil fuel extraction; cases from Pennsylvania and North Dakota in the USA and the Niger Delta in Africa clearly show that the intensification of oil and gas exploration is directly associated with high rates of spills. As shown in Chapters 4 and 5, oil and gas wastewater spills have had a large impact on the associated water resources, far beyond the volume of the spilled oil or wastewater, and the natural remediation is limited. As much as it is apparent that oil and wastewater spills can severely impact local water resources, it is challenging to convert these values into global water volumes. Nonetheless, the large volumes of wastewater that are annually generated from fossil fuel exploration and production (Figure 6.12) imply a major risk to water resources since leaking or spills of even a small percentage of these wastewaters can have profound effects on water quality, and therefore impaired water volume. In short, water contamination and the loss of clean water due to the extraction, processing, and use of fossil fuels result in massive water contamination and the loss of vast water volume, in addition to direct water use, and therefore the impaired water should be included in the budget of global water use for fossil fuels. Thus, if countries and subnational regions continue to exploit and produce coal, oil, and gas, policymakers and populations should be aware of the risk of contamination that goes along with the intensification of fossil fuel exploration and production.

6.4 The Emission of Greenhouse Gases from Fossil Fuels

6.4.1 The Emission of Carbon Dioxide and Water Vapor

The increase of man-made greenhouse gases (GHGs) (i.e., gases that absorb and radiate heat) in the atmosphere has been associated with global warming and climate change. Since the early 1990s, the National Oceanic and Atmospheric Administration (NOAA[779]) has monitored the combined effect of long-lived GHGs on the Earth's surface temperature. The GHGs in the atmosphere are composed of carbon dioxide (66%), methane (16%), nitrous oxide (6%), and other gases (e.g., chlorofluorocarbons; CFCs), in which the majority derive from the combustion of fossil fuels[779]. The concentration of carbon dioxide in the atmosphere has increased by ~1.5-fold since preindustrial levels (~280 ppm) and has risen to a concentration of 419 ppm[780] (updated to May 2021; Figure 6.15). The rise of carbon dioxide has been associated with an increase in global temperature of ~1°C, with estimates that suggest that the doubling of preindustrial carbon dioxide levels will likely cause the global average surface temperature to rise between 1.5° and 4.5°C[5, 781]. Climate scientists have argued that to avoid this increase in surface temperature, countries need to collectively act to reduce the concentration of carbon dioxide to 350ppm[782].

Most of the GHG emissions derive from the burning of fossil fuels, most notably coal, hydrocarbon gas liquids, natural gas, and petroleum. The EIA[783] estimates that out of 6,677 million tons of carbon dioxide equivalent, 75% comes from fossil fuel combustion, in which petroleum, natural gas, and coal contribute 46%, 33%, and 21%, respectively. The predominance of petroleum CO_2 emissions is associated with the transportation sector, while natural gas and coal

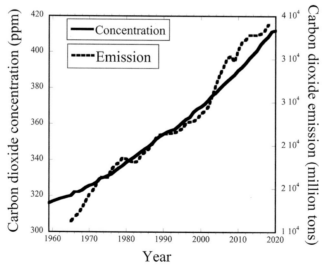

Figure 6.15 The increase of the annual average carbon dioxide concentration in the atmosphere as measured at Mauna Loa, Hawaii (solid line) and global carbon dioxide emission over time (dotted line). Carbon dioxide concentration data are from NOAA Global Monitoring Laboratory[780] and emission data are from BP global dataset[10].

contributions are derived from power generation. Since the combustion of natural gas produces less CO_2 for the same amount of heat produced from burning coal, natural gas contributed in 2019 only 38% of total CO_2 emission from electric power in the USA, while coal contributed 60%, in spite of a larger proportion of natural gas for electricity production[783]. On a global scale, in 2019 annual global carbon dioxide emissions reached 34.2 billion tons (Figure 6.15), of which China contributed 29%, the USA 15%, Europe 12%, and India 7%[10]. Carbon dioxide emissions are directly linked to global warming; humans have emitted about 600 billion tons of carbon, which has resulted in an increase of the average surface temperature of the Earth by 1.3°C. It has been suggested that the emission of one trillion tons of carbon (3.67 trillion tons of carbon dioxide) increases the average global temperature by 1.7 ± 0.4°C, with differential impacts on different regions such as the polar region (temperature increase by 3.6°C per one trillion tons of carbon) and Western North America (2.4°C). Continued emission (Figure 6.15) and accumulation of carbon in the atmosphere are expected to further increase global temperature by 2°C in less than 30 years[89, 784–786].

In addition to carbon dioxide, water vapor is also an important GHG with high heat-trapping properties and thus also contributes to climate change[89]. The combustion of natural gas, petroleum, and coal generate water[787], which is emitted to the atmosphere and should be considered as part of the global water balance induced by the utilization of fossil fuels. Figure 6.16 presents the different water

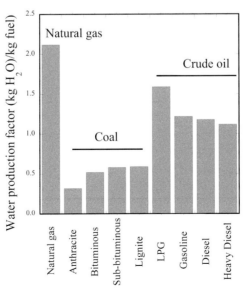

Figure 6.16 Water production factors for stoichiometric combustion of fuels. Data from Belmont et al. (2017)[787].

production capacities of different fossil fuels in which natural gas has the higher water production capacity, followed by natural gas liquids (NGL), petroleum, and coals[787]. Coal has the lowest water generation capacity with values of 0.32 kg H_2O/kg fuel for anthracite to 0.59 kg H_2O/kg fuel for sub-bituminous and lignite coals (Figure 6.16). Based on these water generation capacity values, Grubert and Sanders (2018[29]) estimate that the total water production from coal combustion in the USA (~850 million ton) was 440 million cubic m in 2014. By using the water generation capacity reported for coal[787], the global coal production of 8.1 billion tons in 2019[10] would generate combustion water of 4.1 BCM. As shown in Chapter 3, several volatile contaminants are enriched in coals (Table 3.2), which are emitted to the atmosphere during coal combustion. While some of these elements would be retained in fly ash, highly volatile elements (e.g. mercury, fluoride) with low boiling points may be emitted in the gas phase from the smoke stack. Consequently, the release of combustion water is associated with the fugitive emission of contaminants to the atmosphere, in addition to the increase of water vapor. Combustion of petroleum generates higher water generation capacity, with estimated combustion water-to-fuel weight ratios of 1.12 for heavy diesel, 1.22 for gasoline, and 1.59 for LPG (unit is kgH_2O per kg petroleum)[787] (Figure 6.16). Grubert and Sanders (2018)[29] estimate a volume of 1.1 BCM for the annual combustion water generated from petroleum combustion in the USA. Based on global petroleum consumption, 4.5 billion metric tons of crude oil was consumed in 2019[10], which is equivalent to water combustion of 5.4 BCM. Finally, natural gas has the highest water generation capacity of 2.1 kgH_2O per kg natural gas (Figure 6.16). Grubert and Sanders (2018)[29] estimated a volume of 1.1 BCM for the annual combustion water generated from natural gas combustion in the USA. Based on a global natural consumption of 3,920 BCM of natural gas in 2019[10], we estimate consumption of 2.75 billion metric tons of natural gas (assuming natural gas density of 0.7 kg/m^3), and thus a combustion water volume of 5.8 BCM (using the ratio of 2.1 kgH_2O per kg natural gas[787]). Overall, we estimate that the total global water combustion volume generated from fossil fuels in 2019 was 15.3 BCM per year.

6.4.2 The Fugitive Emission of Methane from Fossil Fuels

As shown in this book, the transition from coal to natural gas that followed the "shale revolution" has substantially reduced the water footprint of electricity generation in the USA (see Section 5.5.3). Yet, the rise of shale gas and tight oil exploration has been associated with fugitive emissions of methane to the atmosphere that have contributed to accumulation of greenhouse gases and global warming, which in turn has had an indirect water footprint through affecting water availability by contributing to reductions of precipitation in arid and semiarid areas. Methane is the second most important human-influenced greenhouse gas in terms of climate change after carbon dioxide[788]. In spite of its relatively low concentration in the atmosphere (~1.9 ppm in 2020) as compared to carbon dioxide (416.4 ppm in 2020)[789], it accounts for ~20% of the radiative effect in the lower atmosphere of anthropogenic greenhouse gas[790]. Since the beginning of the industrial era, methane concentrations in the atmosphere have increased from about 700 ppb in 1750–1,876 ppb in 2020[788, 789]. The National Oceanic and Atmospheric Administration (NOAA) records the global monthly mean methane concentration in the atmosphere[789]. While atmospheric methane concentration was constant between 2000 and 2006, since 2007 the NOAA finds that the methane concentration has increased at a rate of ~8 ppb per year (Figure 6.17), which is equivalent to a net emissions increase of ~25 tera-grams CH_4 per year[791].

Traditionally, methane in the atmosphere was thought to derive from both naturally occurring sources such as methanogenesis in wetlands and fermentation in the soil as well as anthropogenic sources, including paddy rice fields, livestock production systems, biomass burning, anaerobic decomposition of organic waste in landfills, and fugitive emissions during extraction and transport of fossil fuels. Methane is removed from the atmosphere through oxidation within the troposphere, in which chemical reactions with hydroxyl radicals (OH) convert

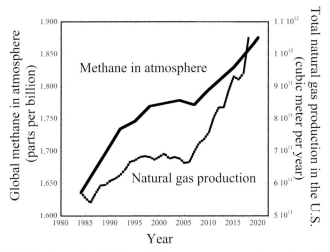

Figure 6.17 Variations of global monthly mean concentrations of methane in the atmosphere as recorded by the NOAA[789] as compared to total natural gas production in the USA. Data on natural gas production in the USA were retrieved from EIA[574]. Since 2006, the global methane concentration in the atmosphere has increased at a rate of 7.9 ppb per year, which was paralleled to the increase of shale gas production and overall natural gas production in the USA.

methane to carbon dioxide[792]. The methane concentrations in the atmosphere reflect the balance between input from the emission of the different sources and removal through oxidation.

Yet, the continuous increase in methane concentrations since 2007 is also associated with the increase in ethane[793], which derives exclusively from oil and gas fugitive emissions, in particular in wet oil fields with large ethane to methane (C2–C1) ratios[794], as well as the rise of shale gas and overall natural gas production in the USA (Figure 5.6). Evidence for methane leaking from shale gas and oil wells has fueled debates concerning the role of the oil and gas industry as a source of anthropogenic methane in the atmosphere[793]. In addition to stray gas contamination of water resources (see Section 5.6), numerous studies have documented methane leaking to the atmosphere. These studies were based on two types of investigation. The first was field on-ground measurements (e.g., mobile laboratory, camera) of methane leaking from shale gas and oil well pads, as well as different stages of life cycles of natural gas and oil exploration[795–801], defined as "bottom-up research." The second was regional observations through monitoring airborne methane and ethane from aircraft flights over oil and gas producing basins[794, 802–807], defined as "top-down research"[808]. Despite a lack of full agreement on the magnitude of methane emissions from the shale gas and tight oil operations, most of these studies have, nevertheless, underscored the large contribution of unconventional oil and gas development to the atmospheric methane budget. Similarly, an analysis of 589,175 operator reports between 2014 and 2018 found annual methane emissions to average 22.1 giga-gram per year from 62,483 wells, including conventional and unconventional wells from Pennsylvania[809]. Integrating the data from these two different types of observations, Alvarez et al. (2018)[808] estimated that in 2015 the total methane atmospheric fugitive flux across the USA was 13 ± 2 tera-grams per year (18.6 BCM), which was 2% of the total natural gas production in the USA during that year. The "bottom up" integrated data showed methane leaking of 7.6 tera-grams per year at the oil and gas production sites, which was double the US Environmental Protection Agency estimate (3.5 tera-grams per year), and combined with other life-cycle stages of processing, transporting, and distribution, the combined leakage ends up being much larger than previously thought[808].

Changes in the concentrations of atmospheric methane over time (Figure 6.17) have stirred a debate in the scientific community on the sources and mechanisms of methane accumulation in the atmosphere; some researchers have suggested that if the stabilization period that occurred between 2000 and 2006 reflects a steady state (input = output), then the renewed growth after 2007 reflects additional sources, like fugitive emission from unconventional oil and gas. In contrast, others have argued that the stabilization period results from temporarily changes in flux rates and/or changes in the oxidation and removal intensity of methane in the atmosphere, and as such, the long-term increase in methane concentration reflects a continuous rise of anthropogenic sources that are not necessarily related to the rise of unconventional oil and gas exploration[793]. Given the relatively short half-life of methane in the atmosphere (9.1 years relative to carbon dioxide of 100 years), combined with the wide range of carbon isotope ratios in the different methane sources (i.e., biogenic from natural sources versus thermogenic from oil and gas emission; see Section 5.2.2), carbon isotopes variations in atmospheric methane could be used to delineate the sources of methane emission to the atmosphere. Thus, the δ^{13}C-CH$_4$ of atmospheric methane may reflect the isotope balance between the integrated contribution of δ^{13}C-CH$_4$ from the different source and an isotope fractionation of 4–6‰ that is associated with the oxidation of methane in the atmosphere, shifting the δ^{13}C-CH$_4$ toward higher (positive) values in the residual atmospheric methane[810]. The δ^{13}C-CH$_4$ in the atmosphere increased from the mid-1980s to 1997, followed by a stabilization period between 1997 to 2008, and since 2009 decreased (data up to 2015 with δ^{13}C-CH$_4$=–47.38‰; data from Schaefer et al. 2016[811]; Figure 6.18). This trend, particularly the decrease in δ^{13}C-CH$_4$ values in the atmosphere since 2009, has raised an additional debate concerning the source of methane in the atmosphere: on one hand, the rise of methane since 2009 was concurrent with an increase in atmospheric ethane[791], which ultimately was derived from fossil fuel emissions that are consistent with the evidence for large methane-leaking from unconventional oil and gas operations, as

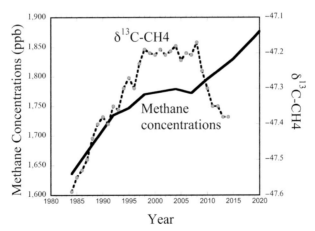

Figure 6.18 Variations of global monthly mean concentrations of methane in the atmosphere as recorded by the NOAA[789]and stable isotope ratio of carbon in atmospheric methane (δ^{13}C-CH$_4$) reported by Schaefer et al. (2016)[811]. The rise in methane since 2007 has been associated with a decrease in δ^{13}C-CH$_4$, which triggered the debate on the role of thermogenic methane with positive δ^{13}C-CH$_4$ values relative to biogenic methane with negative δ^{13}C-CH$_4$ values.

well as the overall increase in natural gas production in the USA and other countries (Figure 5.6). On the other hand, if thermogenic methane was the predominant source, one would also expect to see an increase of δ^{13}C-CH$_4$ toward positive values, and yet the δ^{13}C-CH$_4$ in the atmosphere decreased during this time period toward negative δ^{13}C-CH$_4$ values that characterize biogenic methane sources (Figure 6.18).

The only exception from the general distinction between biogenic methane sources and thermogenic methane with respectively negative and positive δ^{13}C-CH$_4$ fingerprints is the contribution of biomass burning emissions with positive δ^{13}C-CH$_4$ values (Figure 6.19). Changes in biomass burning emissions with positive δ^{13}C-CH$_4$ would therefore also change the isotope composition of the atmospheric methane, regardless of the contribution of methane from fossil fuels. Based on satellite measurements, Worden et al. (2017)[791] have shown a systematic decrease of biomass burning emissions in 2008, after several years (2001–2007) of stable biomass emission. The reduction of biomass emission would in turn reduce the emission of positive δ^{13}C-CH$_4$, and thus could explain the shift in overall δ^{13}C-CH$_4$ of atmospheric methane toward negative values, in spite of the increase in fossil fuel contribution. Changes in the biomass emission provide an explanation to the apparent conflict so that the overall increase of methane and ethane concentrations were derived from increasing contributions of methane from fossil

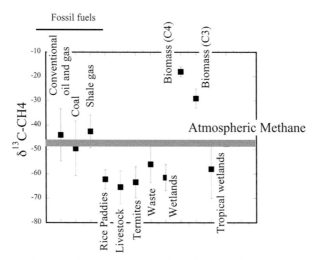

Figure 6.19 Variations of stable isotope ratios of carbon in methane from potential atmospheric sources as compared to atmospheric methane. Note that both naturally occurring and anthropogenic methane sources of biogenic origin (except for biomass) are expected to have negative δ^{13}C-CH$_4$ values relative to thermogenic methane from different sources of fossil fuels with positive δ^{13}C-CH$_4$ values relative to the carbon isotope ratio of atmospheric methane. Data from[793, 811, 816, 817]

fuels, but the decrease of $\delta^{13}C$-CH_4 was due to a smaller contribution of ^{13}C-rich biomass[791]. The revised estimate suggests fossil fuels contribute between 12 and 19 tera-grams methane per year to the recent atmospheric methane increase[791]. This estimate is consistent with Alvarez et al.'s (2018)[808] estimate for unconventional oil and gas contribution of 13 ± 2 tera-grams methane per year for the USA alone. Likewise, Turner et al. (2019)[793] have shown that the concentrations of hydroxyl in the atmosphere during the methane stabilization period were higher than the concentrations post-2007 when methane concentrations started to rise, thus reflecting a higher removal intensity during the stabilization period, rather than a decrease in the contribution of methane sources to the atmosphere.

The slowdown of carbon dioxide emissions from coal combustion increases the impact of methane emissions and accumulation in the atmosphere[812]. Global methane emissions were estimated as 540–568 methane tera-gram per year for the 2003–2012 decade[812], and 550–594 tera-gram per year for the 2008–2017 decade[788], from which ~60% (359 tera-gram per year) is attributed to anthropogenic sources. For the 2008–2017 period, fossil fuel emissions account for 30% of the total anthropogenic fluxes, from which coal mining (11.5% of total anthropogenic fluxes) and the oil and gas industry (22%, 80 methane tera-gram per year based on bottom-up measurements) are the major sources[788]. Through measurements of ^{14}C in methane in ice cores dated to the preindustrial era, Hmiel et al. (2020)[813] have shown that natural geological methane emissions to the atmosphere (1.6–5 tera-gram methane per year) were an order of magnitude lower than previously considered (4–60 tera-gram methane per year), and therefore the anthropogenic methane emissions in modern times has been underestimated by about 25–40% of recent estimates of anthropogenic fossil fuel contributions to the atmosphere[813].

In addition to the input from current oil and gas operations, methane emissions from abandoned oil and gas wells have contributed a substantial source of methane to the atmosphere. Through measurements of methane and other hydrocarbons from abandoned oil and gas wells, Kang et al. (2016)[814] have demonstrated large methane emissions from the oil and gas legacy in Pennsylvania, estimated to 0.04–0.07 tera-gram methane per year from 470,000 to 750,000 abandoned wells in Pennsylvania alone[814]. Similarly, the decline of the coalbed methane industry in the Power River basin in Wyoming left thousands of abandoned wells, from which methane is still emitted to the atmosphere without any monitoring or restrictions[710]. The installation of millions of unconventional oil and gas wells across the USA presents an additional climate change risk for methane emission. While active and operated wells are somehow monitored, abandoned oil and gas wells are not, and therefore present massive potential for unmonitored methane emission without the ability to control or mitigate the methane emission. From a climate policy standpoint, in August 2020 the Trump Administration announced the rollback of a 2016 regulation that required oil and gas companies to detect and repair methane leaks[815]. This rollback cemented the Trump Administration's disdain for global efforts to address the climate crisis, which increasingly have called for curbing methane emissions. However, the Biden electoral victory in November 2020 coupled with the Democrats retaking the Senate allowed the new administration to jumpstart an ambitious climate agenda; most notably, the Democrats in the Senate used for the first time the Congressional Review Act to overturn the Trump Administration's last-ditch effort to overturn the Obama methane rules. Following the House vote in June, President Biden on June 30, 2021 signed legislation to restore the rule limiting methane leaks from oil and gas operations.

6.4.3 Gas Flaring

One of the reasons why natural gas was considered to be an "unwanted by-product" (Stanislaw and Yergin 1993, 90[818]) of crude oil production was the difficulty in capturing the gas, especially in remote locations, and moving it to market; in the absence of gas infrastructure and the lack of market access, companies in the USA[819] and across the world have opted to flare it so as to burn off the methane. In some areas, such as the Niger Delta, flaring has been so pervasive that it has been observed from outer space[820]. Globally in 2018, the World Bank estimated that 145 BCMs of gas was flared and has increased owing to the flaring of gas that has accompanied shale oil production in the USA[821].

Gas flaring (Figure 6.20) is the combustion of natural gas associated with crude oil extraction, distribution, and refinement, and occurs in on-shore drilling sites, off-shore platform production fields, transport ships, port facilities, storage tanks, along distribution pipelines, and refinement facilities. Gas flaring has accompanied conventional oil exploration for decades, and continues to be associated with unconventional tight oil exploration. Since many crude oil reservoirs contain natural gas, the extraction of oil releases large volumes of natural gas that can be hazardous if not tapped or removed by flaring. Thus, in areas where pipelines and gas transformation infrastructure are absent, the untapped natural gas is frequently combusted through flaring that converts methane to carbon dioxide along with other contaminants that are emitted to the atmosphere. The International Energy Agency (IEA) estimated a global natural gas flaring of the magnitude of 145 BCM in 2018, from which the Middle East (29.6%), Eurasia (17.9%), and North America (13.4%) were the major sources (Figure 6.20)[822]. The World Bank used satellite data to show further increase in global gas flaring in 2019, estimated as 150 BCMs, which is equivalent to the total annual gas consumption of Sub-

Figure 6.20 Flaring of unconventional oil well in the Bakken Basin of North Dakota. (A black-and-white version of this figure will appear in some formats. For the colour version, refer to the plate section.)

Saharan Africa, and reflecting flaring increases mainly in the USA (up by 23%), Venezuela (up by 16%), and Russia (up by 9%)[823].

This global gas flaring phenomenon presents major environmental and social issues; the untapped flared gas could be used for the benefit of local communities, particularly marginalized communities living near oil production sites that have been severely affected by the legacy of oil exploration[824–826], as well as utilized as an energy source for remediation of the quality of the oil wastewater that can be a major water source in water-scarce areas (e.g., the Permian Basin in New Mexico and Texas[827, 828]). Yet, at the same time, gas flaring releases carbon dioxide, estimated as an annual flux 275 million ton per year[822], and methane that both contribute to the accumulation of greenhouse gases in the atmosphere, as well as the emission of other contaminants such as nitrogen and sulfur dioxide gases and hydrocarbons that affect the atmosphere and the quality of water resources in the flaring areas[829–831].

The global volume of flared gas has fluctuated since 2007; between 2010 and 2013, flaring declined from 154 BCM per year to 137–140 BCM and then rose again to 150 BCM per year in 2019, tracking an increase in global oil production (Figure 6.21)[832–833]. The World Bank's Global Gas Flaring Initiative, however, which seeks to encourage countries and companies to reduce gas flaring through

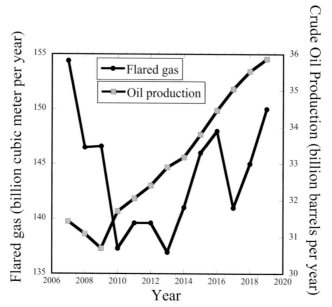

Figure 6.21 Variations of the volume of global flared gas (BCM per year) as compared to global oil production (billion barrels per year) between 2007 and 2011. Data retrieved from Soltanieh et al. (2016)[832], World Bank[833], and BP global dataset[10]

raising awareness, found in its 2020 report that globally "gas flaring has increased to levels last seen in 2009"[833]. According to the World Bank report, of the 150 BCM that was flared globally in 2019, the breakdown from the major emitters was the following: Russia (23.2 BCM; 15.5% of global flared gas emission), Iraq (17.9 BCM, 11.9%), the USA (17.3 BCM, 11.5%), and Iran (13.8 BCM, 9.2%)[833] (Figure 6.22). While Nigeria has attracted lots of international attention for flaring about 10–40% of its associated gas[826], the World Bank ranked Nigeria only seventh among the countries that generate flared gas (Figure 6.22). The USA has joined the ranks of one of the countries that generates the most flared gas owing to the rise of unconventional oil and gas exploration in the early twenty-first century. Based on the World Bank data, the volume of flared gas in the USA increased by a rate of 3.9 BCM per year since 2017, up to 17.3 BCM per year in 2019 (Figure 6.23). The unconventional Permian, Bakken, and Eagle Ford basins have become the major oil producing basins with more than 83% of total US oil production. The absence of infrastructure to bring associated natural gas to market in North Dakota in the early years of the oil boom in the Bakken contributed to the flaring of natural gas[558].

The unconventional oil basins in the USA are rich in gas; the average volume of gas-to-oil ratios in the Bakken and Permian basins wells are 250 and 300 (in cubic

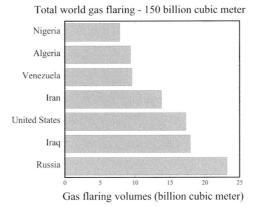

Figure 6.22 Global distribution of natural gas flaring (BCM) in 2019. Data from the World Bank Report (2020)[833]

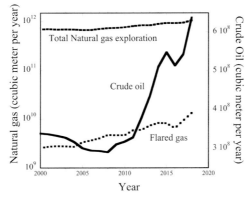

Figure 6.23 Variations of the volume of natural gas flared compared to the annual volumes of the total natural gas (left in logarithmic scale) and crude oil (right) exploration in the USA. Note the rise of crude oil exploration from unconventional oil since 2011–2012 and the increase of flared gas volume. Data from US Department of Energy, Office of Fossil Energy[834] and EIA[835]

m units)[24], and natural gas has been increasingly flared with oil productions in these basins. Data from the US Department of Energy[834] and EIA[835] show increasing flare gas volumes between 2013 and 2018 up to 4.3 BCM per year (Figure 6.23). In the Permian Basin, gas flaring increased from 8.8 million cubic m per day during 2014–2017 by threefold to 26.4 million cubic m per day (9.6 BCM per year) by late 2019, which is equivalent to 12 kg of carbon dioxide per barrel of oil production[836]. Satellite data from the Eagle Ford Shale region of south Texas have identified 43,887 distinct oil and gas flares from 2012 to 2016, with a peak in activity in 2014 and an estimated 4.5 BCM of total gas volume flared over the study period[837]. Overall, the total volume of natural gas that was flared and vented

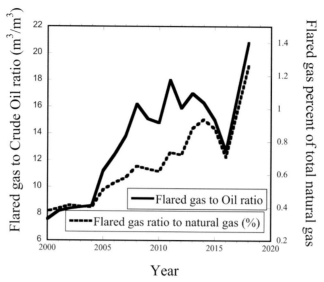

Figure 6.24 Variations of the ratios of flared gas-to-crude oil production (left) and percentage of the total natural gas production (right) in the USA. Note the steady rise of flared gas specifically with the sharp rise of crude oil production. Data from US Department of Energy, Office of Fossil Energy[834] and EIA[835]

in 2018 in the USA was estimated at 13.3 BCM (Figure 6.24)[15], which is equivalent to 1.3% of the total natural gas production in the USA. The flared gas-to-crude oil ratios and the percentage of the annual natural gas production in the USA have increased over time (Figure 6.24), reflecting the increasing role of unconventional crude oil exploration in natural gas flaring. While the US EPA assumes a high (98%) efficiency rate, surveys of hundreds of flare stacks across the Permian Basin have shown that about 11% of the flares are malfunctioning, and at least 7% of Permian gas sent to flares is escaping directly as methane and other pollutants to the atmosphere[838]. The efficiency of the combustion of the natural gas during flaring is therefore an important factor for the atmospheric emission and environmental contamination.

In addition to the emission of greenhouse gases to the atmosphere and large thermal radiation and noise levels, gas flaring releases hazardous air pollutants such as particulate matter, hydrocarbons other than methane, volatile organic compounds (e.g., BTEX), Polycyclic Aromatic Hydrocarbons (PAHs), nitrogen oxides (NOx), sulfur oxides (SOx), and soot (black carbon)[830, 831, 839–841], which have been shown to directly affect the health of local communities. Such health impacts are just some of the environmental injustices experienced by communities living near oil and gas production sites[824–826, 841]. The release of sulfur oxides (SOx) and nitrogen oxides (NOx) to the atmosphere and their dissociation to nitrate and sulfate trigger the formation of acid (low-pH) rain[830, 839], which can have devastating ecological and water quality impacts upon acidification of surface water and shallow groundwater. The formation of a gas flaring plume, often at high temperatures, coupled with the height of the stack, enables some of the pollutants to escape further into the free troposphere resulting in long-range transport of flaring contaminants[840]. The formation of acid rainwater enriched in nitrate and sulfate[842] has been demonstrated in the Niger Delta. Gas flaring has been a major contributor to air pollution across the Niger Delta, with air pollutant concentrations exceeding WHO limits in some locations over certain time periods. Due to the predominant south-westerly wind, concentrations of air pollutants were found also in areas with little flaring activity, affecting 20 million people who are exposed to high flare-associated air pollution[843]. Relatively high sulfate, bicarbonate, and nitrate contents were detected in rain water samples within the radius of 20 km from gas-flaring stations, and found to decrease away from the gas-flaring stations[844]. The enrichment sulfate, bicarbonate, and nitrate combined with low-pH in rainwater was also reflected in surface water and shallow groundwater in the vicinity of the flaring sites[844]. The impact of acid rain and contaminant emissions was detected in surface water and groundwater located near flaring sites across the Niger Delta, Nigeria, with higher salinity and concentrations of heavy metals (lead, iron, zinc, and chromium) and organic contaminants (PAHs) in groundwater located near flaring sites relative to concentrations in groundwater located away from flaring sites[826, 832, 842, 844–848]. The concentrations of PAHs were ~fivefold in groundwater near flaring sites as compared to background levels[845]. The low-pH of rivers and shallow groundwater in the Niger Delta (as low-pH as 5.1) triggers mobilization of heavy metals that resulted in elevated lead, barium, cadmium, and selenium concentrations above the World Health Organization (WHO) maximum permissible limits[846]. Overall, multiple studies have demonstrated the impact of gas flaring on the vegetation and water quality resources in the Niger Delta region with changes in the water chemistry, higher salinity, low-pH, and heavy metals and reduction in soil quality that led to reduction in agricultural productivity[826]. Given the overall uncontrolled pollution and the low quality of water resources in Nigeria[849], the emissions of pollutants and formation of acid rain from large-scale gas flaring further exacerbates the already dire environmental conditions and poor water quality of one of the largest African states. Tapping this volume of flared gas to provide an energy source could have major economic and social benefits, including using this energy source for water treatment. While quantification of the magnitude of water pollution from gas flaring is not available, it is clear that gas flaring induces water contamination and therefore adds to the impaired water intensity of oil production, transport, and refinement.

In many parts of the world flaring has been convenient for dealing with associated gas from oil production, yet owing to public pressure to reduce flaring, governments are taking steps to reduce it. Beginning in 2005 a court case in Nigeria ruled that companies should cease flaring; yet, despite such cases and anti-gas-flaring laws that date back to the mid-1980s[850], enforcement has remained elusive[820]. The environmental damage from flaring and broader oil production in the Niger Delta has been the subject of scrutiny for years, resulting in an environmental assessment in Ogoniland carried out by United Nation Environment Programme (UNEP)[92]. Other countries such as Azerbaijan have sought to reduce flaring by requiring companies to include domestic market obligations in their contracts to capture the gas for local markets[91]. Overall, by being able to capture the gas for local markets, this can help improve energy access in parts of the world where it is low.

At the federal level in the USA, when it comes to regulating flaring and venting, this occurs through the EPA setting standards for air quality under the Clean Air Act (CAA). Yet, for the most part, EPA allows states to develop and implement the necessary regulations to meet federal standards, and thus, regulating venting and flaring takes place at the state level[819]. Where the federal government may try to regulate oil and gas production, including flaring, is on federal lands. Indeed, as mentioned above, the Obama Administration sought to require oil and gas companies to capture leaked methane from escaping into the atmosphere from oil and gas operations on federal and tribal land[851] to help cut greenhouse gas emissions from the oil and gas industries[851]. Under the Trump Administration, progress toward reducing venting and flaring depended on the ability of states to introduce their own regulations or for local ordinances to limit operators (as demonstrated by the state of Colorado's November 2020 landmark ruling to end the practice of routine flaring and venting of natural gas at drilling sites[852]), or on the voluntary actions/commitments taken by the companies themselves.

The devolution of regulatory authority to the states is exemplified by the DOE's Office of Fossil Energy having to compile fact sheets from the 32 oil and gas producing states in the USA to capture the different state-level natural gas flaring and venting regulations[853]. Depending on the state, it could be the department concerned with natural resources, or the state's oil and natural gas commission, the state environmental protection agency, or air-quality board[819] that regulates venting and flaring. The two states with the highest levels of oil and gas drilling and production accompanied by gas flaring are Texas and North Dakota. The Texas Railroad Commission (TRRC) – the entity with jurisdiction for issuing permits for flaring in Texas – has allowed operators to flare gas while drilling a well and up to ten days after a well's completion (DOE 2019, 21[819]). The DOE found that flaring has "increased significantly" since 2010 in Texas owing to the development of tight oil plays in the Permian Basin and the Eagle Ford play (DOE 2019, 21[819]). Yet, in the Permian Basin, with increased public awareness about flaring, it was notable instead that the oil and gas company BP said in 2019 that it would not install new wells unless they had access to a gas pipeline so as to prevent flaring[854]. In contrast, in Colorado, where the Colorado Oil and Gas Conservation Commission (COGCC) is responsible for regulating the state's oil and gas development, the state in 2014 began to regulate methane emissions from oil and gas drilling so as to reduce its greenhouse gas emissions (DOE 2019, 33[819]).

6.5 The Energy Use for Water

Globally, water demand has grown. Since 1900, global water withdrawals and consumption have increased by over 1,000%[855], up to an estimated 4,000 BCM[747–749] (Figure 6.1). The different stages of water utilization, including pumping, transport, treatment of raw water, distribution, domestic, industrial and agricultural use, wastewater treatment, wastewater disposal and reuse (Figure 6.25), all require energy. The energy intensity is defined as a unit of electricity utilized per volume of water and expressed as *kilowatt-hours per cubic m* (kWh/m^3) and varies considerably for the different water extraction and treatment technologies (Table 6.3). In the USA, it was estimated that the energy generated for the water sector in 2010 was 12.3 quads (12.98 exajoules), which was 12.6% of the total national primary energy consumption. About 5.4 quads (5.7 exajoules) of this primary energy were used to generate electricity for pumping, treating, heating, cooling, and pressurizing water in the USA[856]. Of the total electricity generated in

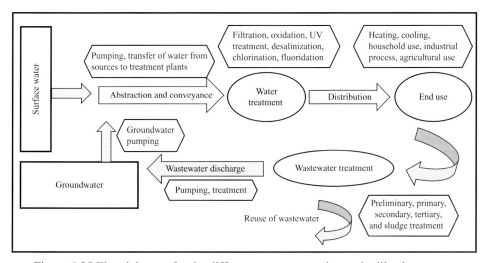

Figure 6.25 Electricity use for the different water processing and utilization stages.

Table 6.3. *Variations of energy intensity values of different water utilization and treatment**

Water use	Energy intensity (kWh/m^3)
Surface water supply	0.0002–1.74
Ground water pumping	0.37–1.44
Desalination (SWRO)	1.0–8.5
Desalination (brackish water RO)	1.0–2.5
Desalination (MSF)	20–30
Desalination (MED)	15–22
Wastewater treatment	0.38–1.12
Wastewater recycling	0.18–0.63
Wastewater recycling via RO desalination	1.0–3.8

* Data from Wakeel et al. (2016)[767] and Shahzad et al. (2017)[771].

the USA, the water sector consumes 1.5–2% for public water and wastewater services. Likewise, the estimated public water supply and wastewater treatment facilities accounted for 4% of the total electricity consumed in the UK[857]. The International Energy Agency (IEA, 2017)[45] estimated that in 2014 about 120 million tons of oil equivalent (Mtoe) of energy was used worldwide for the water sector, from which 60% of that energy was consumed in the form of electricity, corresponding to a global demand of around 820 terawatt-hours, which is equivalent to 4% of global electricity consumption. The residual 40% (48 Mtoe) was used as thermal energy, half in diesel pumps to extract groundwater for irrigation, and half used for desalination, mainly in the form of natural gas, mostly in the Middle East and North Africa[45]. Of the global electricity consumed in 2014 for the water sector, 40% was used for groundwater and surface water extraction, 25% for wastewater treatment, 20% for distribution, and 5% for desalination (for 2014, see section below; Figure 6.26)[45]. The wastewater treatment has a larger electricity consumption in developed countries relative to developing and emerging countries[45].

In the USA, for example, there are 15,000 wastewater treatment plants, which consume about 40% of national electricity consumption in the water sector. Each of the wastewater treatment stages (i.e. preliminary, primary, secondary, tertiary, sludge treatment) has a very different energy consumption. A survey of 600 wastewater treatment plants across several countries has revealed that the magnitude of energy consumption decreases with the increase of the population and flow rate of the plant (i.e. the "dilution factor," which is the ratio of the flow rate to the population that utilizes the wastewater treatment plant), with a value range of 1–4 kWh/m^3 for small wastewater treatment plants to <1 kWh/m^3 for large wastewater treatment plants[858]. The average energy consumption for wastewater treatment varies between 0.25 kWh/m^3 in China to 0.67 kWh/m^3 in

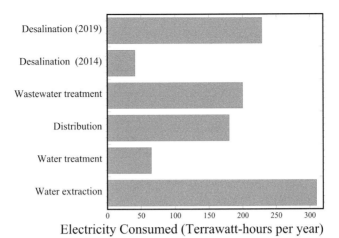

Figure 6.26 Global electricity consumed for the water sector. Electricity data for the different water use components, including desalination in 2014, were from IEA (2017)[45]. Data for desalination in 2019 were calculated based on the global distribution of desalted water generated from the different desalination technologies (see Figure 6.28)

Germany[767]. On a global scale, it was estimated that over 50% of the electricity consumption is allocated for secondary treatment in wastewater treatment plants, 16% for pumping, 14% for sludge treatment, 10% for tertiary treatment, and 8% for primary treatment[45]. Since it is estimated that 35% of the global municipal wastewater is not collected, mostly in developing countries, one would expect that future development would require additional electricity utilized for wastewater treatment. Furthermore, global water scarcity has prompted many countries to reuse domestic wastewater for irrigation, and therefore, the tertiary wastewater treatment is expected to grow and further increase the electricity utilization of global wastewater treatment.

While in 2014 desalination consisted of only 5% of the global electricity used for the water sector, the continued rapid global rise in desalination has increased this electricity footprint. Since 1990, global desalination has increased by 1,200% and reached to 95.4 million cubic m per day (that is equivalent to ~35 BCM per year; Figure 6.27) from 15,906 operational desalination plants[770]. The majority of the desalination plants (48%) were installed in the Middle East and North Africa, with Saudi Arabia (15.5%), the United Arab Emirates (10.1%), and Kuwait (3.7%) as the major producers, followed by East Asia and the Pacific (18%) and North America (12%)[770]. The power consumption for desalination has decreased significantly over the years; for example, the power consumption for the reverse osmosis (RO) stage in seawater RO (SWRO) plants has reduced from 5 to 15 kWh/m^3 during 1970–1990 to about 1 kWh/m^3 (for seawater salinity of 35 g/liter seawater at

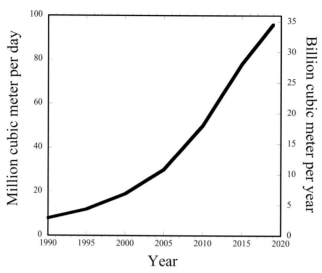

Figure 6.27 Changes in the volume of global desalination. Data from Jones et al. (2019)[770].

50% recovery) in modern SWRO plants[859]. The overall energy consumption directly for RO processes combined with pumping, pre-treatment of the feed water, brine discharge, and electric power used in the desalination plant is estimated as 2–4 kWh/m^3 [860]. Yet thermal desalination technologies such as Multi-stage Flash Distillation (MSF) and Multi-effect Distillation (MED) require much larger electricity footprints, whereas brackish water desalination consumes less energy than SWRO desalination (Table 6.3). The RO desalination is the most common desalination technology, accounting for 84% of the total number of operational desalination plants and 69% of the volume of the total global desalinated water[770]. The relative proportion between global SWRO (33% of the global desalinated water), seawater-MSF (18%), seawater-MED (5%), brackish water RO (19%), river water RO (8.4%), and wastewater RO (5.2%) reported in Jones et al. (2019)[770] (Figure 6.28) is used to estimate the magnitude of electricity production based on the median values of the electricity intensity of the different desalination technologies reported in Table 6.3. Integrating the electricity intensity values with the volume of the desalted water from the different desalination technologies yields an annual electricity of ~230 TWh, which is equivalent to ~28% of the global electricity consumed in 2019 for the water sector, and is much higher than the previous IEA estimate for 2014 (5%)[45]. This reflects the rapid rise of global desalination over time; between 2014 and 2019 the volume of global desalted water increased by ~30% (Figure 6.27). The relationships between the volume of the global desalted water and the electricity consumption of the different desalination technologies are presented in

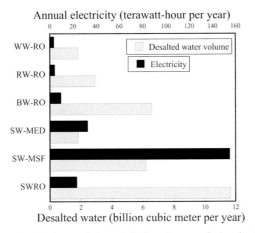

Figure 6.28 The distribution of the global volume of desalted water and the associated utilized electricity sorted by the desalination technologies. Data on the volume of the global desalted water were from Jones et al. (2019)[770] while the associated electricity consumption values were calculated based on the electricity intensity (median) values of the different desalination technologies reported in Table 6.3.

Figure 6.28 and show that while SWRO desalination generates the majority (33%) of global desalinated water, it consumes only 10% of the global electricity used for desalination. In contrast, seawater-MSF and seawater-MED desalination generate 23% of global desalinated water, but consume 82% of the global electricity used for desalination (Figure 6.28). All of the seawater-MSF and seawater-MED desalination plants are located in the Gulf states and, therefore, we posit that future development of SWRO desalination in other locations will not significantly increase the electricity footprint of water desalination. In sum, due to the development of the membrane technology and increase in the efficiency of RO desalination plants, the electricity footprint for RO desalination has substantially decreased. It was estimated that ~2,000 kW per year is needed to desalinate seawater to supply water for one household, which is equivalent to the electricity used in a household's refrigerator[860].

The removal of salts during the desalination process and the generation of potable water create wastewater in the form of brines. The recovery ratio (i.e. the fraction of the feed water volume that is converted to potable water) depends on the quality of the feed water with lower recovery ratio with higher salinity as well as the desalination technology. For example, SWRO desalination has a lower recovery ratio (0.42) compared to brackish water RO desalination (0.65) and river water RO desalination (0.85). Yet, thermal desalination technologies (seawater-MSF = 0.22; seawater MED = 0.25) have typically much lower recovery ratios than membrane technologies, which implies much higher brine volume compared to RO desalination technologies. Based on the relative proportions of desalination technologies and associated recovery ratios, Jones et al. (2019)[770] estimated global brine production of 141.5 million cubic m per day, or 51.7 BCM per year, with most brine production (55%) in Saudi Arabia, UAE, Kuwait, and Qatar[770]. The large scale of brine discharge to the ocean can potentially induce ecological effects given the high salinity and metal enrichment in the brine. Yet over a decade of research of the water quality of the marine environment off-shore of the Mediterranean coast of Israel, where large volumes of brines (up to 187 million cubic of m per year) have been discharged from several large SWRO desalination plants along the Mediterranean seashore, has shown minimal water quality and ecological impacts on the ambient marine environment. Temporal high salinity areas with 5% higher salinity than the background seawater were identified in some restricted areas near brine outfalls but without any other detected water quality issues[860].

6.5.1 Desalination and Water Quality

RO desalination enables selective transport of non-charged ions and isotopes through the membrane, which modifies the chemical and isotope compositions of the permeates (i.e., the treated water) relative to the feed water and generates a new type of "man-made" water with a distinguished geochemical composition that is different from natural water[861]. Non-charged ions such as boric acid (boron species at aquatic pH <8) or arsenite (arsenic species under reduced condition at pH <8), as well as oxygen and hydrogen isotopes, are not rejected by the RO membrane and transferred into the permeates. In cases of high boron (e.g., seawater with boron concentration of 4.7 mg/L) and arsenic (e.g., reduced water) in the feed water, this could result in boron-rich and arsenic-rich permeates that would limit the ability of their use. Since reuse of domestic treated wastewater is becoming a critical and important water source for irrigation, particularly in water-scarce areas (e.g., Israel with 60% of the wastewater reuse for the agriculture sector), high concentrations of boron originated from RO desalted water in domestic wastewater could limit the ability to reuse the wastewater for the agriculture sector, particularly for boron-sensitive crops[862, 863]. Consequently, the configuration of the desalination plants in Israel was modified to include both low-pH (removal of the bulk charged ions) and high-pH (converting all boric acid to the charged species borate under high-pH and its removal) modules, which resulted in better performance of boron removal[864–867]. Likewise, additional post-treatment operations for arsenic removal are needed for cases where the feed water is originated from reduced arsenic-rich groundwater (e.g., the Atlantic coast of North Carolina[868]).

One of the critical aspects of the changes of the permeate chemistry is the high efficiency of the removal of alkaline-earth calcium and magnesium from the RO membrane, resulting in low concentrations of calcium, magnesium, and bicarbonate in low-pH permeates. The high corrosivity of the low-pH permeates requires post-RO treatment aiming to increase the mineralization of the effluents, mostly though interaction with calcium carbonate rock materials. The low calcium and magnesium in permeate also poses human health risks. Since 1960, numerous epidemiological studies have consistently reported chronic health effects in populations consuming naturally occurring low mineral drinking water, as well as artificial softened water with low magnesium and calcium. These studies have shown a significant association between cardiovascular disease and drinking water chemistry, demonstrating the protective role of magnesium in drinking water and the inverse association between magnesium and cardiovascular mortality[869–874] [875–877]. In Israel, uptake of low-magnesium desalinated seawater was associated with a significant decrease in serum magnesium concentrations[878, 879], as well as high rates of heart disease[880, 881]. The human health risk from long-term utilization of desalinated water and magnesium deficiency has triggered a debate in Israel regarding installation of more aggressive post-treatment that includes dolomite rocks that would provide magnesium to the post-treated RO permeate[882–885]. Low-mineral water intake can result also in fluoride deficiency in children and associated health effects, as demonstrated in retarded height growth and increased dental caries in exposed school children in China[886].

Pumping saline groundwater for desalination in coastal aquifers can change the hydraulic dynamics between saline water and low-saline groundwater along the salt–water interface. Selective pumping of the saline groundwater would reduce the saline pressure and shift the interface seawards, with intensification of freshening (i.e., low-saline water flow into saline aquifer)[887] and associated reverse base-exchange reaction (retention of calcium and release of sodium and boron[88, 888]). Along the Atlantic coast of North Carolina, pumping of saline groundwater for desalination along the Outer Bank was not associated with changes in groundwater salinity over time, reflecting continued freshening conditions and maintaining the saline–fresh water interface away from the coast[868]. In contrast, over-pumping of low-saline groundwater could shift the saline–fresh water interface inland, causing groundwater salinization[88]. Overall, given the rise of global RO desalination (Figure 6.27), the volume of RO permeates is increasing and in some cases replacing the naturally occurring meteoric water, generating a new type of man-made potable water. In addition to the chemical fractionation (e.g., selective rejection of double charges cations over a single charge, preferential permeation of smaller ions), differential isotopes permeation through the RO membrane control the water chemistry of RO permeates. While the stable isotope compositions of oxygen and hydrogen in RO desalinated water are identical to the original feed water, boron isotope fractionation is expected due to the selective rejection of the ^{10}B-rich borate species. Consequently, man-made RO desalinated water is expected to have stable isotope ratios identical to seawater (δ^{18}O, δ^2H ~0‰) that are different from naturally occurring meteoric water (δ^{18}O, δ^2H <0‰) as well as a distinctively high δ^{11}B signature that is different from common meteoric water[861].

6.6 Clean Future and the Legacy of Fossil Fuels

In addition to the direct contribution to GHG fugitive emissions and global warming, this concluding chapter has underscored the high-water footprint of fossil fuel exploration and the impact on water quality, which then in turn further reduces water availability (Figure 6.4). Over the last decade, the transition from coal to natural gas as the major energy source for electricity in the USA has reduced overall water withdrawal and consumption, with a reduction of 40 m^3 in water withdrawn and one cubic m for water consumption for every MWh of electricity that has been generated with natural gas instead of coal[234]. As shown in the book, the impaired water intensity of coal mining is far higher than natural gas (e.g., global impact of acid mine drainage, Table 6.3), and therefore policies to reduce coal mining and combustion would lead to much higher levels of water saving and reductions in water contamination. Yet, in spite of the reduction of coal mining and combustion in the USA, and temporary reduction in China, coal production on a global scale has not declined, and in 2019, coal mining reached its second highest level after the 2013 peak (Figure 3.3). Much of this reliance on coal, especially in parts of Asia (e.g., Japan, South Korea, and Taiwan) and Europe (e.g., Poland), has to do with maintaining energy security. For example, for Japan, coal has been critical to its energy security following the shutdown of its nuclear capacity in the aftermath of the Fukushima accident in 2011[889]. Coal has also been a central focus of China's energy security, dating back to the deterioration in relations with the Soviet Union in the 1960s that affected the decline of imports of oil products[889]. If anything, the last few decades of rising temperatures accompanied by extreme weather events ranging from floods to droughts have shown the importance of taking global action to reduce dependence on fossil fuels as one step in addressing climate change.

A country's energy security must then also consider other variables such as economic growth, environmental impact, and health benefits. From a water perspective, we have shown that the transition away from coal to natural gas has some benefits for the environment and water saving. These transitions come with social costs that must be addressed to sustain communities' livelihoods. In particular, as coal mines shut down and workers are unemployed, they may need to relocate and be retooled, as has been the case in Germany and China where commitments have been made to undergo energy transitions. Yet, these transitions disrupt local communities and livelihoods where coal, in particular, has been the major source of employment for decades. In 2016, the Chinese government, for example, announced plans to lay off 1.8 million workers in the coal and steel industries, equivalent to 15% of the workforce, of which 1.3 million would be from the coal sector[890]. In 2019, when the German government announced that it would shut down all of its coal-fired power plants by 2038, it also committed economic aid to the coal-mining states that would be affected[891]. Other countries such as India will likely need to grapple with similar questions of energy transitions given the deterioration in air quality caused not just by coal-powered plants, but also automobile emissions, in cities like Delhi[892].

For countries to meet their climate commitments to reduce greenhouse gas emissions, energy security will thus require adding more renewable sources to their energy portfolio along with introducing more energy-efficient transportation, for example. As such, countries such as China have increased the number of electric vehicles on the road as a way to reduce greenhouse gas emissions[889], as well as experimenting with carbon capture and storage technologies. Even more significant, China has invested heavily in its wind and solar sectors to meet its ambitious greenhouse gas emission reduction targets while also promoting economic development[893]. In the USA there is a growing movement toward consuming energy from renewable sources. For the first time since 1885, the USA consumed more energy in 2020 from renewable sources than it did from coal, owing to greater use of wind and solar in the power sector and the closing of coal-powered plants and the increasing use of natural gas[894].

When it comes to the water–energy nexus, alternative renewable energy sources such as solar and wind have much lower water footprints with several orders of magnitude lower water intensities as compared to fossil fuels (Figure 6.29). Therefore, substituting fossil fuels with renewable energy sources is likely to reduce GHG emissions along with water use and contamination. The negative loop of rising energy consumption that increased the intensification of water use and contamination and in turn result in the need for more water transfers and treatment, and ultimately more energy utilization (Figure 6.4), could then break down with the alternative use of renewable energy sources. While the renewable energy sector holds promise for countries to meet their energy challenges and tackle the climate crisis, it is imperative to recognize that the legacy of fossil fuel utilization will have a long shadow on countries' economies and environments owing to the widespread impact on water quality. For coal, the legacy is the continued formation of acid mine drainage through oxidation of sulfur minerals in the exposed rocks from abandoned coal mines. In

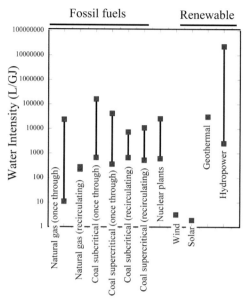

Figure 6.29 Summary of the life-cycle water withdrawal (upper range) and water consumption (lower range) intensity values (L/GJ) associated with different water cooling combinations of fossil fuels (natural gas, coal) and nuclear as compared to renewable energy sources (geothermal, solar, and wind). The water intensity values of renewable energy sources are several orders of magnitude lower than those of fossil fuels. Data for fossil fuels were from Kondash et al. (2019)[234] while data for nuclear and renewable energy sources were from Grubert and Sanders (2018)[29]

addition, CCRs disposed in impoundments and landfills pose major environmental and human health risks due to the continuous mobilization of contaminants from disposed or spilled CCRs to the aquatic phase upon interaction with water. Evidence for unmonitored coal ash spills into lakes near coal ash plants in North Carolina suggests that CCRs disposal is not restricted to designated sites[269] but also occurs in lakes near coal plants and contaminates water resources (e.g., high arsenic and boron in pore water from lakes associated with CCR impoundments[280]) and bioaccumulate into the ecological system[895]. For the oil and gas industry, abandoned oil and gas wells pose major risks to groundwater contamination and methane fugitive emission to the atmosphere[461, 462, 809, 814, 896]. It has been estimated that in the USA about 3.2 million abandoned oil and gas wells emitted 281 kilotons of methane in 2018[896]. In Canada, it is estimated that 313,000 abandoned wells emitted 10.1 kilotons of methane in 2018[896]. Furthermore, oil and brine spills generate contamination that can affect the quality of water resources for decades if not longer (e.g., accumulation of radioactive elements on sediments and soils at spill sites).

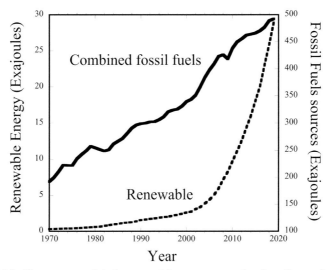

Figure 6.30 Changes to global renewable energy production (in exajoules, left y-coordinate) and the combined fossil fuel production (right y-coordinate) during the last 50 years. While global fossil fuels increased during the last decade by 12%, renewable energy production increased by about 200%.

Indeed, water quality of watersheds and shallow aquifers associated with exploration sites will continue to be affected from the legacy of waste solids and wastewater generated from fossil fuel exploration.

By 2019, the global production of renewable energy sources reached 29 exajoules, which reflects a rapid increase with a rate of 2.4 exajoules per year over the previous decade, a 66% increase (Figure 6.30). Yet this increase has coincided with a parallel increase in the combined fossil fuel utilization (Figure 6.30). While global fossil fuels increased during the last decade by 12%, renewable energy production increased by about 200%, and yet consists of only 5% of the global energy production, whereas fossil fuels consisted of 84% in 2019 (data from BP global dataset[10]).

This conclusion was written as the world is in the midst of the greatest collective challenge in the twenty-first century – the COVID-19 pandemic. At first sight it appears that lockdowns and a pandemic-induced economic collapse in 2020 have led to a major slowdown in the energy sector. Unless governments take action to address the climate crisis in conjunction with the pandemic, this dip in energy demand will be fleeting. Although the price of oil initially dropped and there has been minimal new development and exploration for oil and gas resources globally, the long-term effect of this global pandemic is unknown. As economies begin to reopen and the transportation sector recovers, there has been an increasing demand for fossil fuel, and countries will shift back to the "business as usual" model that will lead to the same negative impacts on water quantity and quality degradation outlined in this book (Figure 6.4).

If countries will accelerate the transition from fossil fuels to renewable energy sources, many of these deleterious impacts on water resources might be reversed. While the COVID-19 pandemic resulted in a reduction of global energy by 5%, a recent IEA report (November 2020[897]) indicated that renewables used for generating electricity grew, mostly in China and the USA, by almost 7% in 2020. The increase of renewable energy sources with combined energy of about 200 GW was predominantly (90%) derived from wind and hydropower, while solar PV growth remained stable[897]. Ultimately, individuals, communities, and governments will need to decide whether they will continue to depend on fossil fuels for their primary energy source or undertake broader energy transitions that will alter global economies while protecting global water resources and the environment.

This chapter further concludes as world leaders gathered in Glasgow for the 26th UN Climate Change Conference, and as before during previous meetings of the conference of the parties to the UN Framework Convention on Climate Change, governments will weigh their commitments to slowing down climate change. Yet, as we show throughout this book, a reliance on fossil fuels not only contributes to global warming, but also affects water availability and quality worldwide. Thus, decarbonization of the energy sector is no longer up for debate, but necessary if the world community also wants to protect its global water resources for future generations.

Chapter 6 Take-Home Messages

- Since fossil fuels comprise 86% of global energy consumption, the energy–water nexus has an important role in the global anthropogenic cycle: increasing water and energy consumption has resulted in over-exploitation and depletion of water resources, which is further exacerbated by large water utilization for energy development, water quality degradation from energy-derived contaminants, as well as global warming and climate change induced by carbon emission by fossil fuel combustion. Water quality degradation requires remediation, and thus the increasing need of energy for water treatment such as desalination further intensifies energy generation, which reactivates the negative-loop cycle.
- Using updated energy production data and water intensity values, this book reevaluates the volumes of global water withdrawal and consumption for the different cycles of energy recovery and processing of fossil fuels. The revised estimate volumes far exceed previous estimates, demonstrating the increasing water footprint and impact of fossil fuels on global water resources.

- Parallel to water use, this book highlights the high volume of global wastewater that is generated from fossil fuel consumption. Given the high levels of contaminants, the release of even small volumes of wastewater through spills, leaks, or inadequate management severely affects the water quality of the receiving water and thus induces a large impaired water intensity.
- As much as water is essential for the recovery of fossil fuels and power generation, electricity is critical for both basic water management and treatment. Increasing global desalination and reuse of domestic wastewater further increase the need for energy for generating clean water.
- In spite of the rapid rise in renewable energy, in 2019 it made up only ~5% of global energy consumption. Substitution of fossil fuels with renewable energy would significantly reduce carbon emissions as well as the large water footprint of fossil fuels.
- The environmental legacy of fossil fuels through abandoned oil and gas wells, leaching of contaminants from oil and wastewater spill sites, acid mine drainage, and coal combustion residues disposal will continue to impact the quality of water resources for decades to come, even in scenarios where fossil fuels are declining.
- While the COVID-19 pandemic provided a pause in fossil fuel consumption, steps are required to ramp up countries' commitments to reducing dependence on fossil fuels and working together through such mechanisms as the Paris Agreement to address climate change, which would then offer opportunities to ameliorate the energy–water quality nexus.

References

1. Steffen, W., Grinevald, J., Crutzen, P., McNeill, J., The Anthropocene: conceptual and historical perspectives. *Philosophical Transactions of the Royal Society A: Mathematical, Physical and Engineering Sciences* **2011**, *369* (1938), 842–867.
2. Nicholson, S., Jinnah, S., *New Earth Politics: Essays from the Anthropocene*. MIT Press: **2016**.
3. Steffen, W., Broadgate, W., Deutsch, L., Gaffney, O., Ludwig, C., The trajectory of the Anthropocene: The Great Acceleration. *The Anthropocene Review* **2015**, *2* (1), 81–98.
4. Steffen, W., Sanderson, A., Jäger, J., Tyson, P. D., Matson, P. A., Moore, B., Oldfield, F., Richardson, K., Schellnhuber, H. J., Turner, B. L., *Global Change and the Earth System: A Planet under Pressure*. Springer: **2005**.
5. Lindsey, R. D., Dahlman, L., Climate Change: Global Temperature. www.climate.gov/news-features/understanding-climate/climate-change-global-temperature (8/8/**2018**).
6. US Department of Energy, Office of Fossil Energy, Our History. www.energy.gov/fe/about-us/our-history (8/8/2018).
7. US Energy Information Administration (EIA), Short-Term Energy Outlook. www.eia.gov/outlooks/steo/report/coal.php (07/**2018**).
8. US Energy Information Administration (EIA), Coal Explained: Coal Imports and Exports. www.eia.gov/energyexplained/coal/imports-and-exports.php (11/7/**2020**).
9. US Energy Information Administration (EIA), US Coal Exports Increased by 61% in 2017 as Exports to Asia More Than Doubled. www.eia.gov/todayinenergy/detail.php?id=35852 (6/22/**2018**).
10. BP, Statistical Review of World Energy. www.bp.com/en/global/corporate/energy-economics/statistical-review-of-world-energy/downloads.html (7/10/2021).
11. BP, Statistical Review of World Energy. www.bp.com/en/global/corporate/energy-economics/statistical-review-of-world-energy/downloads.html (7/10/2021).
12. US Energy Information Administration (EIA), Countries in and around the Middle East Are Adding Coal-Fired Power Plants. www.eia.gov/todayinenergy/detail.php?id=36172 (8/8/**2018**).
13. China Pakistan Economic Corridor (CPEC), CPEC-Energy Priority Projects. http://cpec.gov.pk/energy (8/8/**2018**).
14. US Energy Information Administration (EIA), International Energy Outlook. www.eia.gov/outlooks/ieo/.
15. US Energy Information Administration (EIA), Natural Gas Explained. www.eia.gov/dnav/ng/hist/n9070us2A.htm (8/8/**2018**).
16. Zou, C., Ni, Y., Li, J., Kondash, A., Coyte, R., Lauer, N., Cui, H., Liao, F., Vengosh, A., The water footprint of hydraulic fracturing in Sichuan Basin, China. *Science of the Total Environment* **2018**, *630*, 349–356.
17. Alvarez, R. A., Zavala-Araiza, D., Lyon, D. R., Allen, D. T., Barkley, Z. R., Brandt, A. R., Davis, K. J., Herndon, S. C., Jacob, D. J., Karion, A., Kort, E. A., Lamb, B. K., Lauvaux, T., Maasakkers, J. D., Marchese, A. J., Omara, M., Pacala, S. W., Peischl, J., Robinson, A. L., Shepson, P. B., Sweeney, C., Townsend-Small, A., Wofsy, S. C., Hamburg, S. P., Assessment of methane emissions from the US oil and gas supply chain. *Science* **2018**, *361* (6398), 186–188.
18. Karion, A., Sweeney, C., Kort, E. A., Shepson, P. B., Brewer, A., Cambaliza, M., Conley, S. A., Davis, K., Deng, A., Hardesty, M., Herndon, S. C., Lauvaux, T., Lavoie, T., Lyon, D., Newberger, T., Pétron, G., Rella, C., Smith, M., Wolter, S., Yacovitch, T. I., Tans, P., Aircraft-based estimate of total methane emissions from the Barnett Shale region. *Environmental Science & Technology* **2015**, *49* (13), 8124–8131.

19. Jackson, R. B., Vengosh, A., Carey, J. W., Davies, R. J., Darrah, T. H., O'Sullivan, F., Petron, G., The environmental costs and benefits of fracking. In *Annual Review of Environment and Resources*, Gadgil, A., Liverman, D. M., eds. Annual Reviews: **2014**, Vol. 39, pp. 327–362.

20. Miller, S. M., Wofsy, S. C., Michalak, A. M., Kort, E. A., Andrews, A. E., Biraud, S. C., Dlugokencky, E. J., Eluszkiewicz, J., Fischer, M. L., Janssens-Maenhout, G., Miller, B. R., Miller, J. B., Montzka, S. A., Nehrkorn, T., Sweeney, C., Anthropogenic emissions of methane in the United States. *Proceedings of the National Academy of Sciences* **2013**, *110* (50), 20018–20022.

21. Moore, C. W., Zielinska, B., Pétron, G., Jackson, R. B., Air impacts of increased natural gas acquisition, processing, and use: a critical review. *Environmental Science & Technology* **2014**, *48* (15), 8349–8359.

22. Smith, M. L., Kort, E. A., Karion, A., Sweeney, C., Herndon, S. C., Yacovitch, T. I., Airborne ethane observations in the Barnett Shale: quantification of ethane flux and attribution of methane emissions. *Environmental Science & Technology* **2015**, *49* (13), 8158–8166.

23. Kondash, A., Vengosh, A., Water footprint of hydraulic fracturing. *Environmental Science & Technology Letters* **2015**, *2* (10), 276–280.

24. Kondash, A. J., Lauer, N. E., Vengosh, A., The intensification of the water footprint of hydraulic fracturing. *Science Advances* **2018**, *4* (8).

25. Jackson, R. B., Vengosh, A., Carey, J. W., Davies, R. J., Darrah, T. H., O'Sullivan, F., Pétron, G., The environmental costs and benefits of fracking. *Annual Review of Environment and Resources* **2014**, *39* (1), 327–362.

26. Vengosh, A., Jackson, R. B., Warner, N., Darrah, T. H., Kondash, A., A critical review of the risks to water resources from unconventional shale gas development and hydraulic fracturing in the United States. *Environmental Science & Technology* **2014**, *48* (15), 8334–8348.

27. Vengosh, A., Mitch, W. A., McKenzie, L. M., Environmental and human impacts of unconventional energy development. *Environmental Science & Technology* **2017**, *51* (18), 10271–10273.

28. Gleick, P. H., Water and energy. *Annual Review of Energy and the Environment* **1994**, *19*, 267–299.

29. Grubert, E., Sanders, K. T., Water use in the United States energy system: a national assessment and unit process inventory of water consumption and withdrawals. *Environmental Science & Technology* **2018**, *52* (11), 6695–6703.

30. Grubert, E. A., Water consumption from hydroelectricity in the United States. *Advances in Water Resources* **2016**, *96*, 88–94.

31. Meldrum, J., Nettles-Anderson, S., Heath, G., Macknick, J., Life cycle water use for electricity generation: a review and harmonization of literature estimates. *Environmental Research Letters* **2013**, *8*, 015031.

32. Office of Energy Policy and Systems Analysis, U.S.D.o.E. *Environment Baseline Vol. 4: Energy–Water Nexus*, **2017**, p 93.

33. Sanders, K. T., Critical review: uncharted waters? The future of the electricity–water nexus. *Environmental Science & Technology* **2015**, *49* (1), 51–66.

34. Spang, E. S., Moomaw, W. R., Gallagher, K. S., Kirshen, P. H., Marks, D. H., Multiple metrics for quantifying the intensity of water consumption of energy production. *Environmental Research Letters* **2014**, *9*, 105003.

35. Averyt, K., Macknick, J., Rogers, J., Madden, N., Fisher, J., Meldrum, J., Newmark, R., Water use for electricity in the United States: an analysis of reported and calculated water use information for 2008. *Environmental Research Letters* **2013**, *8*, 015001.

36. Chang, Y., Li, G., Yuan Yao, Y., Zhang, L., Yu, C., Quantifying the water–energy–food nexus: current status and trends. *Energies* **2016**, *9*, 65–82.

37. Weinthal, E., Vengosh, A., Neville, C., The nexus of energy and water quality. In *The Oxford Handbook of Water Politics and Policy,* Conca, K., Weinthal, E., eds. Oxford University Press: **2017**.

38. Endo, A., Tsurita, I., Burnett, K., Orencio, P. M., A review of the current state of research on the water, energy, and food nexus. *Journal of Hydrology: Regional Studies* **2017**, *11*, 20–30.

39. Scott, C. A., Pierce, S. A., Pasqualetti, M. J., Jones, A. L., Montz, B. E., Hoover, J. H., Policy and institutional dimensions of the water–energy nexus. *Energy Policy* **2011**, *39* (10), 6622–6630.

40. Macknick, J., Sattler, S., Averyt, K., Clemmer, S., Rogers, J., The water implications of generating electricity: water use across the United States based on different electricity pathways through 2050. *Environmental Research Letters* **2012**, *7* (4), 045803.

41. Ackerman, F., Fisher, J., Is there a water–energy nexus in electricity generation? Long-term scenarios for the western United States. *Energy Policy* **2013**, *59*, 235–241.

42. Tarroja, B., AghaKouchak, A., Sobhani, R., Feldman, D., Jiang, S., Samuelsen, S., Evaluating options for balancing the water–electricity nexus in California: part 1 – securing water availability. *Science of the Total Environment* **2014**, *497–498*, 697–710.

43. Siddiqi, A., Anadon, L. D., The water–energy nexus in the Middle East and North Africa. *Energy Policy* **2011**, *39* (8), 4529–4540.

44. Crow-Miller, B. L. *Water, Power, and Development in Twenty-First Century China: The Case of the South–North Water Transfer Project.* University of California, Los Angeles: **2013**.

45. International Energy Agency (IEA). *Water Energy Nexus – World Energy Outlook 2016*, International Energy Agency (IEA): **2017**.

46. Dieter, C. A., Maupin, M. A., Caldwell, R. R., Harris, M. A., Ivahnenko, T. I., Lovelace, J. K., Barber, N. L., Linsey, K. S. *Estimated Use of Water in the United States in 2015*, Circular 1441, US Geological Survey, **2018**.

47. Macknick, J., Newmark, R., Heath, G., Hallett, K. C., Operational water consumption and withdrawal factors for electricity generating technologies: a review of existing literature. *Environmental Research Letters* **2012**, *7* (4), 045802.

48. Spang, E. S., Moomaw, W. R., Gallagher, K. S., Kirshen, P. H., Marks, D. H., The water consumption of energy production: an international comparison. *Environmental Research Letters* **2014**, *9*, 105002.

49. Alexey, V., Hal, C., The energy–water nexus: why should we care? *Journal of Contemporary Water Research & Education* **2009**, *143* (1), 17–29.

50. Grubert, E. A., Webber, M. E., Energy for water and water for energy on Maui Island, Hawaii. *Environmental Research Letters* **2015**, *10*, 064009.

51. Pan, S.-Y., Snyder, S. W., Packman, A. I., Lin, Y. J., Chiang, P.-C., Cooling water use in thermoelectric power generation and its associated challenges for addressing water-energy nexus. *Water–Energy Nexus* **2018**, 1, 26–41.

52. Spang, E. S., Loge, F. J., A high-resolution approach to mapping energy flows through water infrastructure systems. *Journal of Industrial Ecology* **2015**, *19* (4), 656–665.

53. McCall, J., Macknick, J., Hillman, D. *Water-Related Power Plant Curtailments: An Overview of Incidents and Contributing Factors* National Renewable Energy Laboratory, US Department of Energy: **2016**, p 32.

54. Raptis, C. E., Vliet, M. T. H. v., Pfister, S., Global thermal pollution of rivers from thermoelectric power plants. *Environmental Research Letters* **2016**, *11* (10), 104011.

55. Goals, U. N. S. D. Take Actions for Sustainable Development Goals. www.un.org/sustainabledevelopment/sustainable-development-goals/.

56. Oki, T., Kanae, S., Global hydrological cycles and world water resources. *Science* **2006**, *313* (5790), 1068–1072.

57. Taylor, R. G., Scanlon, B., Döll, P., Rodell, M., van Beek, R., Wada, Y., Longuevergne, L., Leblanc, M., Famiglietti, J. S., Edmunds, M., Konikow, L., Green, T. R., Chen, J., Taniguchi, M., Bierkens, M. F. P., MacDonald, A., Fan, Y., Maxwell, R. M., Yechieli, Y., Gurdak, J. J., Allen, D. M., Shamsudduha, M., Hiscock, K., Yeh, P. J. F., Holman, I., Treidel, H., Ground water and climate change. *Nature Climate Change* **2012**, *3*, 322.

58. Arnell, N. W., Climate change and global water resources. *Global Environmental Change* **1999**, *9*, S31–S49.

59. Felix, T. P., Petra, D., Stephanie, E., Martina, F., Impact of climate change on renewable groundwater resources: assessing the benefits of avoided greenhouse gas emissions using selected CMIP5 climate projections. *Environmental Research Letters* **2013**, *8* (2), 024023.

60. Roy, S. B., Chen, L., Girvetz, E. H., Maurer, E. P., Mills, W. B., Grieb, T. M., Projecting water withdrawal and supply for future decades in the US under climate change scenarios. *Environmental Science & Technology* **2012**, *46* (5), 2545–2556.

61. Postel, S. L., Daily, G. C., Ehrlich, P. R., Human appropriation of renewable fresh water. *Science* **1996**, *271* (5250), 785–788.

62. Gleick, P. H., Water and conflict: fresh water resources and international security. *International Security* **1993**, *18* (1), 79–112.

63. Wada, Y., van Beek, L. P. H., van Kempen, C. M., Reckman, J. W. T. M., Vasak, S., and Bierkens, M.F.P., Global depletion of groundwater resources. *Geophysical Research Letters* **2010**, *37*, L20402.

64. Vörösmarty, C. J., Green, P., Salisbury, J., Lammers, R. B., Global water resources: vulnerability from climate change and population growth. *Science* **2000**, *289* (5477), 284–288.

65. Petra, D., Vulnerability to the impact of climate change on renewable groundwater resources: a global-scale assessment. *Environmental Research Letters* **2009**, *4* (3), 035006.

66. Vörösmarty, C. J., McIntyre, P. B., Gessner, M. O., Dudgeon, D., Prusevich, A., Green, P., Glidden, S., Bunn, S. E., Sullivan, C. A., Liermann, C. R., Davies, P. M., Global threats to human water security and river biodiversity. *Nature* **2010**, *467*, 555.

67. Gosling, S. N., Arnell, N. W., A global assessment of the impact of climate change on water scarcity. *Climatic Change* **2016**, *134* (3), 371–385.

68. Smith, J. B., Schneider, S. H., Oppenheimer, M., Yohe, G. W., Hare, W., Mastrandrea, M. D., Patwardhan, A., Burton, I., Corfee-Morlot, J., Magadza, C. H. D., Füssel, H.-M., Pittock, A. B., Rahman, A., Suarez, A., van Ypersele, J.-P., Assessing dangerous climate change through an update of the Intergovernmental Panel on Climate Change (IPCC) "reasons for concern". *Proceedings of the National Academy of Sciences* **2009**, *106* (11), 4133–4137.
69. Piao, S., Ciais, P., Huang, Y., Shen, Z., Peng, S., Li, J., Zhou, L., Liu, H., Ma, Y., Ding, Y., Friedlingstein, P., Liu, C., Tan, K., Yu, Y., Zhang, T., Fang, J., The impacts of climate change on water resources and agriculture in China. *Nature* **2010**, *467*, 43.
70. Sowers, J., Vengosh, A., Weinthal, E., Climate change, water resources, and the politics of adaptation in the Middle East and North Africa. *Climatic Change* **2011**, *104* (3), 599–627.
71. Voss, K. A., Famiglietti, J. S., Lo, M., de Linage, C., Rodell, M., and Swenson, S. C., Groundwater depletion in the Middle East from GRACE with implications for transboundary water management in the Tigris–Euphrates–Western Iran region. *Water Resources Research* **2013**, *49* (2), 904–914.
72. Lelieveld, J., Hadjinicolaou, P., Kostopoulou, E., Chenoweth, J., El Maayar, M., Giannakopoulos, C., Hannides, C., Lange, M. A., Tanarhte, M., Tyrlis, E., Xoplaki, E., Climate change and impacts in the Eastern Mediterranean and the Middle East. *Climatic Change* **2012**, *114* (3), 667–687.
73. Kelley, C. P., Mohtadi, S., Cane, M. A., Seager, R., Kushnir, Y., Climate change in the Fertile Crescent and implications of the recent Syrian drought. *Proceedings of the National Academy of Sciences* **2015**, *112* (11), 3241–3246.
74. Evans, J. P., 21st century climate change in the Middle East. *Climatic Change* **2009**, *92* (3), 417–432.
75. Vicuna, S., Dracup, J. A., The evolution of climate change impact studies on hydrology and water resources in California. *Climatic Change* **2007**, *82* (3), 327–350.
76. MacDonald, G. M., Water, climate change, and sustainability in the southwest. *Proceedings of the National Academy of Sciences* **2010**, *107* (50), 21256–21262.
77. Lund, J. R., Jenkins, M. W., Zhu, T., Tanaka, S. K. Climate warming and California's water future. In Bizier, P. and DeBarry, P. (eds), World Water and Environmental Resources Congress 2003, American Society of Civil Engineers, https://doi.org/10.1061/9780784406854.
78. Christensen, N. S., Wood, A. W., Voisin, N., Lettenmaier, D. P., Palmer, R. N., The effects of climate change on the hydrology and water resources of the Colorado River Basin. *Climatic Change* **2004**, *62* (1), 337–363.
79. Famiglietti, J. S., Lo, M., Ho, S. L., Bethune, J., Anderson, K. J., Syed, T. H., Swenson, S. C., de Linage, C. R., Rodell, M., Satellites measure recent rates of groundwater depletion in California's Central Valley. *Geophysical Research Letters* **2011**, *38*, L03403.
80. United Nations Educational, Scientific and Cultural Organization, *The UN World Water Development Report 2014 – Water and Energy*, UNESCO: **2014**, www.unwater.org/publications/world-water-development-report-2014-water-energy/.
81. Mekonnen, M. M., Hoekstra, A. Y., Four billion people facing severe water scarcity. *Science Advances* **2016**, *2* (2), e1500323.
82. Dalin, C., Wada, Y., Kastner, T., Puma, M. J., Groundwater depletion embedded in international food trade. *Nature* **2017**, *543*, 700.
83. Government of India, *Composite Water Management Index*, Niti Aayog, Government of India: **2018**.
84. Asoka, A., Gleeson, T., Wada, Y., Mishra, V., Relative contribution of monsoon precipitation and pumping to changes in groundwater storage in India. *Nature Geoscience* **2017**, *10*, 109.
85. MacDonald, A. M., Bonsor, H. C., Ahmed, K. M., Burgess, W. G., Basharat, M., Calow, R. C., Dixit, A., Foster, S. S. D., Gopal, K., Lapworth, D. J., Lark, R. M., Moench, M., Mukherjee, A., Rao, M. S., Shamsudduha, M., Smith, L., Taylor, R. G., Tucker, J., van Steenbergen, F., Yadav, S. K., Groundwater quality and depletion in the Indo-Gangetic Basin mapped from in situ observations. *Nature Geoscience* **2016**, *9*, 762.
86. Rodell, M., Velicogna, I., Famiglietti, J. S., Satellite-based estimates of groundwater depletion in India. *Nature* **2009**, *460*, 999.
87. Coyte, R. M., Jain, R. C., Srivastava, S. K., Sharma, K. C., Khalil, A., Ma, L., Vengosh, A., Large-scale uranium contamination of groundwater resources in India. *Environmental Science & Technology Letters* **2018**, *5* (6), 341–347.
88. Vengosh, A., Salinization and saline environments. In *Environmental Geochemistry (Volume 9), Treatise in Geochemistry Second Edition*, Sherwood Lollar, B., ed., Vol. 11, pp. 325–378. Elsevier Science: **2014**.
89. Intergovernmental Panel on Climate Change (IPCC), I.P.o.C.C. Climate Change 2013: The Physical Science Basis. www.ipcc.ch/report/ar5/wg1/ (8/8/2018).
90. Allen, L., Cohen, M. J., Abelson, D., Miller, B., Fossil fuels and water quality. In *The World's Water Volume 7*, Gleick, P. H., Allen, L., Christian-Smith, J., Cohen, M. J., Cooley, H., Heberger, M., Morrison, J., Palaniappan, M., Schulte, P., eds. Island Press: **2011**, Vol. 7, pp. 73–96.

91. Jones Luong, P., Weinthal, E., *Oil is Not a Curse: Ownership Structure and Institutions in Soviet Successor States*. Cambridge University Press: **2010**.
92. United Nation Environmental Program, *Environmental Assessment of Ogoniland*, **2011**, https://postconflict .unep.ch/publications/OEA/01_fwd_es_ch01_UNEP_OEA.pdf.
93. Weinthal, E., *State Making and Environmental Cooperation: Linking Domestic and International Politics in Central Asia*. MIT Press: **2002**.
94. Vickers, A., The Energy Policy Act: assessing its impact on utilities. *Journal AWWA* **1993**, *85* (8), 56–62.
95. Freese, B., *Coal: A Human History*. Penguin Books: **2003**.
96. Granitz, E., Klein, B., Monopolization by "raising rivals' costs": the Standard Oil case. *Journal of Law and Economics* **1996**, *39*, 1–47.
97. Yergin, There Will Be Oil. *The Wall Street Journal*, September 17, **2011**, www.wsj.com/articles/ SB10001424053111904060604576572552998674340.
98. Morrison, J., Air Pollution Goes Back Way Further Than You Think. www.smithsonianmag.com/science-nature/air-pollution-goes-back-way-further-you-think-180957716/.
99. Brimblecombe, P., Air pollution in industrializing England. *Journal of the Air Pollution Control Association* **1978**, *28* (2), 115–118.
100. Milici, R. C., Flores, R. M., Stricker, G. D., Coal resources, reserves and peak coal production in the United States. *International Journal of Coal Geology* **2013**, *113*, 109–115.
101. Yuan, J., The future of coal in China. *Resources, Conservation and Recycling* **2018**, *129*, 290–292.
102. Zhang, X., Winchester, N., Zhang, X., The future of coal in China. *Energy Policy* **2017**, *110*, 644–652.
103. Luke, H., Brueckner, M., Emmanouil, N., Unconventional gas development in Australia: a critical review of its social license. *The Extractive Industries and Society* **2018**, *5* (4), 648–662.
104. Zou, C., Zhu, R., Chen, Z.-Q., Ogg, J. G., Wu, S., Dong, D., Qiu, Z., Wang, Y., Wang, L., Lin, S., Cui, J., Su, L., Yang, Z., Organic-matter-rich shales of China. *Earth-Science Review* **2018**, 189, 51–78.
105. Zou, C., Dong, D., Wang, Y., Li, X., Huang, J., Wang, S., Guan, Q., Zhang, C., Wang, H., Liu, H., Bai, W., Liang, F., Lin, W., Zhao, Q., Liu, D., Yang, Z., Liang, P., Sun, S., Qiu, Z., Shale gas in China: characteristics, challenges and prospects (I). *Petroleum Exploration and Development* **2015**, *42* (6), 753–767.
106. Zou, C., Dong, D., Wang, Y., Li, X., Huang, J., Wang, S., Guan, Q., Zhang, C., Wang, H., Liu, H., Bai, W., Liang, F., Lin, W., Zhao, Q., Liu, D., Yang, Z., Liang, P., Sun, S., Qiu, Z., Shale gas in China: characteristics, challenges and prospects (II). *Petroleum Exploration and Development* **2016**, *43* (2), 182–196.
107. US Energy Information Administration (EIA), Energy Units and Calculators Explained. www.eia.gov/energyexplained/index.php?page=about_energy_units (1/19/2019).
108. Global Energy Statistical Yearbook 2018, Lignite. yearbook.enerdata.net/coal-lignite/coal-production-data .html (1/19/2019).
109. Roudi-Fahimi, F., Creel, L., De Souza, R. M. *Finding the Balance: Population and Water Scarcity in the Middle East and North Africa*, Population Reference Bureau Washington, D.C.: **2002**.
110. US Department of Energy, *The Water–Energy Nexus: Challenges and Opportunities*, US Department of Energy: **2014**.
111. Dieter, C. A., Maupin, M. A., Caldwell, R. R., Harris, M. A., Ivahnenko, T. I., Lovelace, J. K., Barber, N. L., Linsey, K. S., *Estimated Use of Water in the United States in 2015*, US Geological Survey: **2018**.
112. US Energy Information Administration, What is US Electricity Generation by Energy Source? www.eia.gov/ tools/faqs/faq.php?id=427&t=3 (4/12/2020).
113. Houser, T., Bordoff, J., Marsters *Can Coal Make a Comeback?*, Columbia, SIPA, Center on Global Energy Policy: **2017**.
114. World Coal Institute, *The Coal Resource: A Comprehensive Overview of Coal*, **2009**, www.scribd.com/ document/18825349/The-Coal-Resource-A-Comprehensive-Overview-of-Coal-World-Coal-Institute.
115. International Energy Agency (IEA), *World Energy Balances: Overview*, **2020**, www.iea.org/reports/world-energy-balances-overview.
116. International Energy Agency (IEA), *Coal Information: Overview*, **2020**, www.iea.org/reports/coal-information-overview.
117. Appunn, K. Coal in Germany, Factsheet. www.cleanenergywire.org/factsheets/coal-germany (22/3/**2020**).
118. Kondash, A. J. *The Water–Energy Nexus for Hydraulic Fracturing*. Duke University PhD Thesis: **2019**.
119. Pan, L., Liu, P., Ma, L., Li, Z., A supply chain based assessment of water issues in the coal industry in China. *Energy Policy* **2012**, *48*, 93–102.
120. Acharya, B. S., Kharel, G., Acid mine drainage from coal mining in the United States – An overview. *Journal of Hydrology* **2020**, *588*, 125061.
121. US Energy Information Administration (EIA), Coal/data. www.eia.gov/coal/data.php (5/10/**2019**).
122. Chu, S., Carbon capture and sequestration. *Science* **2009**, *325*, 1599.

123. Mage, D., Ozolins, G., Peterson, P., Webster, A., Orthofer, R., Vandeweerd, V., Gwynne, M., Urban air pollution in megacities of the world. *Atmospheric Environment* **1996**, *30* (5), 681–686.
124. Ramanathan, V., Feng, Y., Air pollution, greenhouse gases and climate change: global and regional perspectives. *Atmospheric Environment* **2009**, *43* (1), 37–50.
125. Pacyna, E. G., Pacyna, J. M., Steenhuisen, F., Wilson, S., Global anthropogenic mercury emission inventory for 2000. *Atmospheric Environment* **2006**, *40* (22), 4048–4063.
126. Wang, Q., Shen, W., Ma, Z., Estimation of mercury emission from coal combustion in China. *Environmental Science & Technology* **2000**, *34* (13), 2711–2713.
127. Qi, Y., Stern, N., Wu, T., Lu, J., Green, F., China's post-coal growth. *Nature Geoscience* **2016**, *9*, 564–566.
128. Newell, P., Simms, A., Towards a fossil fuel non-proliferation treaty. *Climate Policy* **2020**, *20* (8), 1043–1054.
129. Jakob, M., Steckel, J. C., Jotzo, F., Sovacool, B. K., Cornelsen, L., Chandra, R., Edenhofer, O., Holden, C., Löschel, A., Nace, T., Robins, N., Suedekum, J., Urpelainen, J., The future of coal in a carbon-constrained climate. *Nature Climate Change* **2020**, *10* (8), 704–707.
130. US Energy Information Administration (EIA), Layer Information for Interactive State Maps. www.eia.gov/maps/layer_info-m.php.
131. Meij, R., The fate of trace elements at coal-fired power plants. *Fuel* **1993**, *72* (5), 718.
132. Meij, R., Trace element behavior in coal-fired power plants. *Fuel Processing Technology* **1994**, *39* (1), 199–217.
133. Meij, R., Prediction of environmental quality of by-products of coal-fired power plants: elemental composition and leaching. In *Studies in Environmental Science*, Goumans, J. J. J. M., Senden, G. J., van der Sloot, H. A., eds. Elsevier: **1997**, Vol. 71, pp. 311–325.
134. Goodarzi, F., Swaine, D. J., Chalcophile elements in western Canadian coals. *International Journal of Coal Geology* **1993**, *24* (1), 281–292.
135. Goodarzi, F., Swaine, D. J., The influence of geological factors on the concentration of boron in Australian and Canadian coals. *Chemical Geology* **1994**, *118* (1–4), 301–318.
136. Swaine, D. J., Origin of trace elements in coal. In *Trace Elements in Coal*, Swaine, D. J., ed., pp. 8–26. Butterworth-Heinemann: **1990**.
137. Swaine, D. J., Trace elements in coal and their dispersal during combustion. *Fuel Processing Technology* **1994**, *39* (1–3), 121–137.
138. Dai, S., Ren, D., Chou, C.-L., Finkelman, R. B., Seredin, V. V., Zhou, Y., Geochemistry of trace elements in Chinese coals: a review of abundances, genetic types, impacts on human health, and industrial utilization. *International Journal of Coal Geology* **2012**, *94*, 3–21.
139. Ketris, M. P., Yudovich, Y. E., Estimations of Clarkes for Carbonaceous biolithes: world averages for trace element contents in black shales and coals. *International Journal of Coal Geology* **2009**, *78* (2), 135–148.
140. Yudovich, Y. E., Ketris, M. P., Arsenic in coal: a review. *International Journal of Coal Geology* **2005**, *61* (3), 141–196.
141. Yudovich, Y. E., Ketris, M. P., Mercury in coal: a review: part 1. Geochemistry. *International Journal of Coal Geology* **2005**, *62* (3), 107–134.
142. Yudovich, Y. E., Ketris, M. P., Mercury in coal: a review: part 2. Coal use and environmental problems. *International Journal of Coal Geology* **2005**, *62* (3), 135–165.
143. Yudovich, Y. E., Ketris, M. P., Chlorine in coal: a review. *International Journal of Coal Geology* **2006**, *67* (1), 127–144.
144. Yudovich, Y. E., Ketris, M. P., Selenium in coal: a review. *International Journal of Coal Geology* **2006**, *67* (1), 112–126.
145. Goldschmidt, V. M., Rare elements in coal ashes. *Ind. Eng. Chem.* **1935**, *27*, 1100–1102.
146. Finkelman, R. B., Trace and minor elements in coal. In *Organic Geochemistry*, Engel, M. H., Macko, S., eds. Plenum: **1993**, pp. 593–607.
147. Rudnick, R. L., Gao, S., 4.1 – Composition of the continental crust. In *Treatise on Geochemistry (Second Edition)*, Holland, H. D., Turekian, K. K., eds. Elsevier: **2014**, pp. 1–51.
148. Lauer, N., Vengosh, A., Dai, S., Naturally occurring radioactive materials in coals and coal ash in China. *Abstracts of Papers of the American Chemical Society* **2018**, *255*.
149. Duan, P., Wang, W., Sang, S., Qian, F., Shao, P., Zhao, X., Partitioning of hazardous elements during preparation of high-uranium coal from Rongyang, Guizhou, China. *Journal of Geochemical Exploration* **2018**, *185*, 81–92.
150. Liu, P., Luo, X., Wen, M., Zhang, J., Zheng, C., Gao, W., Ouyang, F., Geoelectrochemical anomaly prospecting for uranium deposits in southeastern China. *Applied Geochemistry* **2018**, *97*, 226–237.

151. Sun, Y., Qi, G., Lei, X., Xu, H., Wang, Y., Extraction of Uranium in Bottom Ash Derived from High-Germanium Coals. *Procedia Environmental Sciences* **2016**, *31*, 589–597.
152. Yang, J., Concentration and distribution of uranium in Chinese coals. *Energy* **2007**, *32* (3), 203–212.
153. Zhang, Y., Shi, M., Wang, J., Yao, J., Cao, Y., Romero, C. E., Pan, W.-p., Occurrence of uranium in Chinese coals and its emissions from coal-fired power plants. *Fuel* **2016**, *166*, 404–409.
154. Dai, S., Finkelman, R. B., Coal as a promising source of critical elements: progress and future prospects. *International Journal of Coal Geology* **2018**, *186*, 155–164.
155. Finkelman, R. B., Palmer, C. A., Wang, P., Quantification of the modes of occurrence of 42 elements in coal. *International Journal of Coal Geology* **2018**, *185*, 138–160.
156. Havelcová, M., Machovič, V., Mizera, J., Sýkorová, I., Borecká, L., Kopecký, L., A multi-instrumental geochemical study of anomalous uranium enrichment in coal. *Journal of Environmental Radioactivity* **2014**, *137*, 52–63.
157. Nxumalo, V., Kramers, J., Mongwaketsi, N., Przybyłowicz, W. J., Micro-PIXE characterisation of uranium occurrence in the coal zones and the mudstones of the Springbok Flats Basin, South Africa. *Nuclear Instruments and Methods in Physics Research Section B: Beam Interactions with Materials and Atoms* **2017**, *404*, 114–120.
158. Chen, G. Q., Li, J. S., Chen, B., Wen, C., Yang, Q., Alsaedi, A., Hayat, T., An overview of mercury emissions by global fuel combustion: The impact of international trade. *Renewable & Sustainable Energy Reviews* **2016**, *65*, 345–355.
159. Oberschelp, C., Pfister, S., Raptis, C. E., Hellweg, S., Global emission hotspots of coal power generation. *Nature Sustainability* **2019**, *2* (2), 113–121.
160. Pacyna, J. M., Travnikov, O., De Simone, F., Hedgecock, I. M., Sundseth, K., Pacyna, E. G., Steenhuisen, F., Pirrone, N., Munthe, J., Kindbom, K., Current and future levels of mercury atmospheric pollution on a global scale. *Atmospheric Chemistry and Physics* **2016**, *16* (19), 12495–12511.
161. Washburn, S. J., Blum, J. D., Johnson, M. W., Tomes, J. M., Carnell, P. J., Isotopic characterization of mercury in natural gas via analysis of mercury removal unit catalysts. *Acs Earth and Space Chemistry* **2018**, *2* (5), 462–470.
162. Blum, J. D., Johnson, M. W., Recent developments in mercury stable isotope analysis. In *Non-Traditional Stable Isotopes*, Teng, F. Z., Watkins, J., Dauphas, N., eds. Mineralogical Society of America: **2017**, Vol. 82, pp. 733–757.
163. Lefticariu, L., Blum, J. D., Gleason, J. D., Mercury isotopic evidence for multiple mercury sources in coal from the Illinois Basin. *Environmental Science & Technology* **2011**, *45* (4), 1724–1729.
164. Sherman, L. S., Blum, J. D., Keeler, G. J., Demers, J. D., Dvonch, J. T., Investigation of local mercury deposition from a coal-fired power plant using mercury isotopes. *Environmental Science & Technology* **2012**, *46* (1), 382–390.
165. Sunderland, E. M., Driscoll, C. T., Jr., Hammitt, J. K., Grandjean, P., Evans, J. S., Blum, J. D., Chen, C. Y., Evers, D. C., Jaffe, D. A., Mason, R. P., Goho, S., Jacobs, W., Benefits of regulating hazardous air pollutants from coal and oil fired utilities in the United States. *Environmental Science & Technology* **2016**, *50* (5), 2117–2120.
166. Schlesinger, W. H., Vengosh, A., Global boron cycle in the Anthropocene. *Global Biogeochemical Cycles* **2016**, *30* (2), 219–230.
167. Schlesinger, W. H., Klein, E. M., Vengosh, A., Global biogeochemical cycle of vanadium. *Proceedings of the National Academy of Sciences of the United States of America* **2017**, *114* (52), E11092–E11100.
168. Lauer, N., Vengosh, A., Dai, S., Naturally occurring radioactive materials in uranium-rich coals and associated coal combustion residues from China. *Environmental Science & Technology* **2017**, *51* (22), 13487–13493.
169. Kondash, A. J., Patino-Echeverri, D., Vengosh, A., Quantification of the water-use reduction associated with the transition from coal to natural gas in the US electricity sector. Environmental Research Letters **2019**, 14, 124028.
170. Zhang, C., Diaz Anadon, L., Life cycle water use of energy production and its environmental impacts in China. *Environmental Science and Technology* **2013**, *47*, 14459–14467.
171. Zhang, C., Anadon, L. D., Mo, H., Zhao, Z., Liu, Z., Water–carbon trade-off in China's coal power industry. *Environmental Science & Technology* **2014**, *48* (19), 11082–11089.
172. US Energy Information Administration (EIA), EIA-7AAnnual Survey of Coal Production and Preparation. www.eia.gov/survey/#eia-7a.
173. Gassert, F., Luck, M. Aqueduct Water Stress Projections: Decadal Projections of Water Supply and Demand Using CMIP5 GCMs. www.wri.org/publication/aqueduct-water-stress-projections-decadal-projections-water-supply-and-demand-using.

174. Zajisz-Zubek, E., Konieczynsky, J., Coal cleaning versus reduction of mercury and other trace elements' emissions from coal combustion processes. *Archives of Environmental Protection* **2014**, *40*, 115–127.

175. Chen, Q., Wang, H., Clean processing and utilization of coal energy. *The Chinese Journal of Process Engineering* **2006**, *6*, 507–511.

176. Pan, L. L., P., Ma, L., Li, Z., A supply chain based assessment of water issues in the coal industry in China. *Energy Policy* **2012**, *48*, 93–102.

177. Jeter, T. S., Sarver, E. A., McNair, H. M., Rezaee, M., 4-MCHM sorption to and desorption from granular activated carbon and raw coal. *Chemosphere* **2016**, *157*, 160–165.

178. Monnot, A. D., Novick, R. M., Paustenbach, D. J., Crude 4-methylcyclohexanemethanol (MCHM) did not cause skin irritation in humans in 48-h patch test. *Cutaneous and Ocular Toxicology* **2017**, *36* (4), 351–355.

179. Paustenbach, D. J., Winans, B., Novick, R. M., Green, S. M., The toxicity of crude 4-methylcyclohexanemethanol (MCHM): review of experimental data and results of predictive models for its constituents and a putative metabolite. *Critical Reviews in Toxicology* **2015**, *45*, 1–55.

180. Sain, A. E., Dietrich, A. M., Smiley, E., Gallagher, D. L., Assessing human exposure and odor detection during showering with crude 4-(methylcyclohexyl)methanol (MCHM) contaminated drinking water. *Science of the Total Environment* **2015**, *538*, 298–305.

181. Cooper, W. J., Responding to crisis: the West Virginia chemical spill. *Environmental Science & Technology* **2014**, *48* (6), 3095–3095.

182. Burrows, J. E., Cravotta, C. A., Peters, S. C., Enhanced Al and Zn removal from coal-mine drainage during rapid oxidation and precipitation of Fe oxides at near-neutral pH. *Applied Geochemistry* **2017**, *78*, 194–210.

183. Burrows, J. E., Peters, S. C., Cravotta, C. A., Temporal geochemical variations in above- and below-drainage coal mine discharge. *Applied Geochemistry* **2015**, *62*, 84–95.

184. Cravotta, C. A., Dissolved metals and associated constituents in abandoned coal-mine discharges, Pennsylvania, USA. Part 2: geochemical controls on constituent concentrations. *Applied Geochemistry* **2008**, *23* (2), 203–226.

185. Cravotta, C. A., Dissolved metals and associated constituents in abandoned coal-mine discharges, Pennsylvania, USA. Part 1: constituent quantities and correlations. *Applied Geochemistry* **2008**, *23* (2), 166–202.

186. Cravotta, C. A., Monitoring, field experiments, and geochemical modeling of Fe(II) oxidation kinetics in a stream dominated by net-alkaline coal-mine drainage, Pennsylvania, USA. *Applied Geochemistry* **2015**, *62*, 96–107.

187. Cravotta, C. A., Brady, K. B. C., Priority pollutants and associated constituents in untreated and treated discharges from coal mining or processing facilities in Pennsylvania, USA. *Applied Geochemistry* **2015**, *62*, 108–130.

188. Blodau, C., A review of acidity generation and consumption in acidic coal mine lakes and their watersheds. *Science of the Total Environment* **2006**, *369*, 307–332.

189. Gray, N. F., Acid mine drainage composition and the implications for its impact on lotic systems. *Water Research* **1998**, *32*, 2122–2134.

190. Gray, N. F., Environmental impact and remediation of acid mine drainage: a management problem. *Environmental Geology* **1997**, *30*, 62–71.

191. Blowes, D. W., Ptacek, C. J., Jambor, J. L., Weisener, C,G., The geochemistry of acid mine drainage. In *Treaties on Geochemistry, Environmental Geochemistry*, Sherwood Lollar, B., ed. Elsevier: **2003**, Vol. 9, pp. 149–204.

192. Ferguson K. D., Erickson, P. M., Pre-mine prediction of acid mine drainage. In *Environmental Management of Solid Waste*, Salomons W., Forstner, U., eds. Springer: **1988**.

193. Pumure, I., Renton, J. J., Smart, R. B., Ultrasonic extraction of arsenic and selenium from rocks associated with mountaintop removal/valley fills coal mining: estimation of bioaccessible concentrations. *Chemosphere* **2010**, *78* (11), 1295–1300.

194. Griffith, M. B., Norton, S. B., Alexander, L. C., Pollard, A. I., LeDuc, S. D., The effects of mountaintop mines and valley fills on the physicochemical quality of stream ecosystems in the central Appalachians: a review. *Science of the Total Environment* **2012**, *417–418*, 1–12.

195. Palmer, M. A., Bernhardt, E. S., Schlesinger, W. H., Eshleman, K. N., Foufoula-Georgiou, E., Hendryx, S., Lemly, A. D., Likens, G. E., Loucks, O. L., Power, M. E., White, P. S., Wilcock, P. R., Mountaintop mining consequences. *Science* **2010**, *327*, 148–149.

196. Pumure, I., Renton, J. J., Smart, R. B., Accelerated aqueous leaching of selenium and arsenic from coal associated rock samples with selenium speciation using ultrasound extraction. *Environmental Geology* **2009**, *56*, 985–991.

197. Vengosh, A., Lindberg, T. T., Merola, B. R., Ruhl, L., Warner, N. R., White, A., Dwyer, G. S., Di Giulio, R. T., Isotopic imprints of mountaintop mining contaminants. *Environmental Science & Technology* **2013**, *47* (17), 10041–10048.
198. Lindberg, T. T., Bernhardt, E. S., Bier, R., Helton, A. M., Merola, R. B., Vengosh, A., Di Giulio, R. T., Cumulative impacts of mountaintop mining on an Appalachian watershed. *Proceedings of the National Academy of Sciences of the United States of America* **2011**, *108* (52), 20929–20934.
199. Ross, M. R. V., Nippgen, F., Hassett, B. A., McGlynn, B. L., Bernhardt, E. S., Pyrite oxidation drives exceptionally high weathering rates and geologic CO2 release in mountaintop-mined landscapes. *Global Biogeochemical Cycles* **2018**, *32* (8), 1182–1194.
200. Nippgen, F., Ross, M. R. V., Bernhardt, E. S., McGlynn, B. L., Creating a more perennial problem? Mountaintop removal coal mining enhances and sustains saline baseflows of Appalachian watersheds. *Environmental Science & Technology* **2017**, *51* (15), 8324–8334.
201. Lutz, B. D., Bernhardt, E. S., Schlesinger, W. H., The environmental price tag on a ton of mountaintop removal coal. *PLoS ONE* **2013**, *8* (9), e73203.
202. Bernhardt, E. S., Palmer, M. A., The environmental costs of mountaintop mining valley fill operations for aquatic ecosystems of the Central Appalachians. In *Year in Ecology and Conservation Biology*, Ostfeld, R. S., Schlesinger, W. H., eds. **2011**, Annals of the New York Academy of Sciences, Vol. 1223, pp. 39–57.
203. Johnson, D. B., Hallberg, K. B., Acid mine drainage remediation options: a review. *Science of the Total Environment* **2005**, *338* (1), 3–14.
204. Wei, X. W., H., Viadero Jr., R. C., Post-reclamation water quality trend in a Mid-Appalachian watershed of abandoned mine lands. *Science of the Total Environment* **2011**, *409*, 941–948.
205. US Fish and Wildlife Service, *Acid Mine Drainage and Effects on Fish Health and Ecology: A Review*. US Fish and Wildlife Service: **2008**.
206. US Environmental Protection Agency, *The Effects of Mountaintop Mines and Valley Fills on Aquatic Ecosystems of the Central Appalachian Coalfields*. US Environmental Protection Agency: **2009**, EPA/600/R-09/138F, 2011.
207. US Environmental Protection Agency. *The Effects of Mountaintop Mines and Valley Fills on Aquatic Ecosystems of the Central Appalachian Coalfields*. US Environmental Protection Agency: **2009**.
208. Schnoor, J. L., Mountaintop mining. *Environmental Science & Technology* **2010**, *44* (23), 8794–8794.
209. Phillips, J. D., Impacts of surface mine valley fills on headwater floods in eastern Kentucky. *Environmental Geology* **2004**, *45* (3), 367–380.
210. Wickham, J. D., Riitters, K. H., Wade, T. G., Coan, M., Homer, C., The effect of Appalachian mountaintop mining on interior forest. *Landscape Ecology* **2007**, *22* (2), 179–187.
211. Palmer, M. A., Bernhardt, E. S., Schlesinger, W. H., Eshleman, K. N., Foufoula-Georgiou, E., Hendryx, M. S., Lemly, A. D., Likens, G. E., Loucks, O. L., Power, M. E., White, P. S., Wilcock, P. R., Mountaintop mining consequences. *Science* **2010**, *327* (5962), 148–149.
212. Pond, G. J., Passmore, M. E., Borsuk, F. A., Reynolds, L., Rose, C. J., Downstream effects of mountaintop coal mining: comparing biological conditions using family- and genus-level macroinvertebrate bioassessment tools. *Journal of the North American Benthological Society* **2008**, *27* (3), 717–737.
213. Fulk, F., Autrey, B., Hutchens, J., Gerritsen, J., Burton, J., Cresswell, C., Jessup, B., *Ecological Assessment of Streams in the Coal Mining Region of West Virginia using Data Collected by US EPA and Environmental Consulting Firms*. National Exposure Research Laboratory, US EPA: **2003**.
214. Hartman, K. J., Kaller, M. D., Howell, J. W., Sweka, J. A., How much do valley fills influence headwater streams? *Hydrobiologia* **2005**, *532*, 91–102.
215. Pond, G. J., Patterns of ephemeroptera taxa loss in Appalachian headwater streams (Kentucky, USA). *Hydrobiologia* **2010**, *641* (1), 185–201.
216. Gilbert, N., Environment mountaintop mining plans close to defeat. *Nature* **2010**, *467* (7319), 1021.
217. Dittman, E. K., Buchwalter, D. B., Manganese bioconcentration in aquatic insects: Mn oxide coatings, molting loss, and Mn(II) thiol scavenging. *Environmental Science & Technology* **2010**, *44* (23), 9182–9188.
218. Pericak, A. A., Thomas, C. J., Kroodsma, D. A., Wasson, M. F., Ross, M. R. V., Clinton, N. E., Campagna, D. J., Franklin, Y., Bernhardt, E. S., Amos, J. F., Mapping the yearly extent of surface coal mining in Central Appalachia using Landsat and Google Earth Engine. *PLoS ONE* **2018**, *13* (7), 15.
219. US Environmental Protection Agency, Improving EPA review of Applachian surface coal mining operation under the Clean Water Act, National Environmental Policy Act, and the Environmental Justice Excecutive Order, April 1, 2010, **2010**.
220. Herlihy, A. T., Kaufmann, P. R., Mitch, M. E., Brown, D. D., Regional estimates of acid mine drainage impact on streams in the mid-Atlantic and Southeastern United States. *Water, Air, and Soil Pollution* **1990**, *50*, 91–107.

221. Das, A., Patel, S. S., Kumar, R., Krishna, K., Dutta, S., Saha, M. C., Sengupta, S., Guha, D., Geochemical sources of metal contamination in a coal mining area in Chhattisgarh, India using lead isotopic ratios. *Chemosphere* **2018**, *197*, 152–164.

222. Neogi, B., Singh, A. K., Pathak, D. D., Chaturvedi, A., Hydrogeochemistry of coal mine water of North Karanpura coalfields, India: implication for solute acquisition processes, dissolved fluxes and water quality assessment. *Environmental Earth Sciences* **2017**, *76* (14), 489.

223. Sahoo, P. K., Tripathy, S., Panigrahi, M. K., Equeenuddin, S. M., Anthropogenic contamination and risk assessment of heavy metals in stream sediments influenced by acid mine drainage from a northeast coalfield, India. *Bulletin of Engineering Geology and the Environment* **2017**, *76* (2), 537–552.

224. Singh, R., Venkatesh, A. S., Syed, T. H., Reddy, A. G. S., Kumar, M., Kurakalva, R. M., Assessment of potentially toxic trace elements contamination in groundwater resources of the coal mining area of the Korba Coalfield, Central India. *Environmental Earth Sciences* **2017**, *76* (16), 566.

225. Tiwari, A. K., De Maio, M., Assessment of sulphate and iron contamination and seasonal variations in the water resources of a Damodar Valley coalfield, India: a case study. *Bulletin of Environmental Contamination and Toxicology* **2018**, *100* (2), 271–279.

226. Sahoo, P. K., Tripathy, S., Panigrahi, M. K., Equeenuddin, S. M., Geochemical characterization of coal and waste rocks from a high sulfur bearing coalfield, India: implication for acid and metal generation. *Journal of Geochemical Exploration* **2014**, *145*, 135–147.

227. Li, X. X., Wu, P., Geochemical characteristics of dissolved rare earth elements in acid mine drainage from abandoned high-As coal mining area, southwestern China. *Environmental Science and Pollution Research* **2017**, *24* (25), 20540–20555.

228. Zhao, Q., Guo, F., Zhang, Y., Ma, S., Jia, X., Meng, W., How sulfate-rich mine drainage affected aquatic ecosystem degradation in northeastern China, and potential ecological risk. *Science of the Total Environment* **2017**, *609*, 1093–1102.

229. Growitz, D. J., Reed, .L.A, Beard, M. M. *Reconnaissance of Mine Drainage in the Coal Fields of Eastern Pennsylvanian*. US Geological Survey: **1985**.

230. Esri, Topographic Map Which Includes Boundaries, Cities, Water Features, Physiographic Features, Parks, Landmarks, Transportation, and Buildings. www.arcgis.com/home/item.html?id=a1dc28de08e6447c8d14085fa15012e1.

231. Qin, Y., Curmi, E., Kopec, G. M., Allwood, J. M., Richards, K. S., China's energy–water nexus – assessment of the energy sector's compliance with the "3 Red Lines" industrial water policy. *Energy Policy* **2015**, *82*, 131–143.

232. Pan, S.-Y., Snyder, S. W., Packman, A. I., Lin, Y. J., Chiang, P.-C., Cooling water use in thermoelectric power generation and its associated challenges for addressing water–energy nexus. *Water–Energy Nexus* **2018**, *1* (1), 26–41.

233. Liao, X., Hall, J. W., Drivers of water use in China's electric power sector from 2000 to 2015. *Environmental Research Letters* **2018**, *13* (9), 094010.

234. Kondash, A. J., Patino-Echeverri, D., Vengosh, A., Quantification of the water-use reduction associated with the transition from coal to natural gas in the US electricity sector. *Environmental Research Letters* **2019**, *14* (12).

235. Zhang, C., Zhong, L., Fu, X., Wang, J., Wu, Z., Revealing water stress by the thermal power industry in China based on a high spatial resolution water withdrawal and consumption inventory. *Environmental Science & Technology* **2016**, *50* (4), 1642–1652.

236. Srinivasan, S., Kholod, N., Chaturvedi, V., Ghosh, P. P., Mathur, R., Clarke, L., Evans, M., Hejazi, M., Kanudia, A., Koti, P. N., Liu, B., Parikh, K. S., Ali, M. S., Sharma, K., Water for electricity in India: a multi-model study of future challenges and linkages to climate change mitigation. *Applied Energy* **2018**, *210*, 673–684.

237. World Resources Institute, 40% of India's Thermal Power Plants Are in Water-Scarce Areas, Threatening Shutdowns. www.wri.org/blog/2018/01/40-indias-thermal-power-plants-are-water-scarce-areas-threatening-shutdowns.

238. US Department of Energy, *The Water–Energy Nexus: Challenges and Opportunities*. US Department of Energy: **2014**, www.energy.gov/articles/water-energy-nexus-challenges-and-opportunities.

239. US Geological Survey, A Coal-Fired Thermoelectric Power Plant. www.usgs.gov/special-topic/water-science-school/science/a-coal-fired-thermoelectric-power-plant?qt-science_center_objects=0#qt-science_center_objects.

240. Raptis, C. E., Boucher, J. M., Pfister, S., Assessing the environmental impacts of freshwater thermal pollution from global power generation in LCA. *Science of the Total Environment* **2017**, *580*, 1014–1026.

241. Raptis, C. E., Pfister, S., Global freshwater thermal emissions from steam-electric power plants with once-through cooling systems. *Energy* **2016**, *97*, 46–57.

242. Vallero, D. A., Thermal pollution. In *Waste (Second Edition)*, Letcher, T. M., Vallero, D. A., eds. Academic Press: **2019**, pp. 381–404.

243. Langford, T., *Ecological Effects of Thermal Discharges*. Elsevier: **1990**.

244. Verones, F., Hanafiah, M. M., Pfister, S., Huijbregts, M. A. J., Pelletier, G. J., Koehler, A., Characterization factors for thermal pollution in freshwater aquatic environments. *Environmental Science & Technology* **2010**, *44* (24), 9364–9369.

245. Raptis, C. E., van Vliet, M. T. H., Pfister, S., Global thermal pollution of rivers from thermoelectric power plants. *Environmental Research Letters* **2016**, *11* (10), 104011.

246. American Coal Ash Association, Production and Use of Coal Combustion Products in the US www.acaa-usa .org/Portals/9/Files/PDFs/ReferenceLibrary/ARTBA-final-forecast.compressed.pdf.

247. US Environmental Protection Agency, US Coal Ash Basics. www.epa.gov/coalash/coal-ash-basics.

248. Yao, Z. T., Ji, X. S., Sarker, P. K., Tang, J. H., Ge, L. Q., Xia, M. S., Xi, Y. Q., A comprehensive review on the applications of coal fly ash. *Earth-Science Reviews* **2015**, *141*, 105–121.

249. Gollakota, A. R. K., Volli, V., Shu, C.-M., Progressive utilisation prospects of coal fly ash: a review. *Science of the Total Environment* **2019**, *672*, 951–989.

250. Harris, D., Heidrich, C., Feuerborn, J., Global aspects on Coal Combustion Products. www.coaltrans.com/insights/article/global-aspects-on-coal-combustion-products.

251. Ma, S.-H., Xu, M.-D., Qiqige, Wang, X-H., Zhou, X, Challenges and developments in the utilization of fly ash in China. *International Journal of Environmental Science and Development* **2017**, *8*, 781–785.

252. Luo, Y., Wu, Y., Ma, S., Zheng, S., Zhang, Y., Chu, P. K., Utilization of coal fly ash in China: a mini-review on challenges and future directions. *Environmental Science and Pollution Research* **2020**, *28*, 18727–18740.

253. Li, J., Zhuang, X., Querol, X., Font, O., Moreno, N., A review on the applications of coal combustion products in China. *International Geology Review* **2018**, *60* (5–6), 671–716.

254. US Environmental Protection Agency, Disposal of Coal Combustion Residuals from Electric Utilities Rulemakings. www.epa.gov/coalash/coal-ash-rule.

255. Clarke, L. B., The fate of trace elements during coal combustion and gasification: an overview. *Fuel* **1993**, *72* (6), 731–736.

256. Meij, R., The fate of trace elements at coal-fired power plants. *Fuel* **1993**, *72*, 718.

257. Meij, R., te Winkel, B. H., Trace elements in world steam coal and their behaviour in Dutch coal-fired power stations: a review. *International Journal of Coal Geology* **2009**, *77*, 289–293.

258. Noda, N., Ito, S., The release and behavior of mercury, selenium, and boron in coal combustion. *Powder Technology* **2008**, *180* (1–2), 227–231.

259. Noda, N., Ito, S., Nunome, Y., Ueki, Y., Yoshiie, R., Naruse, I., Volatilization characteristics of boron compounds during coal combustion. *Proceedings of the Combustion Institute* **2013**, *34* (2), 2831–2838.

260. Goodarzi, F., Swaine, D. J., Behavior of boron in coal during natural and industrial combustion processes. *Energy Sources* **1993**, *15* (4), 609–622.

261. Kashiwakura, S., Takahashi, T., Nagasaka, T., Vaporization behavior of boron from standard coals in the early stage of combustion. *Fuel* **2011**, *90*, 1408–1415.

262. Swaine, D. J., Trace elements in coal and their dispersal during combustion. *Fuel Processing Technology* **1994**, *39*, 121–137.

263. Lauer, N. E., Hower, J. C., Hsu-Kim, H., Taggart, R. K., Vengosh, A., Naturally occurring radioactive materials in coals and coal combustion residuals in the United States. *Environmental Science & Technology* **2015**, *49* (18), 11227–11233.

264. Ruhl, L., Vengosh, A., Dwyer, G. S., Hsu-Kim, H., Deonarine, A., Bergin, M., Kravchenko, J., Survey of the potential environmental and health impacts in the immediate aftermath of the coal ash spill in Kingston, Tennessee. *Environmental Science & Technology* **2009**, *43* (16), 6326–6333.

265. Ruhl, L. S., Dwyer, G. S., Hsu-Kim, H., Hower, J. C., Vengosh, A., Boron and strontium isotopic characterization of coal combustion residuals: validation of new environmental tracers. *Environmental Science & Technology* **2014**, *48* (24), 14790–14798.

266. Schwartz, G. E., Hower, J. C., Phillips, A. L., Rivera, N., Vengosh, A., Hsu-Kim, H., Ranking coal ash materials for their potential to leach arsenic and selenium: relative importance of ash chemistry and site biogeochemistry. *Environmental Engineering Science* **2018**, *35* (7), 728–738.

267. Schwartz, G. E., Redfern, L. K., Ikuma, K., Gunsch, C. K., Ruhl, L. S., Vengosh, A., Hsu-Kim, H., Impacts of coal ash on methylmercury production and the methylating microbial community in anaerobic sediment slurries. *Environmental Science-Processes & Impacts* **2016**, *18* (11), 1427–1439.

268. Schwartz, G. E., Rivera, N., Lee, S.-W., Harrington, J. M., Hower, J. C., Levine, K. E., Vengosh, A., Hsu-Kim, H., Leaching potential and redox transformations of arsenic and selenium in sediment microcosms with fly ash. *Applied Geochemistry* **2016**, *67*, 177–185.

269. Vengosh, A., Cowan, E. A., Coyte, R. M., Kondash, A. J., Wang, Z., Brandt, J. E., Dwyer, G. S., Evidence for unmonitored coal ash spills in Sutton Lake, North Carolina: Implications for contamination of lake ecosystems. *The Science of the Total Environment* **2019**, *686*, 1090–1103.

270. Izquierdo, M. Q., X., Leaching behaviour of elements from coal combustion fly ash: an overview. *International Journal of Coal Geology* **2012**, *94*, 54–66.

271. Hassett, D. J. P.-H., D. F., Heebink, L. V., Leaching of CCBs: observations from over 25 years of research. *Fuel* **2005**, *84*, 1378–1383.

272. Kosson, D. S., F., Kariher, P., Turner, L. H., Delapp, R., Seignette, P. *Characterization of Coal Combustion Residues from Electric Utilities – Leaching and Characterization Data.* US Environmental Protection Agency, EPA/600/R-09/151: **2009**.

273. Thorneloe, S. A., Kosson, D. S., Sanchez, F., Garrabrants, A. C., Helms, G., Evaluating the fate of metals in air pollution control residues from coal-fired power plants. *Environmental Science & Technology* **2010**, *44* (19), 7351–7356.

274. Kosson, D. S., Garrabrants, A. C., DeLapp, R., van der Sloot, H. A., pH-dependent leaching of constituents of potential concern from concrete materials containing coal combustion fly ash. *Chemosphere* **2014**, *103*, 140–147.

275. Kosson, D. S., van der Sloot, H. A., Sanchez, F., Garrabrants, A. C., An integrated framework for evaluating leaching in waste management and utilization of secondary materials. *Environment Science and Technology* **2002**, *19*, 159–204.

276. Ruhl, L., Vengosh, A., Dwyer, G. S., Hsu-Kim, H., Deonarine, A., Environmental impacts of the coal ash spill in Kingston, Tennessee: an 18-month survey. *Environmental Science & Technology* **2010**, *44* (24), 9272–9278.

277. Bartov, G., Deonarine, A., Johnson, T. M., Ruhl, L., Vengosh, A., Hsu-Kim, H., Environmental impacts of the Tennessee Valley Authority Kingston coal ash spill. 1. Source apportionment using mercury stable isotopes. *Environmental Science & Technology* **2013**, *47* (4), 2092–2099.

278. Deonarine, A., Bartov, G., Johnson, T. M., Ruhl, L., Vengosh, A., Hsu-Kim, H., Environmental impacts of the Tennessee Valley Authority Kingston coal ash spill. 2. Effect of coal ash on methylmercury in historically contaminated river sediments. *Environmental Science & Technology* **2013**, *47* (4), 2100–2108.

279. Liu, Y.-T., Chen, T.-Y., Mackebee, W. G., Ruhl, L., Vengosh, A., Hsu-kim, H., Selenium speciation in coal ash spilled at the Tennessee Valley Authority Kingston site. *Environmental Science & Technology* **2013**, *47* (24), 14001–14009.

280. Ruhl, L., Vengosh, A., Dwyer, G. S., Hsu-Kim, H., Schwartz, G., Romanski, A., Smith, S. D., The impact of coal combustion residue effluent on water resources: a North Carolina example. *Environmental Science & Technology* **2012**, *46* (21), 12226–12233.

281. Harkness, J. S., Sulkin, B., Vengosh, A., Evidence for coal ash ponds leaking in the southeastern United States. *Environmental Science & Technology* **2016**, *50* (12), 6583–6592.

282. Environmental Integrity Project, *Coal's Poisonous Legacy Groundwater Contaminated by Coal Ash across the U.S.* Environmental Integrity Project: **2019**, https://environmentalintegrity.org/wp-content/uploads/2019/03/National-Coal-Ash-Report-Revised-7.11.19.pdf.

283. Ailun, Y., Hanhua, Z., Kang, R., Miaohan, S., Xingmin, Z., Hongyuan, T., Xu, H., Fei, L. *The True Cost of Coal – An Investigation into Coal Ash in China*, Greenpeace: **2010**, www.greenpeace.org/usa/research/dow-inspection-report/.

284. North Carolina Department of Environmental Quality, Well Test Information for Residents near Duke Energy Coal Ash Impoundments. https://deq.nc.gov/about/divisions/water-resources/water-resources-hot-topics/dwr-coal-ash-regulation/well-test-information-for-residents-near-duke-energy-coal-ash-impoundments.

285. Vengosh, A., Coyte, R., Karr, J., Harkness, J. S., Kondash, A. J., Ruhl, L. S., Merola, R. B., Dwyer, G. S., Origin of hexavalent chromium in drinking water wells from the Piedmont aquifers of North Carolina. *Environmental Science & Technology Letters* **2016**, *3* (12), 409–414.

286. Coyte, R. M., McKinley, K. L., Jiang, S., Karr, J., Dwyer, G. S., Keyworth, A. J., Davis, C. C., Kondash, A. J., Vengosh, A., Occurrence and distribution of hexavalent chromium in groundwater from North Carolina, USA. *Science of the Total Environment* **2020**, *711*, 135135.

287. Coyte, R. M., Vengosh, A., Factors controlling the risks of co-occurrence of the redox-sensitive elements of arsenic, chromium, vanadium, and uranium in groundwater from the Eastern United States. *Environmental Science & Technology* **2020**, *54* (7), 4367–4375.

288. Pagenkopf, G. K., Connolly, J. M., Retention of boron by coal ash. *Environmental Science & Technology* **1982**, *16* (9), 609–613.

289. US Geological Survey, The National Coal Resources Data System (NCRDS). https://energy.usgs.gov/Tools/NationalCoalResourcesDataSystem.aspx.

290. Sakata, M., Natsumi, M., Tani, Y., Isotopic evidence of boron in precipitation originating from coal burning in Asian continent. *Geochemical Journal* **2010**, *44* (2), 113–123.
291. Takahashi, T., Kashiwakura, S., Kanehashi, K., Hayashi, S., Nagasaka, T., Analysis of atomic scale chemical environments of boron in coal by B-11 solid state NMR. *Environmental Science & Technology* **2011**, *45* (3), 890–895.
292. Williams, L. B., Hervig, R. L., Boron isotope composition of coals: a potential tracer of organic contaminated fluids. *Applied Geochemistry* **2004**, *19* (10), 1625–1636.
293. Chen, S., Gui, H., Isotopic characteristics of D, O-18, C-13(dic), O-18(dic), Sr-87/Sr-86 and their application in coal mine water: a case study. *Water Practice and Technology* **2017**, *12* (1), 97–103.
294. Qu, S., Wang, G., Shi, Z., Xu, Q., Guo, Y., Ma, L., Sheng, Y., Using stable isotopes (delta D, delta O-18, delta S-34 and Sr-87/Sr-86) to identify sources of water in abandoned mines in the Fengfeng coal mining district, northern China. *Hydrogeology Journal* **2018**, *26* (5), 1443–1453.
295. Hamel, B. L., Stewart, B. W., Kim, A. G., Tracing the interaction of acid mine drainage with coal utilization byproducts in a grouted mine: strontium isotope study of the inactive Omega Coal Mine, West Virginia (USA). *Applied Geochemistry* **2010**, *25* (2), 212–223.
296. Hurst, R. W., Davis, T. E., Strontium isotopes as tracers of airborne fly-ash from coal-fired power-plants. *Environmental Geology* **1981**, *3* (6), 363–367.
297. Hurst, R. W., Davis, T. E., Elseewi, A. A., Strontium isotopes as tracers of coal combustion residue in the environment. *Engineering Geology* **1991**, *30* (1), 59–77.
298. Mattigod, S. V., Rai, D., Fruchter, J. S., Strontium isotopic characterization of soils and coal ashes. *Applied Geochemistry* **1990**, *5* (3), 361–365.
299. Spivak-Birndorf, L. J., Stewart, B. W., Capo, R. C., Chapman, E. C., Schroeder, K. T., Brubaker, T. M., Strontium isotope study of coal utilization by-products interacting with environmental waters. *Journal of Environmental Quality* **2012**, *41* (1), 144–154.
300. Widory, D., Liu, X., Dong, S., Isotopes as tracers of sources of lead and strontium in aerosols (TSP & PM2.5) in Beijing. *Atmospheric Environment* **2010**, *44* (30), 3679–3687.
301. US Environmental Protection Agency, *Technical Development Document for the Effluent Limitations Guidelines and Standards for the Steam Electric Power Generating Point Source Category.* US Environmental Protection Agency: **2015**, www.epa.gov/sites/production/files/2015-10/documents/steam-electric-tdd_10-21-15.pdf.
302. US Environmental Protection Agency, National Recommended Water Quality Criteria – Aquatic Life Criteria Table. www.epa.gov/wqc/national-recommended-water-quality-criteria-aquatic-life-criteria-table.
303. US Environmental Protection Agency, Aquatic Life Criterion – Selenium. www.epa.gov/wqc/aquatic-life-criterion-selenium.
304. Gingerich, D. B., Zhao, Y., Mauter, M. S., Environmentally significant shifts in trace element emissions from coal plants complying with the 1990 Clean Air Act Amendments. *Energy Policy* **2019**, *132*, 1206–1215.
305. US Environmental Protection Agency, *Acid Rain and Related Programs: 15 Years of Results.* US Environmental Protection Agency: **2010**.
306. US Environmental Protection Agency, *Technical Development Document for the Effluent Limitations Guidelinesand Standards for the Steam Electric Power Generating Point Source Category.* US Environmental Protection Agency: **2015**.
307. Gingerich, D. B., Grol, E., Mauter, M. S., Fundamental challenges and engineering opportunities in flue gas desulfurization wastewater treatment at coal fired power plants. *Environmental Science-Water Research & Technology* **2018**, *4* (7), 909–925.
308. Gingerich, D. B., Sun, X., Behrer, A. P., Azevedo, I. L., Mauter, M. S., Spatially resolved air–water emissions tradeoffs improve regulatory impact analyses for electricity generation. *Proceedings of the National Academy of Sciences of the United States of America* **2017**, *114* (8), 1862–1867.
309. Sun, X., Gingerich, D. B., Azevedo, I. L., Mauter, M. S., Trace element mass flow rates from US coal fired power plants. *Environmental Science & Technology* **2019**, *53* (10), 5585–5595.
310. US Environmental Protection Agency, Overview of the Cross-State Air Pollution Rule (CSAPR). www.epa.gov/csapr/overview-cross-state-air-pollution-rule-csapr.
311. US Environmental Protection Agency, *National Emission Standards for Hazardous Air Pollutants from Coal- and Oil-Fired Electric Utility Steam In Federal Register.* US Environmental Protection Agency: **2012**, Vol. EPA-HQ-OAR-2009-0234, EPA-HQ-OAR-2011-0044, FRL-9611-4.
312. US Environmental Protection Agency, Mercury and Air Toxics Standards (MATS). www.epa.gov/mats.
313. Young, D. S., S., Senior, C., Meinhardt, S. *Reducing Operating Costs and Risks of Hg Control with Fuel Additives* Advancing Cleaner Energy (ADA): **2016**.

314. Heebink, L. V., Pflughoeft-Hassett, D. F., Hassett, D. J., Effects of mercury emission control technologies using halogens on coal combustion product chemical properties. *Journal of Environmental Monitoring* **2010**, *12* (3), 608–613.

315. Liu, S.-H., Yan, N.-Q., Liu, Z.-R., Qu, Z., Wang, P., Chang, S.-G., Miller, C., Using bromine gas to enhance mercury removal from flue gas of coal-fired power plants. *Environmental Science & Technology* **2007**, *41* (4), 1405–1412.

316. Qu, Z., Yan, N., Liu, P., Chi, Y., Jia, J., Bromine chloride as an oxidant to improve elemental mercury removal from coal-fired flue gas. *Environmental Science & Technology* **2009**, *43* (22), 8610–8615.

317. Rupp, E. C., Wilcox, J., Mercury chemistry of brominated activated carbons – packed-bed breakthrough experiments. *Fuel* **2014**, *117*, 351–353.

318. Sasmaz, E., Kirchofer, A., Jew, A. D., Saha, A., Abram, D., Jaramillo, T. F., Wilcox, J., Mercury chemistry on brominated activated carbon. *Fuel* **2012**, *99*, 188–196.

319. Wilcox, J., Okano, T., Ab initio-based mercury oxidation kinetics via bromine at postcombustion flue gas conditions. *Energy & Fuels* **2011**, *25* (4), 1348–1356.

320. Zhou, Q., Duan, Y.-F., Hong, Y.-G., Zhu, C., She, M., Zhang, J., Wei, H.-Q., Experimental and kinetic studies of gas-phase mercury adsorption by raw and bromine modified activated carbon. *Fuel Processing Technology* **2015**, *134*, 325–332.

321. Wang, S., Zhang, Y., Gu, Y., Wang, J., Yu, X., Wang, T., Sun, Z., liu, Z., Romero, C. E., Pan, W.-p., Coupling of bromide and on-line mechanical modified fly ash for mercury removal at a 1000 MW coal-fired power plant. *Fuel* **2019**, *247*, 179–186.

322. Zhang, Y., Zhang, Z., Liu, Z., Norris, P., Pan, W.-p., Study on the mercury captured by mechanochemical and bromide surface modification of coal fly ash. *Fuel* **2017**, *200*, 427–434.

323. Zhao, S., Pudasainee, D., Duan, Y., Gupta, R., Liu, M., Lu, J., A review on mercury in coal combustion process: content and occurrence forms in coal, transformation, sampling methods, emission and control technologies. *Progress in Energy and Combustion Science* **2019**, *73*, 26–64.

324. Cadwallader, A., VanBriesen, J. M., Temporal and spatial changes in bromine incorporation into drinking water-disinfection by-products in Pennsylvania. *Journal of Environmental Engineering* **2019**, *145* (3).

325. Good, K. D., VanBriesen, J. M., Current and potential future bromide loads from coal-fired power plants in the Allegheny River Basin and their effects on downstream concentrations. *Environmental Science & Technology* **2016**, *50* (17), 9078–9088.

326. Good, K. D., VanBriesen, J. M., Power plant bromide discharges and downstream drinking water systems in Pennsylvania. *Environmental Science & Technology* **2017**, *51* (20), 11829–11838.

327. Good, K. D., VanBriesen, J. M., Coal-fired power plant wet flue gas desulfurization bromide discharges to US watersheds and their contributions to drinking water sources. *Environmental Science & Technology* **2019**, *53* (1), 213–223.

328. US Environmental Protection Agency, Steam Electric Power Generating Effluent Guidelines – 2015 Final Rule. www.epa.gov/eg/steam-electric-power-generating-effluent-guidelines-2015-final-rule (40 CFR Part 423).

329. Tabuchi, H. Republicans move to block rule on coal mining near streams. www.nytimes.com/2017/02/02/business/energy-environment/senate-coal-regulations.html.

330. Eilperin, J. D., B., Muyskens, J. Trump rolled back more than 125 environmental safeguards. Here's how. www.washingtonpost.com/graphics/2020/climate-environment/trump-climate-environment-protections/.

331. Verbong, G., Loorbach, D, *Governing the Energy Transition: Reality, Illusion or Necessity?* Routledge: **2012**.

332. Van de Graaf, T., , Sovacool, B. K., Ghosh, A., Kern, F., Klare, M. T., States, markets, and institutions: integrating international political economy and global energy politics. In *The Palgrave Handbook of the International Political Economy of Energy*, Van de Graaf, T., Sovacool, B. K., Ghosh, A., Kern, F., Klare, M. T., eds. Springer Nature: **2016**, pp. 3–34.

333. Skovgaard, J., van Asselt, H., *The Politics of Fossil Fuel Subsidies and Their Reform*. Cambridge University Press: **2018**.

334. US Energy Information Administration (EIA), Power Sector Coal Demand Has Fallen in Nearly Every State since 2007. www.eia.gov/todayinenergy/detail.php?id=26012.

335. Institute for Energy Economics and Financial Analysis, April is Shaping Up to be Momentous in Transition from Coal to Renewables. https://ieefa.org/ieefa-u-s-april-is-shaping-up-to-be-momentous-in-transition-from-coal-to-renewables/.

336. Appunn, K. Germany Bids Farewell to Domestic Hard Coal Mining. www.cleanenergywire.org/news/germany-bids-farewell-domestic-hard-coal-mining.

337. Whitley, S., van der Burg, L., Reforming fossil fuel subsidies: the art of the possible. In *The Politics of Fossil Fuel Subsidies and Their Reform*, Skovgaard, J., van Asselt, H., eds. Cambridge University Press: **2018**.

338. Ebinger, C. K. *India's Energy and Climate Policy: Can India Meet the Challenges of Industrialization and Climate Change?* Brookings: **2016**.

339. US Energy Information Administration (EIA), India's Coal Industry in Flux as Government Sets Ambitious Coal Production Targets. www.eia.gov/todayinenergy/detail.php?id=22652.

340. Bagirov, S., *Oil of Zerbaijan: Revenues, Expernses, and Risks*. Yeni Nesil Publishing House: **2007**.

341. Jones Luong, P. J., Weinthal, E., *Oil Is Not a Curse*. Cambridge University Press: **2010**.

342. UN Audiovisual Library of International Law, Permanent Sovereignty over Natural Resources General Assembly Resolution 1803 (XVII). https://legal.un.org/avl/ha/ga_1803/ga_1803.html.

343. Ross, M. L., *The Oil Curse: How Petroleum Wealth Shapes the Development of Nations*. Princeton University Press: **2012**.

344. US Energy Information Administration (EIA), International Energy Statistics. www.eia.gov/beta/international/ (1/19/**2019**).

345. Mai-Duc, C. The 1969 Santa Barbara oil spill that changed oil and gas exploration forever. www.latimes.com/local/lanow/la-me-ln-santa-barbara-oil-spill-1969-20150520-htmlstory.html.

346. Leahy, S. Exxon Valdez changed the oil industry forever – but new threats emerge. www.nationalgeographic.com/environment/2019/03/oil-spills-30-years-after-exxon-valdez/.

347. Birkland, T. A., In the wake of the Exxon Valdez: how environmental disasters influence policy. *Environment Science and Policy for Sustainable Development* **1998**, *40*, 4–32.

348. Graham, B. R., W. K., Beinecke, F., Boesch, D. F., Garcia, T. D., Murray, C. A., Ulmer, F., *Deep Water: The Gulf Oil Disaster and the Future of Offshore Drilling*. National Commission on the BP Deepwater Horizon Oil Spill and Offshore Drilling: **2011**.

349. Lyall, S., *At BP, a History of Boldness and Costly Blunders*. The New York Times: **2010**.

350. Xavier Sala-i-Martin, X. Subramanian, A., Addressing the natural resource curse: an illustration from Nigeria. *Journal of African Economies* **2013**, *22* (4), 570–615.

351. Adebayo, B., Nigeria overtakes India in extreme poverty ranking. CNN, **2018**, www.cnn.com/2018/06/26/africa/nigeria-overtakes-india-extreme-poverty-intl/index.html.

352. Mähler, A., An inescapable curse? Resource management, violent conflict, and peacebuilding in the Niger Delta. In *High-Value Natural Resources and Post-Conflict Peacebuilding*, Päivi Lujala, P., Rustad, S. A., eds. Taylor and Francis Group: **2012**.

353. Nossiter, A. Far from Gulf, a spill scourge 5 decades old. www.nytimes.com/2010/06/17/world/africa/17nigeria.html?src=mv.

354. Andrews-Speed, P. *China's Energy Policymaking Processes and Their Consequences* The National Bureau of Asian Research (NBR): **2014**.

355. US Energy Information Administration (EIA), *Key World Energy Statistics*. US Energy Information Administration (EIA): **2019**.

356. US Environmental Protection Agency, Types of Petroleum Oils. https://archive.epa.gov/emergencies/content/learning/web/html/oiltypes.html.

357. S&P Global Platts, *Specifications Guide Americas Crude Oil*. **2021**, www.spglobal.com/platts/plattscontent/_assets/_files/en/our-methodology/methodology-specifications/americas-crude-methodology.pdf.

358. Bush, J. L., Helander, D. P., Empirical prediction of recovery rate in waterflooding depleted sands. *Journal of Petroleum Technology* **1968**, *20*, 933–943.

359. Wu, M., Mintz, M., Wang, M., Arora, S. *Consumptive Water Use in the Production of Ethanol and Petroleum Gasoline*. Argonne National Laboratory, US Department of Energy: **2009**, p 76.

360. Kondash, A. J., Albright, E., Vengosh, A., Quantity of flowback and produced waters from unconventional oil and gas exploration. *Science of the Total Environment* **2017**, *574*, 314–321.

361. US Energy Information Administration (EIA), Crude Oil Production (USA). www.eia.gov/dnav/pet/pet_crd_crpdn_adc_mbbl_m.htm.

362. US Geological Survey, Geological Survey National Produced Waters Geochemical Database v2.3. www.sciencebase.gov/catalog/item/59d25d63e4b05fe04cc235f9.

363. US Energy Information Administration (EIA), *Technically Recoverable Shale Oil and Shale Gas Resources: An Assessment of 137 Shale Formations in 41 Countries Outside the United States*. US Energy Information Administration (EIA): **2013**.

364. US Energy Information Administration (EIA), Maps: Oil and Gas Exploration, Resources, and Production. www.eia.gov/maps/maps.htm.

365. Wu, M., Mintz, M., Wang, M., Arora, S., Chiu, Y-W., Xu, H. *Consumptive Water Use in the Production of Ethanol and Petroleum Gasoline – 2018 Update*. Argonne National Laboratory: **2018**.

366. Neville, K. J., Baka, J., Gamper-Rabindran, S., Bakker, K., Andreasson, S., Vengosh, A., Lin, A., Singh, J. N., Weinthal, E., Debating unconventional energy: social, political, and economic implications. *Annual Review of Environment and Resources*, **2017**, *42*, 241–266.

367. Scanlon, B. R., Reedy, R. C., Nicot, J. P., Comparison of water use for hydraulic fracturing for unconventional oil and gas versus conventional oil. *Environmental Science & Technology* **2014**, *48* (20), 12386–12393.

368. Allen, E. W., Process water treatment in Canada's oil sands industry: I. target pollutants and treatment objectives. *Journal Environment Engineering Science* **2008**, *7*, 123–138.

369. Ali, B., Kumar, A., Development of life cycle water footprints for oil sands-based transportation fuel production. *Energy* **2017**, *131*, 41–49.

370. Sun, P., Elgowainy, A., Wang, M., Han, J., Henderson, R. J., Estimation of US refinery water consumption and allocation to refinery products. *Fuel* **2018**, *221*, 542–557.

371. Zara Khatib, Z., Verbeek, P., Water to value – produced water management for sustainable field development of mature and green fields. *Journal of Petroleum Technology* **2003**, *55*, 26–28.

372. Veil, J. *US Produced Water Volumes and Management.* Groundwater Protection Council: **2015**.

373. Clark, C. E., Veil, J. A. *Produced Water Volumes and Management in the United States.* Argonne National Laboratory: **2009**.

374. Neff, J., Lee, K., DeBloi, E. M., Produced water: overview of composition, fates, and effects. In *Produced Water: Environmental Risks and Advances in Mitigation Technologies*, Lee, K., Neff, J., eds. Springer: **2011**.

375. Veil, J. *US Produced Water Volumes and Management Practices in 2012.* Ground Water Protection Council: **2015**.

376. Foulger, G. R., Wilson, M. P., Gluyas, J. G., Julian, B. R., Davies, R. J., Global review of human-induced earthquakes. *Earth-Science Reviews* **2018**, *178*, 438–514.

377. Frohlich, C., Brunt, M., Two-year survey of earthquakes and injection/production wells in the Eagle Ford Shale, Texas, prior to the MW4.8 20 October 2011 earthquake. *Earth and Planetary Science Letters* **2013**, *379*, 56–63.

378. Roach, T., Oklahoma earthquakes and the price of oil. *Energy Policy* **2018**, *121*, 365–373.

379. Torres, L., Yadav, O. P., Khan, E., A review on risk assessment techniques for hydraulic fracturing water and produced water management implemented in onshore unconventional oil and gas production. *Science of the Total Environment* **2016**, *539*, 478–493.

380. Atkinson, G. M., Eaton, D. W., Ghofrani, H., Walker, D., Cheadle, B., Schultz, R., Shcherbakov, R., Tiampo, K., Gu, J., Harrington, R. M., Liu, Y., van der Baan, M., Kao, H., Hydraulic fracturing and seismicity in the Western Canada Sedimentary Basin. *Seismological Research Letters* **2016**, *87* (3), 631–647.

381. Rubinstein, J., Unconventional oil and gas and induced earthquakes. *Abstracts of Papers of the American Chemical Society* **2016**, *251*.

382. Gomez Alba, S., Vargas, C. A., Zang, A., Evidencing the relationship between injected volume of water and maximum expected magnitude during the Puerto Gaitan (Colombia) earthquake sequence from 2013 to 2015. *Geophysical Journal International* **2020**, *220* (1), 335–344.

383. Ogwari, P. O., DeShon, H. R., Hornbach, M. J., The Dallas-Fort Worth Airport earthquake sequence: seismicity beyond injection period. *Journal of Geophysical Research-Solid Earth* **2018**, *123* (1), 553–563.

384. Schimmel, M., Liu, W., Worrell, E., Facilitating sustainable geo-resources exploitation: a review of environmental and geological risks of fluid injection into hydrocarbon reservoirs. *Earth-Science Reviews* **2019**, *194*, 455–471.

385. Stewart, F. L., Ingelson, A., Regulating energy innovation: US responses to hydraulic fracturing, wastewater injection and induced seismicity. *Journal of Energy & Natural Resources Law* **2017**, *35* (2), 109–146.

386. Wisen, J., Chesnaux, R., Wendling, G., Werring, J., Barbecot, F., Baudron, P., Assessing the potential of cross-contamination from oil and gas hydraulic fracturing: a case study in northeastern British Columbia, Canada. *Journal of Environmental Management* **2019**, *246*, 275–282.

387. Elsworth, D., Spiers, C. J., Niemeijer, A. R., Understanding induced seismicity. *Science* **2016**, *354* (6318), 1380–1381.

388. Guglielmi, Y., Cappa, F., Avouac, J.-P., Henry, P., Elsworth, D., Seismicity triggered by fluid injection-induced aseismic slip. *Science* **2015**, *348* (6240), 1224–1226.

389. Collins, A. G., Origin of oilfield waters. In *Developments in Petroleum Science*, Collins, A. G., ed. Elsevier: **1975**, Vol. 1, pp. 193–252.

390. Connolly, C. A., Walter, L. M., Baadsgaard, H., Longstaffe, F. J., Origin and evolution of formation waters, Alberta Basin, Western Canada Sedimentary Basin. II. Isotope systematics and water mixing. *Applied Geochemistry* **1990**, *5* (4), 397–413.

391. Connolly, C. A., Walter, L. M., Baadsgaard, H., Longstaffe, F. J., Origin and evolution of formation waters, Alberta Basin, Western Canada sedimentary Basin. I. Chemistry. *Applied Geochemistry* **1990**, *5* (4), 375–395.

392. Egeberg, P. K., Aagaard, P., Origin and evolution of formation waters from oil fields on the Norwegian Shelf. *Applied Geochemistry* **1989**, *4* (2), 131–142.

393. Fontes, J. C., Matray, J. M., Geochemistry and origin of formation brines from the Paris Basin, France: 1. Brines associated with Triassic salts. *Chemical Geology* **1993**, *109* (1), 149–175.

394. Kesler, S. E., Martini, A. M., Appold, M. S., Walter, L. M., huston, T. J., furman, F. C., Na-Cl-Br systematics of fluid inclusions from Mississippi Valley-type deposits, Appalachian Basin: constraints on solute origin and migration paths. *Geochimica et Cosmochimica Acta* **1996**, *60* (2), 225–233.

395. Kharaka, Y. K., Geochemistry of oilfield waters. *Earth-Science Reviews* **1977**, *13* (1), 77–78.

396. McNutt, R. H., Frape, S. K., Dollar, P., A strontium, oxygen and hydrogen isotopic composition of brines, Michigan and Appalachian basins, Ontario and Michigan. *Applied Geochemistry* **1987**, *2* (5), 495–505.

397. Moran, J. E., Fehn, U., Hanor, J. S., Determination of source ages and migration patterns of brines from the US Gulf Coast Basin using 129I. *Geochimica et Cosmochimica Acta* **1995**, *59* (24), 5055–5069.

398. Russell, C. W., Cowart, J. B., Russell, G. S., Strontium isotopes in brines and associated rocks from Cretaceous strata in the Mississippi Salt Dome Basin (southeastern Mississippi, U.S.A.). *Chemical Geology* **1988**, *74* (1), 153–171.

399. Spencer, R. J., Origin of CaCl brines in Devonian formations, Western Canada Sedimentary Basin. *Applied Geochemistry* **1987**, *2* (4), 373–384.

400. Surdam, R. C., MacGowan, D. B., Oilfield waters and sandstone diagenesis. *Applied Geochemistry* **1987**, *2* (5), 613–619.

401. Vengosh, A., Chivas, A. R., Starinsky, A., Kolodny, Y., Baozhen, Z., Pengxi, Z., Chemical and boron isotope compositions of non-marine brines from the Qaidam Basin, Qinghai, China. *Chemical Geology* **1995**, *120* (1), 135–154.

402. Wilson, T. P., Long, D. T., Geochemistry and isotope chemistry of Michigan Basin brines: Devonian formations. *Applied Geochemistry* **1993**, *8* (1), 81–100.

403. Bein, A., Dutton, A. R., Origin, distribution, and movement of brine in the Permian Basin (U.S.A.): a model for displacement of connate brine. *GSA Bulletin* **1993**, *105*, 695–707.

404. Warner, N. R., Christie, C. A., Jackson, R. B., Vengosh, A., Impacts of shale gas wastewater disposal on water quality in western Pennsylvania. *Environmental Science & Technology* **2013**, *47* (20), 11849–11857.

405. Warner, N. R., Darrah, T. H., Jackson, R. B., Millot, R., Kloppmann, W., Vengosh, A., New tracers identify hydraulic fracturing fluids and accidental releases from oil and gas operations. *Environmental Science & Technology* **2014**, *48* (21), 12552–12560.

406. Vengosh, A., Kondash, A., Harkness, J., Lauer, N., Warner, N., Darrah, T. H., The geochemistry of hydraulic fracturing fluids. In *15th Water–Rock Interaction International Symposium, Wri-15*, Marques, J. M., Chambel, A., eds. Procedia Earth and Planetary Science: **2017**, Vol. 17, pp. 21–24.

407. Rowan, E. L., Engle, M. A., Kraemer, T. F., Schroeder, K. T., Hammack, R. W., Doughten, M. W., Geochemical and isotopic evolution of water produced from Middle Devonian Marcellus shale gas wells, Appalachian Basin, Pennsylvania. *Aapg Bulletin* **2015**, *99* (2), 181–206.

408. Rowan, E. L. E., M. A., Kirby, C. S., Kraemer, T. F., *Radium Content of Oil- and Gas-Field Produced Waters in the Northern Appalachian Basin (USA) – Summary and Discussion of Data.* US Geological Survey: **2011**.

409. Lauer, N. E., Harkness, J. S., Vengosh, A., Brine spills associated with unconventional oil development in North Dakota. *Environmental Science & Technology* **2016**, *50* (10), 5389–5397.

410. Harkness, J. S., Darrah, T. H., Warner, N. R., Whyte, C. J., Moore, M. T., Millot, R., Kloppmann, W., Jackson, R. B., Vengosh, A., The geochemistry of naturally occurring methane and saline groundwater in an area of unconventional shale gas development. *Geochimica et Cosmochimica Acta* **2017**, *208*, 302–334.

411. Harkness, J. S., Dwyer, G. S., Warner, N. R., Parker, K. M., Mitch, W. A., Vengosh, A., Iodide, bromide, and ammonium in hydraulic fracturing and oil and gas wastewaters: environmental implications. *Environmental Science & Technology* **2015**, *49* (3), 1955–1963.

412. Haluszczak, L. O., Rose, A. W., Kump, L. R., Geochemical evaluation of flowback brine from Marcellus gas wells in Pennsylvania, USA. *Applied Geochemistry* **2013**, *28*, 55–61.

413. Barbot, E., Vidic, N. S., Gregory, K. B., Vidic, R. D., Spatial and temporal correlation of water quality parameters of produced waters from Devonian-Age shale following hydraulic fracturing. *Environmental Science & Technology* **2013**, *47* (6), 2562–2569.

414. Stringfellow, W., Camarillo, M. K., Flowback versus first-flush: new information on the geochemistry of produced water from mandatory reporting. *Environmental Science Processes & Impacts* **2019**, *21*, 370–383.

415. Collins, A. G., *Geochemistry of Oilfield Waters*. Elsevier: **1975**.
416. Veil, J. A., Puder, M. G., Elcock, D., Redweik, R. J., Jr. *A White Paper Describing Produced Water from Production of Crude Oil, Natural Gas, and Coal Bed Methane*. Argonne National Lab., US Department of Energy: **2004**.
417. Amusan, F. O., The environmental impact of oilfield formation water on a freshwater stream in Nigeria. *Journal of Applied Sciences and Environmental Management* **2003**, *7*, 61–66.
418. Varonka, M. S., Gallegos, T. J., Bates, A. L., Doolan, C., Orem, W. H., Organic compounds in produced waters from the Bakken Formation and Three Forks Formation in the Williston Basin, North Dakota. *Heliyon* **2020**, *6* (3), e03590–e03590.
419. Wang, X., Goual, L., Colberg, P. J. S., Characterization and treatment of dissolved organic matter from oilfield produced waters. *Journal of Hazardous Materials* **2012**, *217–218*, 164–170.
420. Ekins, P., Vanner, R., Firebrace, J. *Management of Produced Water on Offshore Oil Installation: A Comparative Assessment Using Flow Analysis*. Policy Studies Institute (PSI): **2005**.
421. Bakke, T., Klungsøyr, J., Sanni, S., Environmental impacts of produced water and drilling waste discharges from the Norwegian offshore petroleum industry. *Marine Environmental Reserach* **2013**, *92*, 154–169.
422. Chapman, E. C., Capo, R. C., Stewart, B. W., Kirby, C. S., Hammack, R. W., Schroeder, K. T., Edenborn, H. M., Geochemical and strontium isotope characterization of produced waters from Marcellus Shale natural gas extraction. *Environmental Science & Technology* **2012**, *46* (6), 3545–3553.
423. Neymark, L. A., Premo, W. R., Emsbo, P., Combined radiogenic (Sr-87/Sr-86, U-234/U-238) and stable (delta Sr-88) isotope systematics as tracers of anthropogenic groundwater contamination within the Williston Basin, USA. *Applied Geochemistry* **2018**, *96*, 11–23.
424. Shrestha, N., Chilkoor, G., Wilder, J., Gadhamshetty, V., Stone, J. J., Potential water resource impacts of hydraulic fracturing from unconventional oil production in the Bakken Shale. *Water Research* **2017**, *108*, 1–24.
425. Thyne, G., Brady, P., Evaluation of formation water chemistry and scale prediction: Bakken Shale. *Applied Geochemistry* **2016**, *75*, 107–113.
426. Wang, H., Lu, L., Chen, X., Bian, Y., Ren, Z. J., Geochemical and microbial characterizations of flowback and produced water in three shale oil and gas plays in the central and western United States. *Water Research* **2019**, *164*, 114942.
427. McMahon, P. B., Kulongoski, J. T., Vengosh, A., Cozzarelli, I. M., Landon, M. K., Kharaka, Y. K., Gillespie, J. M., Davis, T. A., Regional patterns in the geochemistry of oil-field water, southern San Joaquin Valley, California, USA. *Applied Geochemistry* **2018**, *98*, 127–140.
428. Barry, P. H., Kulongoski, J. T., Landon, M. K., Tyne, R. L., Gillespie, J. M., Stephens, M. J., Hillegonds, D. J., Byrne, D. J., Ballentine, C. J., Tracing enhanced oil recovery signatures in casing gases from the Lost Hills oil field using noble gases. *Earth and Planetary Science Letters* **2018**, *496*, 57–67.
429. Everett, R., Gillespie, J., Stephens, M. J., Shimabukuro, D. H., Ducart, A., Gans, K., Metzger, L., *Geochemical and Geophysical Data for Wells in the Fruitvale and Rosedale Ranch Oil and Gas Fields, Kern County, California, USA*. United States Geological Survey: **2018**.
430. Wright, M. T., McMahon, P. B., Landon, M. K., Kulongoski, J. T., Groundwater quality of a public supply aquifer in proximity to oil development, Fruitvale oil field, Bakersfield, California. *Applied Geochemistry* **2019**, *106*, 82–95.
431. Parker, K. M., Zeng, T., Harkness, J., Vengosh, A., Mitch, W. A., Enhanced formation of disinfection byproducts in shale gas wastewater-impacted drinking water supplies. *Environmental Science & Technology* **2014**, *48* (19), 11161–11169.
432. Engle, M. A., Rowan, E. L., Geochemical evolution of produced waters from hydraulic fracturing of the Marcellus Shale, northern Appalachian Basin: a multivariate compositional data analysis approach. *International Journal of Coal Geology* **2014**, *126*, 45–56.
433. Dresel, E. P., Rose, A. W. *Chemistry and Origin of Oil and Gas Well Brines in Western Pennsylvania*. Pennsylvania Geological Survery: **2010**.
434. Ni, Y. Y., Zou, C. N., Cui, H. Y., Li, J., Lauer, N. E., Harkness, J. S., Kondash, A. J., Coyte, R. M., Dwyer, G. S., Liu, D., Dong, D. Z., Liao, F. R., Vengosh, A., Origin of flowback and produced waters from Sichuan Basin, China. *Environmental Science & Technology* **2018**, *52* (24), 14519–14527.
435. Macpherson, G. L., Lithium in fluids from Paleozoic-aged reservoirs, Appalachian Plateau region, USA. *Applied Geochemistry* **2015**, *60*, 72–77.
436. Macpherson, G. L., Capo, R. C., Stewart, B. W., Phan, T. T., Schroeder, K., Hammack, R. W., Temperature-dependent Li isotope ratios in Appalachian Plateau and Gulf Coast Sedimentary Basin saline water. *Geofluids* **2014**, *14* (4), 419–429.

437. Pfister, S., Capo, R. C., Stewart, B. W., Macpherson, G. L., Phan, T. T., Gardiner, J. B., Diehl, J. R., Lopano, C. L., Hakala, J. A., Geochemical and lithium isotope tracking of dissolved solid sources in Permian Basin carbonate reservoir and overlying aquifer waters at an enhanced oil recovery site, northwest Texas, USA. *Applied Geochemistry* **2017**, *87*, 122–135.

438. Köster, M. H., Williams, L. B., Kudejova, P., Gilg, H. A., The boron isotope geochemistry of smectites from sodium, magnesium and calcium bentonite deposits. *Chemical Geology* **2019**, *510*, 166–187.

439. Williams, L. B., Crawford Elliott, W., Hervig, R. L., Tracing hydrocarbons in gas shale using lithium and boron isotopes: Denver Basin USA, Wattenberg Gas Field. *Chemical Geology* **2015**, *417*, 404–413.

440. Williams, L. B., Hervig, R. L., Lithium and boron isotopes in illite-smectite: the importance of crystal size. *Geochimica et Cosmochimica Acta* **2005**, *69* (24), 5705–5716.

441. Williams, L. B., Hervig, R. L., Hutcheon, I., Boron isotope geochemistry during diagenesis. Part II. Applications to organic-rich sediments. *Geochimica et Cosmochimica Acta* **2001**, *65* (11), 1783–1794.

442. Williams, L. B., Hervig, R. L., Wieser, M. E., Hutcheon, I., The influence of organic matter on the boron isotope geochemistry of the Gulf Coast Sedimentary Basin, USA. *Chemical Geology* **2001**, *174* (4), 445–461.

443. Al-Masri, M. S., Spatial and monthly variations of radium isotopes in produced water during oil production. *Applied Radiation and Isotopes* **2006**, *64* (5), 615–623.

444. Kraemer, T. F., Reid, D. F., The occurrence and behavior of radium in saline formation water of the US Gulf Coast region. *Chemical Geology* **1984**, *46* (2), 153–174.

445. Mathews, M., Gotkowitz, M., Ginder-Vogel, M., Effect of geochemical conditions on radium mobility in discrete intervals within the Midwestern Cambrian-Ordovician aquifer system. *Applied Geochemistry* **2018**, *97*, 238–246.

446. Omar, M., Ali, H. M., Abu, M. P., Kontol, K. M., Ahmad, Z., Ahmad, S. H. S. S., Sulaiman, I., Hamzah, R., Distribution of radium in oil and gas industry wastes from Malaysia. *Applied Radiation and Isotopes* **2004**, *60* (5), 779–782.

447. Stackelberg, P. E., Szabo, Z., Jurgens, B. C., Radium mobility and the age of groundwater in public-drinking-water supplies from the Cambrian-Ordovician aquifer system, north-central USA. *Applied Geochemistry* **2018**, *89*, 34–48.

448. Lauer, N. E., Warner, N. R., Vengosh, A., Sources of radium accumulation in stream sediments near disposal sites in Pennsylvania: implications for disposal of conventional oil and gas wastewater. *Environmental Science & Technology* **2018**, *52* (3), 955–962.

449. Beauchamp, R. O., Bus, J. S., Popp, J. A., Boreiko, C. J., Andjelkovich, D. A., Leber, P., A critical review of the literature on hydrogen sulfide toxicity. *CRC Critical Reviews in Toxicology* **1984**, *13* (1), 25–97.

450. Lauer, N., Vengosh, A., Age dating oil and gas wastewater spills using radium isotopes and their decay products in impacted soil and sediment. *Environmental Science & Technology Letters* **2016**, *3* (5), 205–209.

451. Engle, M. A., Reyes, F. R., Varonka, M. S., Orem, W. H., Ma, L., Ianno, A. J., Schell, T. M., Xu, P., Carroll, K. C., Geochemistry of formation waters from the Wolfcamp and "Cline" shales: Insights into brine origin, reservoir connectivity, and fluid flow in the Permian Basin, USA. *Chemical Geology* **2016**, *425*, 76–92.

452. McDevitt, B., McLaughlin, M., Cravotta, C. A., III, Ajemigbitse, M. A., Van Sice, K. J., Blotevogel, J., Borch, T., Warner, N. R., Emerging investigator series: radium accumulation in carbonate river sediments at oil and gas produced water discharges: implications for beneficial use as disposal management. *Environmental Science-Processes & Impacts* **2019**, *21* (2), 324–338.

453. Tasker, T. L., Burgos, W. D., Ajemigbitse, M. A., Lauer, N. E., Gusa, A. V., Kuatbek, M., May, D., Landis, J. D., Alessi, D. S., Johnsen, A. M., Kaste, J. M., Headrick, K. L., Wilke, F. D. H., McNeal, M., Engle, M., Jubb, A. M., Vidic, R. D., Vengosh, A., Warner, N. R., Accuracy of methods for reporting inorganic element concentrations and radioactivity in oil and gas wastewaters from the Appalachian Basin, US based on an inter-laboratory comparison. *Environmental Science-Processes & Impacts* **2019**, *21* (2), 224–241.

454. Tasker, T. L., Burgos, W. D., Piotrowski, P., Castillo-Meza, L., Blewett, T. A., Ganow, K. B., Stallworth, A., Delompre, P. L. M., Goss, G. G., Fowler, L. B., Vanden Heuvel, J. P., Dorman, F., Warner, N. R., Environmental and human health impacts of spreading oil and gas wastewater on roads. *Environmental Science & Technology* **2018**, *52* (12), 7081–7091.

455. Van Sice, K., Cravotta, C. A., III, McDevitt, B., Tasker, T. L., Landis, J. D., Puhr, J., Warner, N. R., Radium attenuation and mobilization in stream sediments following oil and gas wastewater disposal in western Pennsylvania. *Applied Geochemistry* **2018**, *98*, 393–403.

456. Babatunde, B. B., Sikoki, F. D., Avwiri, G. O., Chad-Umoreh, Y. E., Review of the status of radioactivity profile in the oil and gas producing areas of the Niger Delta region of Nigeria. *Journal of Environmental Radioactivity* **2019**, *202*, 66–73.

457. Moskovchenko, D. V., Babushkin, A. G., Artamonova, G. N., Surface water quality assessment of the Vatinsky Egan River catchment, west Siberia. *Environmental Monitoring and Assessment* **2009**, *148*, 359–368.

458. Moquet, J.-S., Maurice, L., Crave, A., Viers, J., Arevalo, N., Lagane, C., Lavado-Casimiro, W., Guyot, J.-L., Cl and Na fluxes in an Andean Foreland Basin of the Peruvian Amazon: an anthropogenic impact evidence. *Aquatic Geochemistry* **2014**, *20* (6), 613–637.

459. An, Y.-J., Kampbell, D. H., Jeong, S.-W., Jewell, K. P., Masoner, J. R., Impact of geochemical stressors on shallow groundwater quality. *Science of the Total Environment* **2005**, *348* (1), 257–266.

460. Ma, J., Pan, F., He, J., Chen, L., Fu, S., Jia, B., Petroleum pollution and evolution of water quality in the Malian river basin of the Longdong Loess Plateau, northwestern China. *Environmental Earth Science* **2011**, *66*, 1769–1782.

461. Kang, M., Kanno, C. M., Reid, M. C., Zhang, X., Mauzerall, D. L., Celia, M. A., Chen, Y., Onstott, T. C., Direct measurements of methane emissions from abandoned oil and gas wells in Pennsylvania. *Proceedings of the National Academy of Sciences* **2014**, *111* (51), 18173–18177.

462. King, G. E., Valencia, R. L., Environmental risk and well integrity of plugged and abandoned wells. In *SPE Annual Technical Conference and Exhibition*, Society of Petroleum Engineers: **2014**.

463. Harrison, S. S., Evaluating system for ground-water contamination hazards due to gas-well drilling on the glaciated Appalachian Plateau. *GroundWater* **1983**, *21*, 689–700.

464. Harrison, S. S., Contamination of aquifers by overpressuring the annulus of oil and gas wells. *GroundWater* **1985**, *23*, 317–324.

465. Thyne, T. *Review of Phase II Hydrogeologic Study (Prepared for Garfield County)*. **2008**, https://s3.amazonaws.com/propublica/assets/methane/thyne_review.pdf.

466. Warner, N. R., Jackson, R. B., Darrah, T. H., Osborn, S. G., Down, A., Zhao, K., White, A., Vengosh, A., Geochemical evidence for possible natural migration of Marcellus Formation brine to shallow aquifers in Pennsylvania. *Proceedings of the National Academy of Sciences* **2012**, *109* (30), 11961–11966.

467. Darrah, T. H., Vengosh, A., Jackson, R. B., Warner, N. R., Poreda, R. J., Noble gases identify the mechanisms of fugitive gas contamination in drinking-water wells overlying the Marcellus and Barnett shales. *Proceedings of the National Academy of Sciences* **2014**, *111* (39), 14076–14081.

468. Eger, C. K., Vargo, J. S., Prevention: ground water contamination at the Martha Oil Field, Lawrence and Johnson counties, Kentucky. In *Environmental Concerns in the Petroleum Industry*, Testa, M., ed. AAPG: **1989**.

469. McIntosh, J. C., Ferguson, G., Conventional oil – the forgotten part of the water–energy nexus. *Groundwater* **2019**, *57*, 669–677.

470. US General Accounting Office (GAO), *Drinking Water: Safeguards Are Not Preventing Contamination from Injected Oil and Gas Wastes*. US General Accounting Office: **1989**.

471. North Dakota State Government, North Dakota General Statistics. www.dmr.nd.gov/oilgas/stats/statisticsvw.asp.

472. Haghshenas, A., Nasr-El-Din, H. A., Effect of dissolved solids on reuse of produced water at high temperature in hydraulic fracturing jobs. *Journal of Natural Gas Science and Engineering* **2014**, *21*, 316–325.

473. Chang, H., Liu, B.,Yang, B., Yang, X., Guo, C., He, Q., Liang, S., Chen, S., Yang, P., An integrated coagulation-ultrafiltration-nanofiltration process for internal reuse of shale gas flowback and produced water. *Separation and Purification Technology* **2019**, *21*, 310–321.

474. Liang, T., Shao, L., Yao, E., Zuo, J., Liu, X., Zhang, B., Zhou, F., Study on fluid–rock interaction and reuse of flowback fluid for gel fracturing in desert areas. *Geofluids* **2018**, *8948961*.

475. Sun, Y., Wang, D., Tsang, D. C. W., Wang, L., Ok, Y. S., Feng, Y., A critical review of risks, characteristics, and treatment strategies for potentially toxic elements in wastewater from shale gas extraction. *Environmental International* **2019**, *125*, 452–469.

476. Esmaeilirad, N., Terry, C., Kennedy, H., Prior, A., Carlson, K, Recycling fracturing flowback water for use in hydraulic fracturing: influence of organic matter on stability of carboxyl-methyl-cellulose-based fracturing fluids. *SPE Journal* **2016**, *21*, SPE-179723-PA.

477. Menefee, A. H., Ellis, B. R., Wastewater management strategies for sustained shale gas production. *Environmental Research Letters* **2020**, *15* (2), 024001.

478. Liu, D. Li, J., Zou, C., Cui, H., Ni, Y., Liu, J., Wu, W., Zhang, L., Coyte, R., Kondash, A. J., Vengosh, A., Recycling flowback water for hydraulic fracturing in Sichuan Basin, China: implications for gas production, water footprint, and water quality of regenerated flowback water. *Fuel* **2020**, *272*, 117621.

479. Christian-Smith, J., Levy, M. C., Gleick, P. H., Maladaptation to drought: a case report from California, USA. *Sustain Science* **2015**, *10*, 491–501.

480. Kondash, A. J., Redmon, J. H., Lambertini, E., Feinstein, L., Weinthal, E., Cabrales, L., Vengosh, A., The impact of using low-saline oilfield produced water for irrigation on water and soil quality in California. *Science of the Total Environment* **2020**, *733*, 139392.

481. Oetjen, K., Chan, K. E., Gulmark, K., Christensen, J. H., Blotevogel, J., Borch, T., Spear, J. R., Cath, T. Y., Higgins, C. P., Temporal characterization and statistical analysis of flowback and produced waters and their potential for reuse. *Science of the Total Environment* **2018**, *619–620*, 654–664.

482. Miller, H. T., Trivedi, P., Qiu, Y., Sedlacko, E. M., Higgins, C. P., Borch, T., Food crop irrigation with oilfield-produced water suppresses plant immune response. *Environmental Science and Technology Letters* **2019**, *6*, 656–661.

483. Pica, N. E., Carlon, K., Steiner, J. J., Waskom, R., Produced water reuse for irrigation of non-food biofuel crops: effects on switchgrass and rapeseed germination, physiology and biomass yield. *Industrial Crops and Products* **2017**, *100*, 65–76.

484. Sedlacko, E. M., Jahn C E., Heuberger, A. L., Sindt, N. M., Miller, H. M., Borch, T., Blaine, A. C., Cath, T. Y., Higgins, C. P., Potential for beneficial reuse of oil-and-gas-derived produced water in agriculture: physiological and morphological responses in spring wheat (triticum aestivum). *Environ. Toxicology Chemistry* **2019**, *38*, 1756–1769.

485. McLaughlin, M. C., Blotevogel, J., Watson, R. A., Schell, B., Blewett, T. A. Folkerts. E. J., Goss, G. G., Truong, L., Tanguay, R. L., Argueso, J. L., Borch, T., Mutagenicity assessment downstream of oil and gas produced water discharges intended for agricultural beneficial reuse. *Science for Total Environment* **2020**, *715*, 136944.

486. Shariq, L., Health risks associated with arsenic and cadmium uptake in wheat grain irrigated with simulated hydraulic fracturing flowback water. *Journal Environtal Health* **2019**, *81*, E1–E9.

487. Boo, C., Khalil, Y. F., Elimelech, M., Performance evaluation of trimethylamine–carbon dioxide thermolytic draw solution for engineered osmosis. *Journal of Membrane Science* **2015**, *473*, 302–309.

488. Chang, H., Li, T., Liu, B., Vidic, R. D., Elimelech, M., Crittenden, J. C., Potential and implemented membrane-based technologies for the treatment and reuse of flowback and produced water from shale gas and oil plays: a review. *Desalination* **2019**, *455*, 34–57.

489. Shaffer, D. L., Arias Chavez, L. H., Ben-Sasson, M., Romero-Vargas Castrillón, S., Yip, N. Y., Elimelech, M., Desalination and reuse of high-salinity shale gas produced water: drivers, technologies, and future directions. *Environmental Science & Technology* **2013**, *47* (17), 9569–9583.

490. Canada, N. R. Crude oil facts. www.nrcan.gc.ca/science-data/data-analysis/energy-data-analysis/energy-facts/crude-oil-facts/20064#L4.

491. Harkness, J. S., Warner, N. R., Ulrich, A., Millot, R., Kloppmann, W., Ahad, J. M. E., Savard, M. M., Gammon, P., Vengosh, A., Characterization of the boron, lithium, and strontium isotopic variations of oil sands process-affected water in Alberta, Canada. *Applied Geochemistry* **2018**, *90*, 50–62.

492. Renault, S., Lait, C., Zwiazek, J. J., MacKinnon, M., Effect of high salinity tailings waters produced from gypsum treatment of oil sands tailings on plants of the boreal forest. *Environmental Pollution* **1998**, *102* (2), 177–184.

493. Jones, D., Scarlett, A. G., West, C. E., Frank, R. A., Gieleciak, R., Hager, D., Pureveen, J., Tegelaar, E., Rowland, S. J., Elemental and spectroscopic characterization of fractions of an acidic extract of oil sands process water. *Chemosphere* **2013**, *93* (9), 1655–1664.

494. Scarlett, A. G., Reinardy, H. C., Henry, T. B., West, C. E., Frank, R. A., Hewitt, L. M., Rowland, S. J., Acute toxicity of aromatic and non-aromatic fractions of naphthenic acids extracted from oil sands process-affected water to larval zebrafish. *Chemosphere* **2013**, *93* (2), 415–420.

495. Holden, A. A., Haque, S. E., Mayer, K. U., Ulrich, A. C., Biogeochemical processes controlling the mobility of major ions and trace metals in aquitard sediments beneath an oil sand tailing pond: laboratory studies and reactive transport modeling. *Journal of Contaminant Hydrology* **2013**, *151*, 55–67.

496. Gosselin, P., Hrudey, S. E., Naeth, A., Plourde, A., Therrien, R., Van Der Kraak, G., Xu, Z. *Environmental and Health Impactsof Canada's Oil Sands Industry.* Royal Society of Canada: **2010**.

497. Jasechko, S., Gibson, J. J., Jean Birks, S., Yi, Y., Quantifying saline groundwater seepage to surface waters in the Athabasca oil sands region. *Applied Geochemistry* **2012**, *27* (10), 2068–2076.

498. Ahad, J. M. E., Pakdel, H., Savard, M. M., Calderhead, A. I., Gammon, P. R., Rivera, A., Peru, K. M., Headley, J. V., Characterization and quantification of mining-related "naphthenic acids" in groundwater near a major oil sands tailings pond. *Environmental Science & Technology* **2013**, *47* (10), 5023–5030.

499. Frank, R. A., Roy, J. W., Bickerton, G., Rowland, S. J., Headley, J. V., Scarlett, A. G., West, C. E., Peru, K. M., Parrott, J. L., Conly, F. M., Hewitt, L. M., Profiling oil sands mixtures from industrial developments and natural groundwaters for source identification. *Environmental Science & Technology* **2014**, *48* (5), 2660–2670.

500. Headley, J. V., Peru, K. M., Mohamed, M. H., Frank, R. A., Martin, J. W., Hazewinkel, R. R. O., Humphries, D., Gurprasad, N. P., Hewitt, L. M., Muir, D. C. G., Lindeman, D., Strub, R., Young, R. F., Grewer, D. M., Whittal, R. M., Fedorak, P. M., Birkholz, D. A., Hindle, R., Reisdorph, R., Wang, X., Kasperski, K. L., Hamilton, C., Woudneh, M., Wang, G., Loescher, B., Farwell, A., Dixon, D. G., Ross, M., Pereira, A. D. S., King, E., Barrow, M. P., Fahlman, B., Bailey, J., McMartin, D. W., Borchers, C. H., Ryan, C. H., Toor, N. S., Gillis, H. M., Zuin, L., Bickerton, G., McMaster, M., Sverko, E., Shang, D., Wilson, L. D., Wrona, F. J., Chemical fingerprinting of naphthenic acids and oil sands process waters – a review of analytical methods for environmental samples. *Journal of Environmental Science and Health, Part A* **2013**, *48* (10), 1145–1163.
501. Kavanagh, R. J., Burnison, B. K., Frank, R. A., Solomon, K. R., Van Der Kraak, G., Detecting oil sands process-affected waters in the Alberta oil sands region using synchronous fluorescence spectroscopy. *Chemosphere* **2009**, *76* (1), 120–126.
502. Savard, M. M., Ahad, J. M. E., Gammon P., et al., *A Local Test Study Distinguishes Natural from Anthropogenic Groundwater Contaminants near an Athabasca Oil Sands Mining Operation.* Geological Survey of Canada: **2012**.
503. Lari, E., Steinkey, D., Morandi, G., Rasmussen, J. B., Giesy, J. P., Pyle, G. G., Oil sands process-affected water impairs feeding by Daphnia magna. *Chemosphere* **2017**, *175*, 465–472.
504. Lari, E., Wiseman, S., Mohaddes, E., Morandi, G., Alharbi, H., Pyle, G. G., Determining the effect of oil sands process-affected water on grazing behaviour of Daphnia magna, long-term consequences, and mechanism. *Chemosphere* **2016**, *146*, 362–370.
505. Li, C., Fu, L., Stafford, J., Belosevic, M., Gamal El-Din, M., The toxicity of oil sands process-affected water (OSPW): a critical review. *Science of the Total Environment* **2017**, *601–602*, 1785–1802.
506. Lyons, D. D., Morrison, C., Philibert, D. A., Gamal El-Din, M., Tierney, K. B., Growth and recovery of zebrafish embryos after developmental exposure to raw and ozonated oil sands process-affected water. *Chemosphere* **2018**, *206*, 405–413.
507. McQueen, A. D., Kinley, C. M., Hendrikse, M., Gaspari, D. P., Calomeni, A. J., Iwinski, K. J., Castle, J. W., Haakensen, M. C., Peru, K. M., Headley, J. V., Rodgers, J. H., A risk-based approach for identifying constituents of concern in oil sands process-affected water from the Athabasca Oil Sands region. *Chemosphere* **2017**, *173*, 340–350.
508. Miles, S. M., Hofstetter, S., Edwards, T., Dlusskaya, E., Cologgi, D. L., Gänzle, M., Ulrich, A. C., Tolerance and cytotoxicity of naphthenic acids on microorganisms isolated from oil sands process-affected water. *Science of the Total Environment* **2019**, *695*, 133749.
509. Morandi, G. D., Wiseman, S. B., Guan, M., Zhang, X. W., Martin, J. W., Giesy, J. P., Elucidating mechanisms of toxic action of dissolved organic chemicals in oil sands process-affected water (OSPW). *Chemosphere* **2017**, *186*, 893–900.
510. Philibert, D. A., Lyons, D. D., Qin, R., Huang, R., El-Din, M. G., Tierney, K. B., Persistent and transgenerational effects of raw and ozonated oil sands process-affected water exposure on a model vertebrate, the zebrafish. *Science of the Total Environment* **2019**, *693*, 133611.
511. Bauer, A. E., Hewitt, L. M., Parrott, J. L., Bartlett, A. J., Gillis, P. L., Deeth, L. E., Rudy, M. D., Vanderveen, R., Brown, L., Campbell, S. D., Rodrigues, M. R., Farwell, A. J., Dixon, D. G., Frank, R. A., The toxicity of organic fractions from aged oil sands process-affected water to aquatic species. *Science of the Total Environment* **2019**, *669*, 702–710.
512. Gibson, J. J., Yi, Y., Birks, S. J., Isotope-based partitioning of streamflow in the oil sands region, northern Alberta: towards a monitoring strategy for assessing flow sources and water quality controls. *Journal of Hydrology: Regional Studies* **2016**, *5*, 131–148.
513. Fennell, J., Arciszewski, T. J., Current knowledge of seepage from oil sands tailings ponds and its environmental influence in northeastern Alberta. *Science of the Total Environment* **2019**, *686*, 968–985.
514. Schuster, J. K., Harner, T., Su, K., Mihele, C., Eng, A., First results from the oil sands passive air monitoring network for polycyclic aromatic compounds. *Environmental Science & Technology* **2015**, *49* (5), 2991–2998.
515. Hsu, Y.-M., Harner, T., Li, H., Fellin, P., PAH measurements in air in the Athabasca oil sands region. *Environmental Science & Technology* **2015**, *49* (9), 5584–5592.
516. Parajulee, A., Wania, F., Evaluating officially reported polycyclic aromatic hydrocarbon emissions in the Athabasca oil sands region with a multimedia fate model. *Proceedings of the National Academy of Sciences* **2014**, *111* (9), 3344–3349.
517. Ahad, J. M. E., Gammon, P. R., Gobeil, C., Jautzy, J., Krupa, S., Savard, M. M., Studabaker, W. B., Evaporative emissions from tailings ponds are not likely an important source of airborne PAHs in the Athabasca oil sands region. *Proceedings of the National Academy of Sciences* **2014**, *111* (24), E2439–E2439.

518. Landis, M. S., Berryman, S. D., White, E. M., Graney, J. R., Edgerton, E. S., Studabaker, W. B., Use of an epiphytic lichen and a novel geostatistical approach to evaluate spatial and temporal changes in atmospheric deposition in the Athabasca oil sands region, Alberta, Canada. *Science of the Total Environment* **2019**, *692*, 1005–1021.

519. Landis, M. S., Studabaker, W. B., Pancras, J. P., Graney, J. R., White, E. M., Edgerton, E. S., Source apportionment of ambient fine and coarse particulate matter polycyclic aromatic hydrocarbons at the Bertha Ganter-Fort McKay community site in the oil sands region of Alberta, Canada. *Science of the Total Environment* **2019**, *666*, 540–558.

520. Landis, M. S., Studabaker, W. B., Patrick Pancras, J., Graney, J. R., Puckett, K., White, E. M., Edgerton, E. S., Source apportionment of an epiphytic lichen biomonitor to elucidate the sources and spatial distribution of polycyclic aromatic hydrocarbons in the Athabasca Oil Sands Region, Alberta, Canada. *Science of the Total Environment* **2019**, *654*, 1241–1257.

521. Studabaker, W. B., Puckett, K. J., Percy, K. E., Landis, M. S., Determination of polycyclic aromatic hydrocarbons, dibenzothiophene, and alkylated homologs in the lichen Hypogymnia physodes by gas chromatography using single quadrupole mass spectrometry and time-of-flight mass spectrometry. *Journal of Chromatography A* **2017**, *1492*, 106–116.

522. Jautzy, J., Ahad, J. M. E., Gobeil, C., Savard, M. M., Century-long source apportionment of PAHs in Athabasca oil sands region lakes using diagnostic ratios and compound-specific carbon isotope signatures. *Environmental Science & Technology* **2013**, *47* (12), 6155–6163.

523. US Environmental Protection Agency, *Detailed Study of the Petroleum Refining Category – 2019 Report.* US Environmental Protection Agency: **2019**.

524. Al Zarooni, M., Elshorbagy, W., Characterization and assessment of Al Ruwais refinery wastewater. *Journal of Hazardous Materials* **2006**, *136* (3), 398–405.

525. Coelho, A., Castro, A. V., Dezotti, M., Sant'Anna, G. L., Treatment of petroleum refinery sourwater by advanced oxidation processes. *Journal of Hazardous Materials* **2006**, *137* (1), 178–184.

526. Munirasu, S., Haija, M. A., Banat, F., Use of membrane technology for oil field and refinery produced water treatment – a review. *Process Safety and Environmental Protection* **2016**, *100*, 183–202.

527. Wake, H., Oil refineries: a review of their ecological impacts on the aquatic environment. *Estuarine, Coastal and Shelf Science* **2005**, *62* (1), 131–140.

528. Board, T. R., Council, N. R., *Oil in the Sea III: Inputs, Fates, and Effects.* The National Academies Press: **2003**.

529. Chen, J., Zhang, W., Wan, Z., Li, S., Huang, T., Fei, Y., Oil spills from global tankers: status review and future governance. *Journal of Cleaner Production* **2019**, *227*, 20–32.

530. Barron, M. G., Vivian, D. N., Heintz, R. A., Yim, U. H., Long-term ecological impacts from oil spills: comparison of Exxon Valdez, Hebei Spirit, and Deepwater Horizon. *Environmental Science & Technology* **2021**, *54* (11), 6456–6467.

531. Patterson, L. A., Konschnik, K. E., Wiseman, H., Fargione, J., Maloney, K. O., Kiesecker, J., Nicot, J.-P., Baruch-Mordo, S., Entrekin, S., Trainor, A., Saiers, J. E., Unconventional oil and gas spills: risks, mitigation priorities, and state reporting requirements. *Environmental Science & Technology* **2017**, *51* (5), 2563–2573.

532. Duffy, J. J., Peake, E., Mohtadi, M. F., Oil spills on land as potential sources of groundwater contamination. *Environmental International* **1980**, *3*, 107–120.

533. Delin, G. N., Essaid, H. I., Cozzarelli, I. M., Lahvis, M. H., Bekins, B. *Ground Water Contamination by Crude Oil near Bemidji, Minnesota.* US Geological Survey, Fact Sheet 084-98: **1998**.

534. Cozzarelli, I. M., Baedecker, M. J., Eganhouse, R. P., Goerlitz, D. F., The geochemical evolution of low-molecular-weight organic acids derived from the degradation of petroleum contaminants in groundwater. *Geochimica et Cosmochimica Acta* **1994**, *58* (2), 863–877.

535. Bennett, P. C., Siegel, D. I., Baedecker, M. J., Hult, M. F., Crude oil in a shallow aquifer, 1 – Aquifer characterization and hydrogeochemical controls on inorganic solutes. *Applied Geochemistry* **1993**, *8*, 529–549.

536. Dillard, L. A., Essaid, H. I., Herkelrath, W. N., Multiphase flow modeling of a crude-oil spill site with a bimodal permeability distribution. *Water Resources Research* **1997**, *33* (7), 1617–1632.

537. Ng, G. H. C., Bekins, B. A., Cozzarelli, I. M., Baedecker, M. J., Bennett, P. C., Amos, R. T., A mass balance approach to investigating geochemical controls on secondary water quality impacts at a crude oil spill site near Bemidji, MN. *Journal of Contaminant Hydrology* **2014**, *164*, 1–15.

538. Ziegler, B. A., McGuire, J. T., Cozzarelli, I. M., Rates of As and trace-element mobilization caused by Fe reduction in mixed BTEX-ethanol experimental plumes. *Environmental Science & Technology* **2015**, *49* (22), 13179–13189.

539. Fish, F. United States: Petroleum Coke, Total Production. www.factfish.com/statistic-country/united%20states/petroleum%20coke%2C%20total%20production.

540. Nesbitt, J. A., Lindsay, M. B. J., Vanadium geochemistry of oil sands fluid petroleum coke. *Environmental Science & Technology* **2017**, *51* (5), 3102–3109.

541. Schlesinger, W. H., Klein, E. M., Vengosh, A., Global biogeochemical cycle of vanadium. *Proceedings of the National Academy of Sciences* **2017**, *114* (52), E11092–E11100.

542. Wilhelm, S. M., Liang, L., Cussen, D., Kirchgessner, D. A., Mercury in crude oil processed in the United States (2004). *Environmental Science & Technology* **2007**, *41* (13), 4509–4514.

543. Mojammal, A. H. M., Back, S.-K., Seo, Y.-C., Kim, J.-H., Mass balance and behavior of mercury in oil refinery facilities. *Atmospheric Pollution Research* **2019**, *10* (1), 145–151.

544. US Environmental Protection Agency, Mercury Emissions: The Global Context. www.epa.gov/international-cooperation/mercury-emissions-global-context.

545. Barwise, A. J. G., Role of nickel and vanadium in petroleum classification. *Energy & Fuels* **1990**, *4* (6), 647–652.

546. Moreno, T., Querol, X., Alastuey, A., Gibbons, W., Identification of FCC refinery atmospheric pollution events using lanthanoid- and vanadium-bearing aerosols. *Atmospheric Environment* **2008**, *42* (34), 7851–7861.

547. Soldi, T., Riolo, C., Alberti, G., Gallorini, M., Peloso, G. F., Environmental vanadium distribution from an industrial settlement. *Science of the Total Environment* **1996**, *181* (1), 45–50.

548. Mitchell, R. B., *Intentional Oil Pollution at Sea: Environmental Policy and Treaty Compliance*. MIT Press: **1994**.

549. International Maritime Organization, International Convention for the Prevention of Pollution from Ships (MARPOL). www.imo.org/en/About/Conventions/ListOfConventions/Pages/International-Convention-for-the-Prevention-of-Pollution-from-Ships-(MARPOL).aspx.

550. US Environmental Protection Agency, Summary of the Oil Pollution Act. www.epa.gov/laws-regulations/summary-oil-pollution-act.

551. Cohen, S., The use of strategic planning, information, and analysis in environmental policy making and management. In *The Oxford Handbook of US Environmental Policy*, Sheldon Kamieniecki, S., Kraft, M. E., eds. Oxford University Press: **2013**.

552. Büthe, T., Mattli, W., *The New Global Rulers: The Privatization of Regulation in the World Economy*. Princeton University Press: **2011**.

553. Frynas, J. G., Corporate social responsibility or government regulation? Evidence on oil spill prevention. *Ecology and Society* **2012**, *17*, 4.

554. O'Rourke, D., Connolly, S., Just oil? The distribution of environmental and social impacts of oil production and consumption *Annual Review of Environment and Resources* **2003**, *28*, 587–617.

555. Council, G. R. *Produced Water Report: Regulations, Current Practices, and Reserach Needs*. Groundwater Rptection Council: **2019**.

556. Lee, M. EPA may let oil waste in waterways. Is the public at risk? www.eenews.net/stories/1061525917.

557. Yergin, D., *The Prize: The Epic Question for Oil, Money and Power*. Simon & Schuster: **1991**.

558. Bradshaw, M. J., Boersma, T., *Natural Gas*. Polity Press: **2020**.

559. Evans, P. C., Farima, M. F. *The Age of Gas & The Power Networks*. General Electric Company: **2013**.

560. Stern, J. P., *The Future of Russian Gas and Gazprom*. Oxford University Press: **2005**.

561. Lewis, P. A Soviet Project Tempts Europe. www.nytimes.com/1982/05/30/business/a-soviet-project-tempts-europe.html.

562. China National Petroleum Corporation, Central Asia–China Gas Pipeline. www.cnpc.com.cn/en/CentralAsia/CentralAsia_index.shtml.

563. Xu, M., Aizhu, C., Astakhova, O. Landmark Siberian gas to test CNPC's marketing mettle in China's backwaters. www.reuters.com/places/russia/article/us-china-russia-gas-pipeline/landmark-siberian-gas-to-test-cnpcs-marketing-mettle-in-chinas-backwaters-idUSKBN1Y30JH.

564. International Energy Agency (IEA), *WEO-2011 Special Report: Are We Entering a Golden Age?* International Energy Agency (IEA): **2011**.

565. Wu, K., China's energy security: oil and gas. *Energy Policy* **2014**, *73*, 4–11.

566. US Energy Information Administration (EIA), *Technically Recoverable Shale Oil and Shale Gas Resources: China*. US Department of Energy: **2015**.

567. Dong, D., Wang, Y., Li, X., Zou, C., Guan, Q., Zhang, C., Huang, J., Wang, S., Wang, H., Liu, H., Bai, W., Liang, F., Lin, W., Zhao, Q., Liu, D., Qiu, Z., Breakthrough and prospect of shale gas exploration and development in China. *Natural Gas Industry B* **2016**, *3* (1), 12–26.

568. Dong, D., Zou, C., Dai, J., Huang, S., Zheng, J., Gong, J., Wang, Y., Li, X., Guan, Q., Zhang, C., Huang, J., Wang, S., Liu, D., Qiu, Z., Suggestions on the development strategy of shale gas in China. *Journal of Natural Gas Geoscience* **2016**, *1* (6), 413–423.

569. Speight, J. G., *Natural Gas: A Basic Handbook*. Gulf Pub. Co.: **2007**.

570. Speight, J. G., *Handbook of Natural Gas Analysis*. John Wiley & Sons: **2018**.

571. Faramawy, S., Zaki, T., Sakr, A. A. E., Natural gas origin, composition, and processing: a review. *Journal of Natural Gas Science and Engineering* **2016**, *34*, 34–54.

572. Lollar, B. S., Lacrampe-Couloume, G., Voglesonger, K., Onstott, T. C., Pratt, L. M., Slater, G. F., Isotopic signatures of CH4 and higher hydrocarbon gases from Precambrian Shield sites: a model for abiogenic polymerization of hydrocarbons. *Geochimica et Cosmochimica Acta* **2008**, *72* (19), 4778–4795.

573. Etiope, G., Schoell, M., Abiotic gas: atypical but not rare. *Elements* **2014**, *10*, 291–296.

574. US Energy Information Administration (EIA), Coalbed Methane Production. www.eia.gov/dnav/ng/ng_prod_coalbed_s1_a.htm.

575. Schoell, M., The hydrogen and carbon isotopic composition of methane from natural gases of various origins. *Geochimica et Cosmochimica Acta* **1980**, *44* (5), 649–661.

576. Schoell, M., Recent advances in petroleum isotope geochemistry. *Organic Geochemistry* **1984**, *6*, 645–663.

577. Schoell, M., Multiple origins of methane in the Earth. *Chemical Geology* **1988**, *71* (1), 1–10.

578. Whiticar, M. J., Faber, E., Schoell, M., Biogenic methane formation in marine and freshwater environments: CO2 reduction vs. acetate fermentation – isotope evidence. *Geochimica et Cosmochimica Acta* **1986**, *50* (5), 693–709.

579. Milkov, A. V., Etiope, G., Revised genetic diagrams for natural gases based on a global dataset of >20,000 samples. *Organic Geochemistry* **2018**, *125*, 109–120.

580. Martini, A. M., Walter, L. M., Budai, J. M., Ku, T. C. W., Kaiser, C. J., Schoell, M., Genetic and temporal relations between formation waters and biogenic methane: Upper Devonian Antrim Shale, Michigan Basin, USA. *Geochimica et Cosmochimica Acta* **1998**, *62* (10), 1699–1720.

581. Kinnaman, F. S., Valentine, D. L., Tyler, S. C., Carbon and hydrogen isotope fractionation associated with the aerobic microbial oxidation of methane, ethane, propane and butane. *Geochimica et Cosmochimica Acta* **2007**, *71* (2), 271–283.

582. Schoell, M., Genetic characterization of natural gases. *American Association of Petroleum Geologists Bulletin* **1983**, *67*, 2225–2238.

583. Bernard, B., Brooks, J. M., Sackett, W. M., A geochemical model for characterization of hydrocarbon gas sources in marine sediments. In *9th Annual Offshore Technology Conference*. **1977**, pp. 435–438, https://doi.org/10.4043/2934-MS.

584. Milkov, A. V., Faiz, M., Etiope, G., Geochemistry of shale gases from around the world: composition, origins, isotope reversals and rollovers, and implications for the exploration of shale plays. *Organic Geochemistry* **2020**, *143*, 103997.

585. Golding, S. D., Boreham, C. J., Esterle, J. S., Stable isotope geochemistry of coal bed and shale gas and related production waters: a review. *International Journal of Coal Geology* **2013**, *120*, 24–40.

586. Milkov, A. V., Worldwide distribution and significance of secondary microbial methane formed during petroleum biodegradation in conventional reservoirs. *Organic Geochemistry* **2011**, *42* (2), 184–207.

587. Belyadi, H., Fathi, E., Belyadi, F., Hydraulic fracturing fluid systems. In *Hydraulic Fracturing in Unconventional Reservoirs*, Belyadi, H., Fathi, E., Belyadi, F., eds. Gulf Professional Publishing: **2017**, pp. 49–72.

588. US Energy Information Administration (EIA), Oil and Gas Supply Module. www.eia.gov/outlooks/aeo/assumptions/pdf/oilgas.pdf.

589. Liu, P., Feng, Y., Zhao, L., Li, N., Luo, Z., Technical status and challenges of shale gas development in Sichuan Basin, China. *Petroleum* **2015**, *1* (1), 1–7.

590. Ma, X., Xie, J., The progress and prospects of shale gas exploration and development in southern Sichuan Basin, SW China. *Petroleum Exploration and Development* **2018**, *45* (1), 172–182.

591. Wang, S., Shale gas exploitation: status, problems and prospect. *Natural Gas Industry B* **2018**, *5* (1), 60–74.

592. Zhao, Q., Yang, S., Wang, H., Wang, N., Liu, D., Liu, H., Zang, H., Prediction of marine shale gas production in South China based on drilling workload analysis. *Natural Gas Industry B* **2016**, *3* (6), 545–551.

593. US Energy Information Administration (EIA), *The Distribution of US Oil and Natural Gas Wells by Production Rate*. US Department of Energy: **2019**.

594. Gallegos, T. J., Varela, B. A., Haines, S. S., Engle, M. A., Hydraulic fracturing water use variability in the United States and potential environmental implications. *Water Resources Research* **2015**, *51*, 5839–5845.

595. Makhanov, K., Habibi, A., Dehghanpour, H., Kuru, E., Liquid uptake of gas shales: A workflow to estimate water loss during shut-in periods after fracturing operations. *Journal of Unconventional Oil and Gas Resources* **2014**, *7*, 22–32.

596. Lan, Q., Ghanbari, E., Dehghanpour, H., Hawkes, R., Water loss versus soaking time: spontaneous imbibition in tight rocks. *Energy Technology* **2014**, *2*, 1033–1039.

597. Xu, Y., Adefidipe, O. A., Dehghanpour, H., Estimating fracture volume using flowback data from the Horn River Basin: a material balance approach. *Journal of Natural Gas Science Engineering* **2015**, *25*, 253–270.

598. Ghanbari, E. a. D., H., The fate of fracturing water: a field and simulation study. *Fuel* **2016**, *163*, 282–294.

599. Yu, M., Weinthal, E., Patiño-Echeverri, D., Deshusses, M. A., Zou, C. Ni, Y., Vengosh, A., Water availability for shale gas development in Sichuan Basin, China. *Environmental Science & Technology* **2016**, *50*, 2837–2845.

600. Yang, B., Zhang, H., Kang, Y., You, L., She, J., Wang, K., Chen, Z., In situ sequestration of a hydraulic fracturing fluid in Longmaxi Shale gas formation in the Sichuan Basin. *Energy Fuels* **2019**, *33*, 6983–6994.

601. Binazadeh, M., Xu, M., Zolfaghari, A., Dehghanpour, H., Effect of electrostatic interactions on water uptake of gas shales: the interplay of solution ionic strength and electrostatic double layer. *Energy Fuels* **2016**, *30*, 992–1001.

602. Fakcharoenphol, P., Torcuk, M., Kazemi, H. Wu, Y-S., Effect of shut-in time on gas flow rate in hydraulic fractured shale reservoirs. *Journal of Natural Gas Science and Engineering* **2016**, *32*, 109–121.

603. Singh, H., A critical review of water uptake by shales. *Journal of Natural Gas Science and Engineering* **2016**, *34*, 751–766.

604. Fan, K., Li, Y., Elsworth, D., Dong, M., Yin, C., Li, Y., Chen, Z., Three stages of methane adsorption capacity affected by moisture content. *Fuel* **2018**, *231*, 352–360.

605. Zhou, J., Mao, Q., Luo, K. H., Effects of moisture and salinity on methane adsorption in kerogen: a molecular simulation study. *Energy Fuels* **2019**, *33*, 5368–5376.

606. Clark, C. E., Horner, R. M., Harto, C. B., Life cycle water consumption for shale gas and conventional natural gas. *Environmental Science & Technology* **2013**, *47* (20), 11829–11836.

607. Guo, T. L., Discovery and characteristics of the Fuling Shale gas field and its enlightenment and thinking. *Earth Science Frontiers* **2016**, *23*, 29–43.

608. US Energy Information Administration (EIA), The Basics of Underground Natural Gas Storage. www.eia.gov/naturalgas/storage/basics/.

609. US Energy Information Administration (EIA), What is US Electricity Generation by Energy Source? www.eia.gov/tools/faqs/faq.php?id=427&t=3.

610. Stringfellow, W. T., Domen, J. K., Camarillo, M. K., Sandelin, W. L., Borglin, S., Physical, chemical, and biological characteristics of compounds used in hydraulic fracturing. *Journal of Hazardous Materials* **2014**, *275*, 37–54.

611. Chen, H., Carter, K. E., Characterization of the chemicals used in hydraulic fracturing fluids for wells located in the Marcellus Shale Play. *Journal of Environmental Management* **2017**, *200*, 312–324.

612. Kassotis, C. D., Harkness, J. S., Vo, P. H., Vu, D. C., Hoffman, K., Cinnamon, K. M., Cornelius-Green, J. N., Vengosh, A., Lin, C.-H., Tillitt, D. E., Kruse, R. L., McElroy, J. A., Nagel, S. C., Endocrine disrupting activities and geochemistry of water resources associated with unconventional oil and gas activity. *Science of the Total Environment* **2020**, *748*, 142236.

613. Ellsworth, W. L., Injection-induced earthquakes. *Science* **2013**, *341* (6142), 1225942.

614. Dresel, P. E., Rose, A. W., *Chemistry and Origin of Oil and Gas Well Brines in Western Pennsylvania*. Pennsylvania Geological Survey: **2010**.

615. Osborn, S. G., McIntosh, J. C., Chemical and isotopic tracers of the contribution of microbial gas in Devonian organic-rich shales and reservoir sandstones, northern Appalachian Basin. *Applied Geochemistry* **2010**, *25*, 456–471.

616. Osborn, S. G., McIntosh, J. C., Hanor, J., Biddulph, Iodine-129, 87Sr/86Sr, and trace elemental geochemistry of Northern Appalachian Basin brines: evidence for basinal-scale fluid migration and clay mineral diagenesis. *American Journal of Science* **2012**, *312*, 263–287.

617. Darrah, T. H., Jackson, R. B., Vengosh, A., Warner, N. R., Whyte, C. J., Walsh, T. B., Kondash, A. J., Poreda, R. J., The evolution of Devonian hydrocarbon gases in shallow aquifers of the northern Appalachian Basin: Insights from integrating noble gas and hydrocarbon geochemistry. *Geochimica et Cosmochimica Acta* **2015**, *170*, 321–355.

618. Darrah, T. H., Vengosh, A., Jackson, R. B., Warner, N. R., Poreda, R. J., Noble gases identify the mechanisms of fugitive gas contamination in drinking-water wells overlying the Marcellus and Barnett Shales. *Proceedings of the National Academy of Sciences of the United States of America* **2014**, *111* (39), 14076–14081.

619. Darrah, T. H., Jackson, R. B., Vengosh, A., Warner, N. R., Poreda, R. J., Noble gases: a new technique for fugitive gas investigation in groundwater. *Groundwater* **2015**, *53* (1), 23–28.
620. Jackson, R. B., Vengosh, A., Darrah, T. H., Warner, N. R., Down, A., Poreda, R. J., Osborn, S. G., Zhao, K. G., Karr, J. D., Increased stray gas abundance in a subset of drinking water wells near Marcellus Shale gas extraction. *Proceedings of the National Academy of Sciences of the United States of America* **2013**, *110* (28), 11250–11255.
621. Kreuzer, R. L., Darrah, T. H., Grove, B. S., Moore, M. T., Warner, N. R., Eymold, W. K., Whyte, C. J., Mitra, G., Jackson, R. B., Vengosh, A., Poreda, R. J., Structural and hydrogeological controls on hydrocarbon and brine migration into drinking water aquifers in southern New York. *Groundwater* **2018**, *56* (2), 225–244.
622. McIntosh, J. C., Hendry, M. J., Ballentine, C., Haszeldine, R. S., Mayer, B., Etiope, G., Elsner, M., Darrah, T. H., Prinzhofer, A., Osborn, S., Stalker, L., Kuloyo, O., Lu, Z. T., Martini, A., Lollar, B. S., A Critical review of state-of-the-art and emerging approaches to identify fracking-derived gases and associated contaminants in aquifers. *Environmental Science & Technology* **2019**, *53* (3), 1063–1077.
623. Rosenblum, J., Nelson, A. W., Ruyle, B., Schultz, M. K., Ryan, J. N., Linden, K. G., Temporal characterization of flowback and produced water quality from a hydraulically fractured oil and gas well. *Science of the Total Environment* **2017**, *596–597*, 369–377.
624. Rowan, E. L., Engle, M. A., Kraemer, T. F., Schroeder, K. T., Hammack, R. W., Doughten, M. W., Geochemical and isotopic evolution of water produced from Middle Devonian Marcellus Shale gas wells, Appalachian Basin, Pennsylvania. *AAPG Bulletin* **2015**, *99*, 181–206.
625. Balashov, V. N., Engelder, T., Gu, X., Fantle, M. S., Brantley, S. L., A model describing flowback chemistry changes with time after Marcellus Shale hydraulic fracturing. *AAPG Bulletin* **2015**, *99*, 143–154.
626. Stewart, B. W., Chapman, E. C., Capo, R. C., Johnson, J. D., Graney, J. R., Kirby, C. S., Schroeder, K. T., Origin of brines, salts and carbonate from shales of the Marcellus Formation: evidence from geochemical and Sr isotope study of sequentially extracted fluids. *Applied Geochemistry* **2015**, *60*, 78–88.
627. Tieman, Z. G., Stewart, B. W., Capo, R. C., Phan, T. T., Lopano, C. L., Hakala, J. A., Barium isotopes track the source of dissolved solids in produced water from the unconventional Marcellus Shale Gas Play. *Environmental Science & Technology* **2020**, *54* (7), 4275–4285.
628. Gao, J., Zou, C., Li, W., Ni, Y., Liao, F., Yao, L., Sui, J., Vengosh, A., Hydrochemistry of flowback water from Changning Shale gas field and associated shallow groundwater in Southern Sichuan Basin, China: implications for the possible impact of shale gas development on groundwater quality. *Science of the Total Environment* **2020**, *713*, 136591.
629. Vidic, R. D., Brantley, S. L., Vandenbossche, J. M., Yoxtheimer, D., Abad, J. D., Impact of Shale Gas Development on Regional Water Quality. *Science* **2013**, *340* (6134), 1235009.
630. Huang, T., Pang, Z., Li, Z., Li, Y., Hao, Y., A framework to determine sensitive inorganic monitoring indicators for tracing groundwater contamination by produced formation water from shale gas development in the Fuling Gasfield, SW China. *Journal of Hydrology* **2020**, *581*, 124403.
631. Wang, B., Xiong, M., Wang, P., Shi, B., Chemical characterization in hydraulic fracturing flowback and produced water (HF-FPW) of shale gas in Sichuan of China. *Environmental Science and Pollution Research* **2020**, *27* (21), 26532–26542.
632. Jacobs, R. P. W. M., Grant, R. O. H., Kwant J., Marquenie J. M., Mentzer E., The composition of produced water from Shell operated oil and gas production in the North Sea. In *Produced Water*, Ray, J. P., Engelhardt, R., eds. Springer: **1992**.
633. Kondash, A. J., Warner, N. R., Lahav, O., Vengosh, A., Radium and barium removal through blending hydraulic fracturing fluids with acid mine drainage. *Environmental Science & Technology* **2014**, *48* (2), 1334–1342.
634. Balashov, V. N., Engelder, T., Gu, X., Fantle, M. S., Brantley, S. L., A model describing flowback chemistry changes with time after Marcellus Shale hydraulic fracturing. *AAPG Bulletin* **2015**, *99* (1), 143–154.
635. Webster, I. T., Hancock, G. J., Murray, A. S., Modelling the effect of salinity on radium desorption from sediments. *Geochimica et Cosmochimica Acta* **1995**, *59* (12), 2469–2476.
636. Sturchio, N. C., Banner, J. L., Binz, C. M., Heraty, L. B., Musgrove, M., Radium geochemistry of ground waters in Paleozoic carbonate aquifers, midcontinent, USA. *Applied Geochemistry* **2001**, *16* (1), 109–122.
637. Lüning, S., Kolonic, S., Uranium spectral gamma-ray response as a proxy for organic richness in black shales: applicability and limitations. *Journal of Petroleum Geology* **2003**, *26* (2), 153–174.
638. Liu, B., Mastalerz, M., Schieber, J., Teng, J., Association of uranium with macerals in marine black shales: insights from the Upper Devonian New Albany Shale, Illinois Basin. *International Journal of Coal Geology* **2020**, *217*, 103351.

639. Jew, A. D., Besançon, C. J., Roycroft, S. J., Noel, V. S., Bargar, J. R., Brown, G. E., Chemical speciation and stability of uranium in unconventional shales: impact of hydraulic fracture fluid. *Environmental Science & Technology* **2020**, *54* (12), 7320–7329.

640. Wang, G., Jin, Z., Liu, G., Liu, Q., Liu, Z., Wang, H., Liang, X., Jiang, T., Wang, R., Geological implications of gamma ray (GR) anomalies in marine shales: a case study of the Ordovician-Silurian Wufeng-Longmaxi succession in the Sichuan Basin and its periphery, Southwest China. *Journal of Asian Earth Sciences* **2020**, *199*, 104359.

641. Vengosh, A., Hirschfeld, D., Vinson, D., Dwyer, G., Raanan, H., Rimawi, O., Al-Zoubi, A., Akkawi, E., Marie, A., Haquin, G., Zaarur, S., Ganor, J., High naturally occurring radioactivity in fossil groundwater from the Middle East. *Environmental Science & Technology* **2009**, *43* (6), 1769–1775.

642. Tasker, T. L., Warner, N. R., Burgos, W. D., Geochemical and isotope analysis of produced water from the Utica/Point Pleasant Shale, Appalachian Basin. *Environmental Science: Processes & Impacts* **2020**, *22* (5), 1224–1232.

643. Szczuka, A., Parker, K. M., Harvey, C., Hayes, E., Vengosh, A., Mitch, W. A., Regulated and unregulated halogenated disinfection byproduct formation from chlorination of saline groundwater. *Water Research* **2017**, *122*, 633–644.

644. US Environmental Protection Agency, *Detailed Study of the Centralized Waste Treatment Point Source Category for Facilities Managing Oil and Gas Extraction Wastes*. US Environmental Protection Agency: **2018**.

645. Ferrar, K. J., Michanowicz, D. R., Christen, C. L., Mulcahy, N., Malone, S. L., Sharma, R. K., Assessment of effluent contaminants from three facilities discharging Marcellus Shale wastewater to surface waters in Pennsylvania. *Environmental Science & Technology* **2013**, *47* (7), 3472–3481.

646. Landis, M. S., Kamal, A. S., Kovalcik, K. D., Croghan, C., Norris, G. A., Bergdale, A., The impact of commercially treated oil and gas produced water discharges on bromide concentrations and modeled brominated trihalomethane disinfection byproducts at two downstream municipal drinking water plants in the upper Allegheny River, Pennsylvania, USA. *Science of the Total Environment* **2016**, *542*, 505–520.

647. Geeza, T. J., Gillikin, D. P., McDevitt, B., Van Sice, K., Warner, N. R., Accumulation of Marcellus Formation oil and gas wastewater metals in freshwater mussel shells. *Environmental Science & Technology* **2018**, *52* (18), 10883–10892.

648. Brantley, S. L., Yoxtheimer, D., Arjmand, S., Grieve, P., Vidic, R., Pollak, J., Llewellyn, G. T., Abad, J., Simon, C., Water resource impacts during unconventional shale gas development: the Pennsylvania experience. *International Journal of Coal Geology* **2014**, *126*, 140–156.

649. Maloney, K. O., Baruch-Mordo, S., Patterson, L. A., Nicot, J.-P., Entrekin, S. A., Fargione, J. E., Kiesecker, J. M., Konschnik, K. E., Ryan, J. N., Trainor, A. M., Saiers, J. E., Wiseman, H. J., Unconventional oil and gas spills: materials, volumes, and risks to surface waters in four states of the US *Science of the Total Environment* **2017**, *581–582*, 369–377.

650. Rozell, D. J., Reaven, S. J., Water pollution risk associated with natural gas extraction from the Marcellus Shale. *Risk Analysis* **2012**, *32* (8), 1382–1393.

651. Jackson, R. B., Lowry, E. R., Pickle, A., Kang, M., DiGiulio, D., Zhao, K., The depths of hydraulic fracturing and accompanying water use across the United States. *Environmental Science & Technology* **2015**, *49* (15), 8969–8976.

652. DiGiulio, D. C., Jackson, R. B., Impact to underground sources of drinking water and domestic wells from production well stimulation and completion practices in the Pavillion, Wyoming, Field. *Environmental Science & Technology* **2016**, *50* (8), 4524–4536.

653. Drollette, B. D., Hoelzer, K., Warner, N. R., Darrah, T. H., Karatum, O., O'Connor, M. P., Nelson, R. K., Fernandez, L. A., Reddy, C. M., Vengosh, A., Jackson, R. B., Elsner, M., Plata, D. L., Elevated levels of diesel range organic compounds in groundwater near Marcellus gas operations are derived from surface activities. *Proceedings of the National Academy of Sciences* **2015**, *112* (43), 13184–13189.

654. Llewellyn, G. T., Dorman, F., Westland, J. L., Yoxtheimer, D., Grieve, P., Sowers, T., Humston-Fulmer, E., Brantley, S. L., Evaluating a groundwater supply contamination incident attributed to Marcellus Shale gas development. *Proceedings of the National Academy of Sciences* **2015**, *112* (20), 6325–6330.

655. Dilmore, R. M., Sams, J. I., Glosser, D., Carter, K. M., Bain, D. J., Spatial and temporal characteristics of historical oil and gas wells in Pennsylvania: implications for new shale gas resources. *Environmental Science & Technology* **2015**, *49* (20), 12015–12023.

656. King, G. E., Valencia, R. L., Environmental risk and well integrity of plugged and abandoned wells. In *SPE Annual Technical Conference and Exhibition*. Society of Petroleum Engineers: **2014**.

657. Chapman, E. C., Capo, R. C., Stewart, B. W., Hedin, R. S., Weaver, T. J., Edenborn, H. M., Strontium isotope quantification of siderite, brine and acid mine drainage contributions to abandoned gas well discharges in the Appalachian Plateau. *Applied Geochemistry* **2013**, *31*, 109–118.

658. Skuce, M., Longstaffe, F. J., Carter, T. R., Potter, J., Isotopic fingerprinting of groundwaters in southwestern Ontario: Applications to abandoned well remediation. *Applied Geochemistry* **2015**, *58*, 1–13.

659. Illinois Department of Public Health, Methane in Groundwater. www.dph.illinois.gov/topics-services/envir onmental-health-protection/private-water/methane-groundwater.

660. Osborn, S. G., Vengosh, A., Warner, N. R., Jackson, R. B., Methane contamination of drinking water accompanying gas-well drilling and hydraulic fracturing. *Proceedings of the National Academy of Sciences* **2011**, *108* (20), 8172–8176.

661. Sherwood, O. A., Rogers, J. D., Lackey, G., Burke, T. L., Osborn, S. G., Ryan, J. N., Groundwater methane in relation to oil and gas development and shallow coal seams in the Denver-Julesburg Basin of Colorado. *Proceedings of the National Academy of Sciences* **2016**, *113* (30), 8391–8396.

662. Molofsky, L. J., Connor, J. A., Wylie, A. S., Wagner, T., Farhat, S. K., Evaluation of methane sources in groundwater in northeastern Pennsylvania. *Groundwater* **2013**, *51* (3), 333–349.

663. Siegel, D. I., Azzolina, N. A., Smith, B. J., Perry, A. E., Bothun, R. L., Methane concentrations in water wells unrelated to proximity to existing oil and gas wells in northeastern Pennsylvania. *Environmental Science & Technology* **2015**, *49* (7), 4106–4112.

664. Barth-Naftilan, E., Sohng, J., Saiers, J. E., Methane in groundwater before, during, and after hydraulic fracturing of the Marcellus Shale. *Proceedings of the National Academy of Sciences* **2018**, *115* (27), 6970–6975.

665. Nicot, J.-P., Larson, T., Darvari, R., Mickler, P., Slotten, M., Aldridge, J., Uhlman, K., Costley, R., Controls on methane occurrences in shallow aquifers overlying the Haynesville Shale Gas Field, East Texas. *Groundwater* **2017**, *55* (4), 443–454.

666. Nicot, J.-P., Larson, T., Darvari, R., Mickler, P., Uhlman, K., Costley, R., Controls on methane occurrences in aquifers overlying the Eagle Ford Shale Play, South Texas. *Groundwater* **2017**, *55* (4), 455–468.

667. Nicot J. P., M. P., Larson T., Castro M. C., Darvari R., Uhlman K., Costley, R., Methane occurrences in aquifers overlying the Barnett Shale Play with a focus on Parker County, Texas. *GroundWater* **2017**, *55*, 469–481.

668. Warner, N. R., Kresse, T. M., Hays, P. D., Down, A., Karr, J. D., Jackson, R. B., Vengosh, A., Geochemical and isotopic variations in shallow groundwater in areas of the Fayetteville Shale development, north-central Arkansas. *Applied Geochemistry* **2013**, *35*, 207–220.

669. Down, A., Schreglmann, K., Plata, D. L., Elsner, M., Warner, N. R., Vengosh, A., Moore, K., Coleman, D., Jackson, R. B., Pre-drilling background groundwater quality in the Deep River Triassic Basin of central North Carolina, USA. *Applied Geochemistry* **2015**, *60*, 3–13.

670. Chapman, M. J., Gurley, L. N., Fitzgerald, S. A., *Baseline Well Inventory and Groundwater-Quality Data from a Potential Shale Gas Resource Area in Parts of Lee and Chatham Counties, North Carolina, October 2011–August 2012*. US Geological Survey: **2014**.

671. Eckhardt, D. A., Sloto, R. A., *Baseline Groundwater Quality in National Park Units within the Marcellus and Utica Shale Gas Plays, New York, Pennsylvania, and West Virginia, 2011*. US Geological Survey: **2012**.

672. Jackson, R. E., Heagle, D. J., Sampling domestic/farm wells for baseline groundwater quality and fugitive gas. *Hydrogeology Journal* **2016**, *24*, 269–272.

673. Humez, P., Mayer, B., Nightingale, M., Ing, J., Becker, V., Jones, D., Lam, V., An 8-year record of gas geochemistry and isotopic composition of methane during baseline sampling at a groundwater observation well in Alberta (Canada). *Hydrogeology Journal* **2016**, *24* (1), 109–122.

674. Moritz, A., Hélie, J.-F., Pinti, D. L., Larocque, M., Barnetche, D., Retailleau, S., Lefebvre, R., Gélinas, Y., Methane baseline concentrations and sources in shallow aquifers from the shale gas-prone region of the St. Lawrence Lowlands (Quebec, Canada). *Environmental Science & Technology* **2015**, *49* (7), 4765–4771.

675. Lavoie, D., Rivard, C., Lefebvre, R., Séjourné, S., Thériault, R., Duchesne, M. J., Ahad, J. M. E., Wang, B., Benoit, N., Lamontagne, C., The Utica Shale and Gas Play in southern Quebec: geological and hydrogeological syntheses and methodological approaches to groundwater risk evaluation. *International Journal of Coal Geology* **2014**, *126*, 77–91.

676. Rhodes, A. L., Horton, N. J., Establishing baseline water quality for household wells within the Marcellus Shale gas region, Susquehanna County, Pennsylvania, U.S.A. *Applied Geochemistry* **2015**, *60*, 14–28.

677. Montcoudiol, N., Banks, D., Isherwood, C., Gunning, A., Burnside, N., Baseline groundwater monitoring for shale gas extraction: definition of baseline conditions and recommendations from a real site (Wysin, Northern Poland). *Acta Geophysica* **2019**, *67* (1), 365–384.

678. Huang, T., Pang, Z., Tian, J., Li, Y., Yang, S., Luo, L., Methane content and isotopic composition of shallow groundwater: implications for environmental monitoring related to shale gas exploitation. *Journal of Radioanalytical and Nuclear Chemistry* **2017**, *312* (3), 577–585.

679. Jackson, R. E., Gorody, A. W., Mayer, B., Roy, J. W., Ryan, M. C., Van Stempvoort, D. R., Groundwater protection and unconventional gas extraction: the critical need for field-based hydrogeological research. *Groundwater* **2013**, *51* (4), 488–510.

680. Smedley, P. L., Ward, R. S., Bearcock, J. M., Bowes, M. J., Establishing the baseline in groundwater chemistry in connection with shale-gas exploration: Vale of Pickering, UK. *Procedia Earth and Planetary Science* **2017**, *17*, 678–681.

681. Li Z, H. T., Ma B., Long Y., Zhang F., Tian J., Li Y., Pang Z.., Baseline groundwater quality before shale gas development in Xishui, Southwest China: analyses of hydrochemistry and multiple environmental isotopes (2H, 18O, 13C, 87Sr/86Sr, 11B, and noble gas isotopes). *Water* **2020**, *12*, 1741.

682. Bell, R. A., Darling, W. G., Ward, R. S., Basava-Reddi, L., Halwa, L., Manamsa, K., Ó Dochartaigh, B. E., A baseline survey of dissolved methane in aquifers of Great Britain. *Science of the Total Environment* **2017**, *601–602*, 1803–1813.

683. Humez, P., Mayer, B., Ing, J., Nightingale, M., Becker, V., Kingston, A., Akbilgic, O., Taylor, S., Occurrence and origin of methane in groundwater in Alberta (Canada): Gas geochemical and isotopic approaches. *Science of the Total Environment* **2016**, *541*, 1253–1268.

684. Siegel, D. I., Smith, B., Perry, E., Bothun, R., Hollingsworth, M., Pre-drilling water-quality data of groundwater prior to shale gas drilling in the Appalachian Basin: analysis of the Chesapeake Energy Corporation dataset. *Applied Geochemistry* **2015**, *63*, 37–57.

685. Eymold, W. K., Swana, K., Moore, M. T., Whyte, C. J., Harkness, J. S., Talma, S., Murray, R., Moortgat, J. B., Miller, J., Vengosh, A., Darrah, T. H., Hydrocarbon-rich groundwater above shale-gas formations: a Karoo Basin case study. *Groundwater* **2018**, *56* (2), 204–224.

686. Claire Botner, E., Townsend-Small, A., Nash, D. B., Xu, X., Schimmelmann, A., Miller, J. H., Monitoring concentration and isotopic composition of methane in groundwater in the Utica Shale hydraulic fracturing region of Ohio. *Environmental Monitoring and Assessment* **2018**, *190* (6), 322.

687. Harkness, J. S., Swana, K., Eymold, W. K., Miller, J., Murray, R., Talma, S., Whyte, C. J., Moore, M. T., Maletic, E. L., Vengosh, A., Darrah, T. H., Pre-drill groundwater geochemistry in the Karoo Basin, South Africa. *Groundwater* **2018**, *56* (2), 187–203.

688. Schout, G., Hartog, N., Hassanizadeh, S. M., Griffioen, J., Impact of an historic underground gas well blowout on the current methane chemistry in a shallow groundwater system. *Proceedings of the National Academy of Sciences* **2018**, *115* (2), 296–301.

689. Woda, J., Wen, T., Oakley, D., Yoxtheimer, D., Engelder, T., Castro, M. C., Brantley, S. L., Detecting and explaining why aquifers occasionally become degraded near hydraulically fractured shale gas wells. *Proceedings of the National Academy of Sciences* **2018**, *115* (49), 12349–12358.

690. Wen, T., Castro, M. C., Nicot, J.-P., Hall, C. M., Larson, T., Mickler, P., Darvari, R., Methane sources and migration mechanisms in shallow groundwaters in Parker and Hood counties, Texas – a heavy noble gas analysis. *Environmental Science & Technology* **2016**, *50* (21), 12012–12021.

691. Wen, T., Castro, M. C., Nicot, J.-P., Hall, C. M., Pinti, D. L., Mickler, P., Darvari, R., Larson, T., Characterizing the noble gas isotopic composition of the Barnett Shale and Strawn Group and constraining the source of stray gas in the Trinity Aquifer, North-Central Texas. *Environmental Science & Technology* **2017**, *51* (11), 6533–6541.

692. Muehlenbachs, K. *Identifying the Sources of Fugitive Methane Associated with Shale Gas Development.* Resources for the Future: **2013**.

693. Rowe, D., Muehlenbachs, K., Isotopic fingerprints of shallow gases in the Western Canadian Sedimentary Basin: tools for remediation of leaking heavy oil wells. *Organic Geochemistry* **1999**, *30* (8, part 1), 861–871.

694. Lefebvre, R., Mechanisms leading to potential impacts of shale gas development on groundwater quality. *WIREs Water* **2017**, *4* (1), e1188.

695. Jackson, R. B., The integrity of oil and gas wells. *Proceedings of the National Academy of Sciences* **2014**, *111* (30), 10902–10903.

696. Brufatto, C. C., J., Conn, L., Ower, D., From mud to cement – building gas wells. *Oil Field Review* **2003**, *2003*, 62–76.

697. Ingraffea, A. R., Wells, M. T., Santoro, R. L., Shonkoff, S. B. C., Assessment and risk analysis of casing and cement impairment in oil and gas wells in Pennsylvania, 2000–2012. *Proceedings of the National Academy of Sciences* **2014**, *111* (30), 10955–10960.

698. Wen, T., Woda, J., Marcon, V., Niu, X., Li, Z., Brantley, S. L., Exploring how to use groundwater chemistry to identify migration of methane near shale gas wells in the Appalachian Basin. *Environmental Science & Technology* **2019**, *53* (15), 9317–9327.

699. Van Stempvoort, D., Maathuis, H., Jaworski, E., Mayer, B., Rich, K., Oxidation of fugitive methane in ground water linked to bacterial sulfate reduction. *Groundwater* **2005**, *43* (2), 187–199.

700. Fontenot, B. E., Hunt, L. R., Hildenbrand, Z. L., Carlton Jr., D. D., Oka, H., Walton, J. L., Hopkins, D., Osorio, A., Bjorndal, B., Hu, Q. H., Schug, K. A., An evaluation of water quality in private drinking water wells near natural gas extraction sites in the Barnett Shale formation. *Environmental Science & Technology* **2013**, *47* (17), 10032–10040.

701. Hammond, P. A., Wen, T., Brantley, S. L., Engelder, T., Gas well integrity and methane migration: evaluation of published evidence during shale-gas development in the USA. *Hydrogeology Journal* **2020**, *28* (4), 1481–1502.

702. Al-Jubori, A., Johnston, S., Boyer, C., Lambert, S. W., Bustos, O. A., Pashin, J. C., Wray, A., Coalbed methane: clean energy for the world. *Oilfield Review* **2009**, *21*, 4–13.

703. Boger, C., Marshall, J. S., Pilcher, R. C., Worldwide coal mine methane and coalbed methane activities. In *Coal Bed Methane*, Thakur, P., Schatzel, S., Aminian, K., eds. Elsevier: **2014**, pp. 351–407.

704. Flores, R. M., *Coal and Coalbed Gas: Fueling the Future*. Elsevier Science & Technology: **2013**.

705. Li, H., Lau, H. C., Huang, S., China's coalbed methane development: a review of the challenges and opportunities in subsurface and surface engineering. *Journal of Petroleum Science and Engineering* **2018**, *166*, 621–635.

706. National Research Council, *Management and Effects of Coalbed Methane Produced Water in the Western United States*. The National Academies Press: **2010**.

707. Zeng, Q., Wang, Z., McPherson, B. J., McLennan, J. D., Modeling competitive adsorption between methane and water on coals. *Energy & Fuels* **2017**, *31* (10), 10775–10786.

708. Colmenares, L. B., Zoback, M. D., Hydraulic fracturing and wellbore completion of coalbed methane wells in the Powder River Basin, Wyoming: implications for water and gas production. *AAPG Bulletin* **2007**, *91*, 51–67.

709. Ayers, W. B., Kaiser, W. R. *Coalbed Methane in the Upper Cretaceous Fruitland Formation, San Juan Basin, New Mexico and Colorado*. Bureau of Economic Geology, the University of Texas at Austin: **1994**.

710. Bleizeffer, D., Coalbed Methane: Boom, Bust and Hard Lessons. www.wyohistory.org/encyclopedia/coalbed-methane-boom-bust-and-hard-lessons.

711. Cheung, K., Klassen, P., Mayer, B., Goodarzi, F., Aravena, R., Major ion and isotope geochemistry of fluids and gases from coalbed methane and shallow groundwater wells in Alberta, Canada. *Applied Geochemistry* **2010**, *25* (9), 1307–1329.

712. Johnston, C. R., Vance, G. F., Ganjegunte, G. K., Irrigation with coalbed natural gas co-produced water. *Agricultural Water Management* **2008**, *95* (11), 1243–1252.

713. Myers, T., Groundwater management and coal bed methane development in the Powder River Basin of Montana. *Journal of Hydrology* **2009**, *368* (1), 178–193.

714. Snyder, G. T., Riese, W. C. R., Franks, S., Fehn, U., Pelzmann, W. L., Gorody, A. W., Moran, J. E., Origin and history of waters associated with coalbed methane: 129I, 36Cl, and stable isotope results from the Fruitland Formation, CO and NM. *Geochimica et Cosmochimica Acta* **2003**, *67* (23), 4529–4544.

715. Van Voast, W. A., Geochemical signature of formation waters associated with coalbed methane. *AAPG Bulletin* **2003**, *87*, 667–676.

716. Pashin, J. C., Stratigraphy and structure of coalbed methane reservoirs in the United States: an overview. *International Journal of Coal Geology* **1998**, *35* (1), 209–240.

717. Pashin, J. C., Hydrodynamics of coalbed methane reservoirs in the Black Warrior Basin: key to understanding reservoir performance and environmental issues. *Applied Geochemistry* **2007**, *22* (10), 2257–2272.

718. Pashin, J. C., Variable gas saturation in coalbed methane reservoirs of the Black Warrior Basin: implications for exploration and production. *International Journal of Coal Geology* **2010**, *82* (3), 135–146.

719. Campbell, C. E., Pearson, B. N. and Frost, C. D., Strontium isotopes as indicators of aquifer communication in an area of coal-bed natural gas production, Powder River Basin, Wyoming and Montana. *Rocky Mountain Geology* **2008**, *43*, 171–197.

720. Rice, C. A., Ellis, M. S., Bullock, J. H., *Water Co-produced with Coalbed Methane in the Powder River Basin, Wyoming: Preliminary Compositional Data*. US Geolgical Survey: **2000**.

721. Rice, C. A., Flores, R. M., Stricker, G. D., Ellis, M. S., Chemical and stable isotopic evidence for water/rock interaction and biogenic origin of coalbed methane, Fort Union Formation, Powder River Basin, Wyoming and Montana U.S.A. *International Journal of Coal Geology* **2008**, *76* (1), 76–85.

722. Jackson, R. E., Reddy, K. J., Trace element chemistry of coal bed natural gas produced water in the Powder River Basin, Wyoming. *Environmental Science & Technology* **2007**, *41* (17), 5953–5959.

723. Ayers, W. B., Kaiser, W. R., *Coalbed Methane in the Upper Cretaceous Fruitland Formation, San Juan Basin, New Mexico and Colorado*. New Mexico Bureau of Mines and Mineral Resources: **1994**, Vol. 146.

724. Golding, S. D., Boreham, C. J., Esterle, J. S., Stable isotope geochemistry of coal bed and shale gas and related production waters: a review. *International Journal of Coal Geology* **2013**, *120*, 24–40.

725. Frost, C. D., Pearson, B. N., Ogle, K. M., Heffern, E. L., Lyman, R. M., Sr isotope tracing of aquifer interactions in an area of accelerating coal-bed methane production, Powder River Basin, Wyoming *Geology* **2002**, *30*, 923–926.

726. Dahm, K. G., Guerra, K. L., Munakata-Marr, J., Drewes, J. E., Trends in water quality variability for coalbed methane produced water. *Journal of Cleaner Production* **2014**, *84*, 840–848.

727. Cheung, K., Sanei, H., Klassen, P., Mayer, B., Goodarzi, F., Produced fluids and shallow groundwater in coalbed methane (CBM) producing regions of Alberta, Canada: trace element and rare earth element geochemistry. *International Journal of Coal Geology* **2009**, *77* (3), 338–349.

728. Healy, R. W., Rice, C. A., Bartos, T. T., McKinley, M. P., Infiltration from an impoundment for coal-bed natural gas, Powder River Basin, Wyoming: evolution of water and sediment chemistry. *Water Resources Research* **2008**, *44* (6).

729. Davis, W. N., Bramblett, R. G., Zale, A. V., Endicott, C. L., A review of the potential effects of coal bed natural gas development activities on fish assemblages of the Powder River Geologic Basin. *Reviews in Fisheries Science* **2009**, *17* (3), 402–422.

730. Farag, A., Harper, D. D., Senecal, A., Hubert, W. A., Potential effects of coalbed natural gas development on fish and aquatic resources. In *Coalbed Natural Gas: Energy and Environment*, Reddy, K. J., ed. Nova Science Publishers: **2010**, pp. 227–242.

731. Dauwalter, D. C., Wenger, S. J., Gelwicks, K. R., Fesenmyer, K. A., Land use associations with distributions of declining native fishes in the Upper Colorado River Basin. *Transactions of the American Fisheries Society* **2011**, *140* (3), 646–658.

732. Singh, U., Colosi, L. M, Water–energy sustainability synergies and health benefitsas means to motivate potable reuse of coalbed methane-produced waters. *Ambio* **2019**, *48*, 752–768.

733. Li, G.-H., Sjursen, H. P., Characteristics of produced water during coalbed methane (CBM) development and its feasibility as irrigation water in Jincheng, China. *Journal of Coal Science and Engineering (China)* **2013**, *19* (3), 369–374.

734. Lester, J. P., Federalism and state environmental policy. In *Environmental Politics and Policy: Theories and Evidence*, Lester, J. P., ed. Duke University Press: **1977**, pp. 39–62.

735. US Congress, *Energy Policy Act 2005*. U.S. Congress: **2005**.

736. Zirogiannis, N., Alcorn, J., Rupp, J., Carley, S., Graham, J. D., State regulation of unconventional gas development in the U.S.: an empirical evaluation. *Energy Research & Social Science* **2016**, *11*, 142–154.

737. Christopherson, S., Rightor, N., How shale gas extraction affects drilling localities: lessons for regional and city policy makers. *Journal of Town & City Management* **2012**, *2*, 1–20.

738. Richardson, N., Gottlieb, M., Krupnick, A., Wiseman, H., *The State of State Shale Gas Regulation*. Resources for the Future: **2013**.

739. Litzow, E., Neville, K. J., Johnson-King, B., Weinthal, E., Why does industry structure matter for unconventional oil and gas development? Examining revenue sharing outcomes in North Dakota. *Energy Research & Social Science* **2018**, *44*, 371–384.

740. McFeeley, M., *State Hydraulic Fracturing Disclosure Rules and Enforcement: A Comparison*. Natural Resources Defense Council: **2012**.

741. Wiseman, H., *The Private Role in Public Fracturing Disclosure and Regulation*. Harvard Business Law Review Online: **2013**.

742. Pi, G., Dong, X., Dong, C., Guo, J., Ma, Z., The status, obstacles and policy recommendations of shale gas development in China. *Sustainability* **2015**, *7* (3), 2353–2372.

743. Lin, A., Replacing coal with shale gas: could reducing China's regional air pollution lead to more local pollution in rural China? In *The Shale Dilemma: Political, Economic and Scientific Issues behind the "Fracking" Debate in Global Perspective*, Gamper-Rabindran, S., ed. University of Pittsburgh Press: **2017**.

744. UN Department of Economic and Social Affairs Population Dynamics, Population Division World Population Prospects 2019, Online Edition. Rev. 1. https://population.un.org/wpp/Download/Standard/Population/.

745. World Economic Forum, These 12 Charts Show How the World's Population Has Exploded in the Last 200 Years. www.weforum.org/agenda/2019/07/populations-around-world-changed-over-the-years.

746. Gleick, P. a. I., C. *Water, Security, and Conflict*. World Resource Institute: **2018**.

747. Wada, Y., Flörke, M., Hanasaki, N., Eisner, S., Fischer, G., Tramberend, S., Satoh, Y., van Vliet, M. T. H., Yillia, P., Ringler, C., Burek, P., Wiberg, D., Modeling global water use for the 21st century: the Water Futures and Solutions (WFaS) initiative and its approaches. *Geoscientific Model Development* **2016**, *9* (1), 175–222.

748. Wada, Y., van Beek, L. P. H., Wanders, N., Bierkens, M. F. P., Human water consumption intensifies hydrological drought worldwide. *Environmental Research Letters* **2013**, *8* (3), 034036.
749. Wada, Y., Wisser, D., Eisner, S., Flörke, M., Gerten, D., Haddeland, I., Hanasaki, N., Masaki, Y., Portmann, F. T., Stacke, T., Tessler, Z., Schewe, J., Multimodel projections and uncertainties of irrigation water demand under climate change. *Geophysical Research Letters* **2013**, *40* (17), 4626–4632.
750. Fischer, G., Tubiello, F. N., van Velthuizen, H., Wiberg, D. A., Climate change impacts on irrigation water requirements: effects of mitigation, 1990–2080. *Technological Forecasting and Social Change* **2007**, *74* (7), 1083–1107.
751. Somanathan, E., Taming the anarchy: groundwater governance in South Asia. *Indian Growth and Development Review* **2010**, *3* (1), 92–94.
752. Rodell, M., Velicogna, I., Famiglietti, J. S., Satellite-based estimates of groundwater depletion in India. *Nature* **2009**, *460* (7258), 999–1002.
753. Coyte, R. M., Singh, A., Furst, K. E., Mitch, W. A., Vengosh, A., Co-occurrence of geogenic and anthropogenic contaminants in groundwater from Rajasthan, India. *Science of the Total Environment* **2019**, *688*, 1216–1227.
754. Dai, A., Drought under global warming: a review. *WIREs Climate Change* **2011**, *2* (1), 45–65.
755. Haddeland, I., Heinke, J., Biemans, H., Eisner, S., Flörke, M., Hanasaki, N., Konzmann, M., Ludwig, F., Masaki, Y., Schewe, J., Stacke, T., Tessler, Z. D., Wada, Y., Wisser, D., Global water resources affected by human interventions and climate change. *Proceedings of the National Academy of Sciences* **2014**, *111* (9), 3251–3256.
756. Lettenmaier, D. P., Wood, A. W., Palmer, R. N., Wood, E. F., Stakhiv, E. Z., Water resources implications of global warming: a US regional perspective. *Climatic Change* **1999**, *43* (3), 537–579.
757. Buhaug, H., Climate–conflict research: some reflections on the way forward. *WIREs Climate Change* **2015**, *6* (3), 269–275.
758. Buhaug, H., Climate not to blame for African civil wars. *Proceedings of the National Academy of Sciences* **2010**, *107* (38), 16477–16482.
759. Daoudy, M., *The Origins of the Syrian Conflict: Climate Change and Human Security*. Cambridge University Press: **2020**.
760. Alova, G., A global analysis of the progress and failure of electric utilities to adapt their portfolios of power-generation assets to the energy transition. *Nature Energy* **2020**, *5*, 920–927.
761. Schewe, J., Heinke, J., Gerten, D., Haddeland, I., Arnell, N. W., Clark, D. B., Dankers, R., Eisner, S., Fekete, B. M., Colón-González, F. J., Gosling, S. N., Kim, H., Liu, X., Masaki, Y., Portmann, F. T., Satoh, Y., Stacke, T., Tang, Q., Wada, Y., Wisser, D., Albrecht, T., Frieler, K., Piontek, F., Warszawski, L., Kabat, P., Multimodel assessment of water scarcity under climate change. *Proceedings of the National Academy of Sciences* **2014**, *111* (9), 3245–3250.
762. Gosling, S. N., Arnell, N. W., A global assessment of the impact of climate change on water scarcity. *Climatic Change* **2016**, *134*, 371–385.
763. Mekonnen, M. M., Hoekstra, A. Y., Four billion people facing severe water scarcity. *Science Advances* **2016**, *2* (2), e1500323.
764. Barnett, T. P., Adam, J. C., Lettenmaier, D. P., Potential impacts of a warming climate on water availability in snow-dominated regions. *Nature* **2005**, *438* (7066), 303–309.
765. Jiménez Cisneros, B. E., Oki, T., Arnell, N. W., Benito, G., Cogley, J. G., Jiang, D. T., Mwakalila, S. S., Freshwater resources. In *Climate Change 2014: Impacts, Adaptation, and Vulnerability. Part A: Global and Sectoral Aspects. Contribution of Working Group II to the Fifth Assessment Report of the Intergovernmental Panel on Climate Change*. Cambridge University Press: **2014**, pp. 229–269.
766. National Centers for Environmental Information, Global Temperature and Precipitation Maps. www.ncdc.noaa.gov/temp-and-precip/global-maps/.
767. Wakeel, M., Chen, B., Hayat, T., Alsaedi, A., Ahmad, B., Energy consumption for water use cycles in different countries: a review. *Applied Energy* **2016**, *178*, 868–885.
768. Escobar, I. C., Schäfer, A., *Sustainable Water for the Future, Volume 2 – Water Recycling versus Desalination*. Elsevier: **2017**.
769. Shahzad, M. W., Burhan, M., Ghaffour, N., Ng, K. C., A multi evaporator desalination system operated with thermocline energy for future sustainability. *Desalination* **2018**, *435*, 268–277.
770. Jones, E., Qadir, M., van Vliet, M. T. H., Smakhtin, V., Kang, S.-m., The state of desalination and brine production: a global outlook. *Science of the Total Environment* **2019**, *657*, 1343–1356.
771. Shahzad, M. W., Burhan, M., Ang, L., Ng, K. C., Energy–water–environment nexus underpinning future desalination sustainability. *Desalination* **2017**, *413*, 52–64.

772. Ritchie, H., Rosser, M. Water Use and Sress. https://ourworldindata.org/water-use-stress#freshwater-with drawals-by-country.
773. International Energy Agency (IEA), *Offshore Energy Outlook*. International Energy Agency (IEA): **2018**.
774. US Energy Information Administration (EIA), Offshore Production Nearly 30% of Global Crude Oil Output in 2015. www.eia.gov/todayinenergy/detail.php?id=28492.
775. UN Economic Commission for Europe, The Convention and Its Achievements. www.unece.org/environ mental-policy/conventions/envlrtapwelcome/the-air-convention-and-its-protocols/the-convention-and-its-achievements.html.
776. European Environment Agency, EN10 – Residues from Combustion of Coal for Energy Production. www.eea .europa.eu/data-and-maps/indicators/en10-residues-from-combustion-of/residues-from-combustion-of-coal.
777. Qian, G., Li, Y., Acid and metalliferous drainage – a global environmental issue. *Journal of Mining and Mechanical Engineering* **2019**, *1*, 1–4.
778. Naidu, G., Ryu, S., Thiruvenkatachari, R., Choi, Y., Jeong, S., Vigneswaran, S., A critical review on remedia-tion, reuse, and resource recovery from acid mine drainage. *Environmental Pollution* **2019**, *247*, 1110–1124.
779. National Oceanic and Atmospheric Administration (NOAA), Climate Change: Annual Greenhouse Gas Index. www.climate.gov/news-features/understanding-climate/climate-change-annual-greenhouse-gas-index.
780. NOAA Global Monitoring Laboratory, Trends in Atmospheric Carbon Dioxide. www.esrl.noaa.gov/gmd/ ccgg/trends/mlo.html.
781. National Centers for Environmental Information, State of the Climate: Global Climate Report for Annual 2019. www.ncdc.noaa.gov/sotc/global/201913.
782. Hansen, J., Sato, M., Kharecha, P., Beerling, D., Berner, R., Masson-Delmotte, V., Pagani, M., Raymo, M., Royer, D. L., Zachos, J. C., Target atmospheric CO: where should humanity aim? *The Open Atmospheric Science Journal* **2008**, *2*, 217–231.
783. US Energy Information Administration (EIA), Energy and the Environment Explained: Where Greenhouse Gases Come From. www.eia.gov/energyexplained/energy-and-the-environment/where-greenhouse-gases-come-from.php.
784. Leduc, M., Matthews, H. D., de Elía, R., Regional estimates of the transient climate response to cumulative CO2 emissions. *Nature Climate Change* **2016**, *6* (5), 474–478.
785. Allen, M. R., Frame, D. J., Huntingford, C., Jones, C. D., Lowe, J. A., Meinshausen, M., Meinshausen, N., Warming caused by cumulative carbon emissions towards the trillionth tonne. *Nature* **2009**, *458* (7242), 1163–1166.
786. Zickfeld, K., MacDougall, A. H., Matthews, H. D., On the proportionality between global temperature change and cumulative CO2 emissions during periods of net negative CO2 emissions. *Environmental Research Letters* **2016**, *11* (5), 055006.
787. Belmont, E. L., Davidson, F. T., Glazer, Y. R., Beagle, E. A., Webber, M. E., Accounting for water formation from hydrocarbon fuel combustion in life cycle analyses. *Environmental Research Letters* **2017**, *12* (9), 094019.
788. Saunois, M., Stavert, A. R., Poulter, B., Bousquet, P., Canadell, J. G., Jackson, R. B., Raymond, P. A., Dlugokencky, E. J., Houweling, S., Patra, P. K., Ciais, P., Arora, V. K., Bastviken, D., Bergamaschi, P., Blake, D. R., Brailsford, G., Bruhwiler, L., Carlson, K. M., Carrol, M., Castaldi, S., Chandra, N., Crevoisier, C., Crill, P. M., Covey, K., Curry, C. L., Etiope, G., Frankenberg, C., Gedney, N., Hegglin, M. I., Höglund-Isaksson, L., Hugelius, G., Ishizawa, M., Ito, A., Janssens-Maenhout, G., Jensen, K. M., Joos, F., Kleinen, T., Krummel, P. B., Langenfelds, R. L., Laruelle, G. G., Liu, L., Machida, T., Maksyutov, S., McDonald, K. C., McNorton, J., Miller, P. A., Melton, J. R., Morino, I., Müller, J., Murguia-Flores, F., Naik, V., Niwa, Y., Noce, S., O'Doherty, S., Parker, R. J., Peng, C., Peng, S., Peters, G. P., Prigent, C., Prinn, R., Ramonet, M., Regnier, P., Riley, W. J., Rosentreter, J. A., Segers, A., Simpson, I. J., Shi, H., Smith, S. J., Steele, L. P., Thornton, B. F., Tian, H., Tohjima, Y., Tubiello, F. N., Tsuruta, A., Viovy, N., Voulgarakis, A., Weber, T. S., van Weele, M., van der Werf, G. R., Weiss, R. F., Worthy, D., Wunch, D., Yin, Y., Yoshida, Y., Zhang, W., Zhang, Z., Zhao, Y., Zheng, B., Zhu, Q., Zhu, Q., Zhuang, Q., The global methane budget 2000–2017. *Earth System Science Data* **2020**, *12*, 1561–1623.
789. Dlugokencky, E. J. Trends in Atmospheric Methane. www.esrl.noaa.gov/gmd/ccgg/trends_ch4/.
790. Ciais, P., Sabine, C., Bala, G., Bopp, L., Brovkin, V., Canadell, J., Chhabra, A., DeFries, R., Galloway, J., M., H., Jones, C., Le Quéré, C., Myneni, R. B., Piao, S., Thornton, P. *Carbon and Other Biogeochemical Cycles*. Cambridge University Press: **2013**.
791. Worden, J. R., Bloom, A. A., Pandey, S., Jiang, Z., Worden, H. M., Walker, T. W., Houweling, S., Röckmann, T., Reduced biomass burning emissions reconcile conflicting estimates of the post-2006 atmos-pheric methane budget. *Nature Communications* **2017**, *8* (1), 2227.

792. Cloy, J. M., Smith, K. A., Greenhouse gas sources and sinks. In *Encyclopedia of the Anthropocene*, Dellasala, D. A., Goldstein, M. I., eds. Elsevier: **2018**, pp. 391–400.

793. Turner, A. J., Frankenberg, C., Kort, E. A., Interpreting contemporary trends in atmospheric methane. *Proceedings of the National Academy of Sciences* **2019**, *116* (8), 2805–2813.

794. Kort, E. A., Smith, M. L., Murray, L. T., Gvakharia, A., Brandt, A. R., Peischl, J., Ryerson, T. B., Sweeney, C., Travis, K., Fugitive emissions from the Bakken Shale illustrate role of shale production in global ethane shift. *Geophysical Research Letters* **2016**, *43* (9), 4617–4623.

795. Lan, X., Talbot, R., Laine, P., Torres, A., Characterizing fugitive methane emissions in the Barnett Shale area using a mobile laboratory. *Environmental Science & Technology* **2015**, *49* (13), 8139–8146.

796. Rella, C. W., Tsai, T. R., Botkin, C. G., Crosson, E. R., Steele, D., Measuring emissions from oil and natural gas well pads using the mobile flux plane technique. *Environmental Science & Technology* **2015**, *49* (7), 4742–4748.

797. Robertson, A. M., Edie, R., Snare, D., Soltis, J., Field, R. A., Burkhart, M. D., Bell, C. S., Zimmerle, D., Murphy, S. M., Variation in methane emission rates from well pads in four oil and gas basins with contrasting production volumes and compositions. *Environmental Science & Technology* **2017**, *51* (15), 8832–8840.

798. Brandt, A. R., Heath, G. A., Cooley, D., Methane leaks from natural gas systems follow extreme distributions. *Environmental Science & Technology* **2016**, *50* (22), 12512–12520.

799. Zavala-Araiza, D., Alvarez, R. A., Lyon, D. R., Allen, D. T., Marchese, A. J., Zimmerle, D. J., Hamburg, S. P., Super-emitters in natural gas infrastructure are caused by abnormal process conditions. *Nature Communications* **2017**, *8* (1), 14012.

800. Allen, D. T., Torres, V. M., Thomas, J., Sullivan, D. W., Harrison, M., Hendler, A., Herndon, S. C., Kolb, C. E., Fraser, M. P., Hill, A. D., Lamb, B. K., Miskimins, J., Sawyer, R. F., Seinfeld, J. H., Measurements of methane emissions at natural gas production sites in the United States. *Proceedings of the National Academy of Sciences* **2013**, *110* (44), 17768–17773.

801. Lamb, B. K., Edburg, S. L., Ferrara, T. W., Howard, T., Harrison, M. R., Kolb, C. E., Townsend-Small, A., Dyck, W., Possolo, A., Whetstone, J. R., Direct measurements show decreasing methane emissions from natural gas local distribution systems in the United States. *Environmental Science & Technology* **2015**, *49* (8), 5161–5169.

802. Peischl, J., Ryerson, T. B., Aikin, K. C., de Gouw, J. A., Gilman, J. B., Holloway, J. S., Lerner, B. M., Nadkarni, R., Neuman, J. A., Nowak, J. B., Trainer, M., Warneke, C., Parrish, D. D., Quantifying atmospheric methane emissions from the Haynesville, Fayetteville, and northeastern Marcellus Shale gas production regions. *Journal of Geophysical Research: Atmospheres* **2015**, *120* (5), 2119–2139.

803. Caulton, D. R., Shepson, P. B., Santoro, R. L., Sparks, J. P., Howarth, R. W., Ingraffea, A. R., Cambaliza, M. O. L., Sweeney, C., Karion, A., Davis, K. J., Stirm, B. H., Montzka, S. A., Miller, B. R., Toward a better understanding and quantification of methane emissions from shale gas development. *Proceedings of the National Academy of Sciences* **2014**, *111* (17), 6237–6242.

804. Smith, M. L., Gvakharia, A., Kort, E. A., Sweeney, C., Conley, S. A., Faloona, I., Newberger, T., Schnell, R., Schwietzke, S., Wolter, S., Airborne quantification of methane emissions over the Four Corners region. *Environmental Science & Technology* **2017**, *51* (10), 5832–5837.

805. Schwietzke, S., Pétron, G., Conley, S., Pickering, C., Mielke-Maday, I., Dlugokencky, E. J., Tans, P. P., Vaughn, T., Bell, C., Zimmerle, D., Wolter, S., King, C. W., White, A. B., Coleman, T., Bianco, L., Schnell, R. C., Improved mechanistic understanding of natural gas methane emissions from spatially resolved aircraft measurements. *Environmental Science & Technology* **2017**, *51* (12), 7286–7294.

806. Barkley, Z. R., Lauvaux, T., Davis, K. J., Deng, A., Miles, N. L., Richardson, S. J., Cao, Y., Sweeney, C., Karion, A., Smith, M., Kort, E. A., Schwietzke, S., Murphy, T., Cervone, G., Martins, D., Maasakkers, J. D., Quantifying methane emissions from natural gas production in north-eastern Pennsylvania. *Atmospheric Chemistry and Physics* **2017**, *17* (22), 13941–13966.

807. Karion, A., Sweeney, C., Pétron, G., Frost, G., Michael Hardesty, R., Kofler, J., Miller, B. R., Newberger, T., Wolter, S., Banta, R., Brewer, A., Dlugokencky, E., Lang, P., Montzka, S. A., Schnell, R., Tans, P., Trainer, M., Zamora, R., Conley, S., Methane emissions estimate from airborne measurements over a western United States natural gas field. *Geophysical Research Letters* **2013**, *40* (16), 4393–4397.

808. Alvarez, R. A., Zavala-Araiza, D., Lyon, D. R., Allen, D. T., Barkley, Z. R., Brandt, A. R., Davis, K. J., Herndon, S. C., Jacob, D. J., Karion, A., Kort, E. A., Lamb, B. K., Lauvaux, T., Maasakkers, J. D., Marchese, A. J., Omara, M., Pacala, S. W., Peischl, J., Robinson, A. L., Shepson, P. B., Sweeney, C., Townsend-Small, A., Wofsy, S. C., Hamburg, S. P., Assessment of methane emissions from the US oil and gas supply chain. *Science* **2018**, *361* (6398), 186–188.

809. Ingraffea, A. R., Wawrzynek, P. A., Santoro, R., Wells, M., Reported methane emissions from active oil and gas wells in Pennsylvania, 2014–2018. *Environmental Science & Technology* **2020**, *54* (9), 5783–5789.

810. Allan, W., Struthers, H., Lowe, D. C., Methane carbon isotope effects caused by atomic chlorine in the marine boundary layer: global model results compared with Southern Hemisphere measurements. *Journal of Geophysical Research: Atmospheres* **2007**, *112* (D4).

811. Schaefer, H., Fletcher, S. E. M., Veidt, C., Lassey, K. R., Brailsford, G. W., Bromley, T. M., Dlugokencky, E. J., Michel, S. E., Miller, J. B., Levin, I., Lowe, D. C., Martin, R. J., Vaughn, B. H., White, J. W. C., A 21st-century shift from fossil-fuel to biogenic methane emissions indicated by $^{13}CH^4$. *Science* **2016**, *352* (6281), 80–84.

812. Saunois, M., Jackson, R. B., Bousquet, P., Poulter, B., Canadell, J. G., The growing role of methane in anthropogenic climate change. *Environmental Research Letters* **2016**, *11* (12), 120207.

813. Hmiel, B., Petrenko, V. V., Dyonisius, M. N., Buizert, C., Smith, A. M., Place, P. F., Harth, C., Beaudette, R., Hua, Q., Yang, B., Vimont, I., Michel, S. E., Severinghaus, J. P., Etheridge, D., Bromley, T., Schmitt, J., Faïn, X., Weiss, R. F., Dlugokencky, E., Preindustrial 14CH4 indicates greater anthropogenic fossil CH4 emissions. *Nature* **2020**, *578* (7795), 409–412.

814. Kang, M., Christian, S., Celia, M. A., Mauzerall, D. L., Bill, M., Miller, A. R., Chen, Y., Conrad, M. E., Darrah, T. H., Jackson, R. B., Identification and characterization of high methane-emitting abandoned oil and gas wells. *Proceedings of the National Academy of Sciences* **2016**, *113* (48), 13636–13641.

815. Davenport, C. Trump Eliminates Major Methane Rule, Even as Leaks Are Worsening. www.nytimes.com/2020/08/13/climate/trump-methane.html.

816. Brownlow, R., Lowry, D., Fisher, R. E., France, J. L., Lanoisellé, M., White, B., Wooster, M. J., Zhang, T., Nisbet, E. G., Isotopic ratios of tropical methane emissions by atmospheric measurement. *Global Biogeochemical Cycles* **2017**, *31* (9), 1408–1419.

817. Sherwood, O. A., Schwietzke, S., Arling, V. A., Etiope, G., Global inventory of gas geochemistry data from fossil fuel, microbial and burning sources, version 2017. *Earth Syst. Sci. Data* **2017**, *9* (2), 639–656.

818. Stanislaw, J., Yergin, D., Oil: reopening the door. *Foreign Affairs* **1993**, *72*, 81–93

819. US Department of Energy, *Natural Gas Flaring and Venting: State and Federal Regulatory Overview, Trends, and Impacts.* US Department of Energy: **2019**.

820. Osofsky, H. M., Climate change and environmental justice: reflections on litigation over oil extraction and rights violations in Nigeria. *Journal of Human Rights and the Environment* **2012**, *1*, 189–210.

821. Collins, R., Adams-Heard, R. Flaring, or Why So Much Gas Is Going Up in Flames. www.bloombergquint.com/quicktakes/flaring-or-why-so-much-gas-is-going-up-in-flames-quicktake.

822. International Energy Agency (IEA), Flaring Emmissions. www.iea.org/reports/flaring-emissions.

823. The World Bank, Global Gas Flaring Jumps to Levels Last Seen in 2009. www.worldbank.org/en/news/press-release/2020/07/21/global-gas-flaring-jumps-to-levels-last-seen-in-2009.

824. Cushing, L. J., Vavra-Musser, K., Chau, K., Franklin, M., Johnston, J. E., Flaring from unconventional oil and gas development and birth outcomes in the Eagle Ford Shale in South Texas. *Environmental Health Perspectives* **2020**, *128* (7), 077003.

825. Johnston, J. E., Chau, K., Franklin, M., Cushing, L., Environmental justice dimensions of oil and gas flaring in South Texas: disproportionate exposure among Hispanic communities. *Environmental Science & Technology* **2020**, *54* (10), 6289–6298.

826. Seiyaboh, E. I., Izah, S. C., A review of impacts of gas flaring on vegetation and water resources in the Niger Delta region of Nigeria. *International Journal of Economy, Energy and Environment* **2017**, *2*, 48–55.

827. Glazer, Y. R., Kjellsson, J. B., Sanders, K. T., Webber, M. E., Potential for using energy from flared gas for on-site hydraulic fracturing wastewater treatment in Texas. *Environmental Science & Technology Letters* **2014**, *1* (7), 300–304.

828. Kar, A., Bahadur, V., Using excess natural gas for reverse osmosis-based flowback water treatment in US shale fields. *Energy* **2020**, *196*, 117145.

829. US Environmental Protection Agency, *Frequent, Routine Flaring May Cause Excessive, Uncontrolled Sulfur Dioxide Releases.* US Environmental Protection Agency (EPA): **2000**.

830. Eman, E. A., Gas flaring in industry: an overview. *Petroleum and Coal* **2015**, *57*, 532–555.

831. Sorrels, J. L., Bradley, K., Randall, D., *Flares.* RTI International: **2019**.

832. Soltanieh, M., Zohrabian, A., Gholipour, M. J., Kalnay, E., A review of global gas flaring and venting and impact on the environment: case study of Iran. *International Journal of Greenhouse Gas Control* **2016**, *49*, 488–509.

833. The World Bank, *Global Gas Flaring Tracker Report.* The World Bank: **2020**.

834. US Department of Energy, *North Dakota Natural Gas Flaring and Venting Regulations.* US Department of Energy: **2019**.

835. US Energy Information Administration (EIA), Natural Gas Venting and Flaring Increased in North Dakota and Texas in 2018. www.eia.gov/todayinenergy/detail.php?id=42195.
836. Rystad Energy, A Downturn Silver Lining: Permian Gas Flaring Has Decreased and Is Expected to Fall Further in 2020. www.rystadenergy.com/newsevents/news/press-releases/a-downturn-silver-lining-permian-gas-flaring-has-decreased-and-is-expected-to-fall-further-in-2020/.
837. Franklin, M., Chau, K., Cushing, L. J., Johnston, J. E., Characterizing flaring from unconventional oil and gas operations in South Texas using satellite observations. *Environmental Science & Technology* **2019**, *53* (4), 2220–2228.
838. Environmental Defense Fund, Helicopter Surveys Indicate Malfunctioning Flares in the Permian Basin Are Releasing at Least 300,000 Metric Tons of Unburned Methane a Year. www.edf.org/media/helicopter-surveys-indicate-malfunctioning-flares-permian-basin-are-releasing-least-300000.
839. Eman, E. A., Environmental pollution and measurement of gas flaring. *International Journal of Scientific Research in Science, Engineering and Technology* **2016**, *2*, 252–262.
840. Fawole, O. G., Cai, X. M., MacKenzie, A. R., Gas flaring and resultant air pollution: a review focusing on black carbon. *Environmental Pollution* **2016**, *216*, 182–197.
841. Mirrezaei, M. A., Orkomi, A. A., Gas flares contribution in total health risk assessment of BTEX in Asalouyeh, Iran. *Process Safety and Environmental Protection* **2020**, *137*, 223–237.
842. Rim-Rukeh, A., Ikiafa, G. O., Okokoyo, P. A., Monitoring air pollutants due to gas flaring using rain water *Global Journal of Environmental Sciences* **2005**, *4*, 123–126
843. Anejionu, O. C. D., Whyatt, J. D., Blackburn, G. A., Price, C. S., Contributions of gas flaring to a global air pollution hotspot: spatial and temporal variations, impacts and alleviation. *Atmospheric Environment* **2015**, *118*, 184–193.
844. Amadi, A. N., Impact of gas-flaring on the quality of rain water, groundwater and surface water in parts of Eastern Niger Delta, Nigeria. *Journal of Geosciences and Geomatics* **2014**, *2*, 114–119.
845. Alani, R., Nwude, D., Joseph, A., Akinrinade, O., Impact of gas flaring on surface and underground water: a case study of Anieze and Okwuibome areas of Delta State, Nigeria. *Environmental Monitoring and Assessment* **2020**, *192* (3), 166.
846. Nwankwo, C. N., Ogagarue D. O., Effects of gas flaring on surface and ground waters in Delta State Nigeria. *Journal of Geology and Mining Research* **2011**, *3*, 131–136.
847. Raimi, M., Ezugwu, S. C., An assessment of trace elements in surface and ground water quality in the Ebocha-Obrikom oil and gas producing area of Rivers State, Nigeria. *International Journal of Scientific & Engineering Research* **2017**, *2*.
848. Emumejaye, K., Effects of gas flaring on surface and ground water in Irri town and environs, Niger-Delta, Nigeria. *IOSR Journal of Environmental Science, Toxicology and Food Technology* **2012**, *1*, 29–33.
849. Ighalo, J. O., Adeniyi, A. G., A comprehensive review of water quality monitoring and assessment in Nigeria. *Chemosphere* **2020**, *260*, 127569.
850. Deutsche Welle, Gas Flaring Continues Scorching Niger Delta. www.dw.com/en/gas-flaring-continues-scorching-niger-delta/a-46088235.
851. Friedman, L. Trump Administration Formally Rolls Back Rule Aimed at Limiting Methane Pollution. www.nytimes.com/2018/09/18/climate/trump-methane-rollback.html.
852. Baltz, T., Colorado in Landmark Rule Moves to End Routine Flaring, Venting. *Bloomberg Law* **2020**.
853. US Department of Energy, State-Level Natural Gas Flaring and Venting Regulations. www.energy.gov/fe/state-level-natural-gas-flaring-and-venting-regulations.
854. Tabuchi, H. Despite Their Promises, Giant Energy Companies Burn Away Vast Amounts of Natural Gas. www.nytimes.com/2019/10/16/climate/natural-gas-flaring-exxon-bp.html.
855. Clayton, R, *Review of Current Knowledge: Desalination for Water Supply*. Foundation for Water Reserach: **2011**, www.fwr.org/desal.pdf.
856. Sanders, K. T., Webber, M. E., Evaluating the energy consumed for water use in the United States. *Environmental Reserach Letters* **2012**, *7*, 034034.
857. Copeland, C., Carter, N. T., *Energy Water Nexus: The Water Sector's Energy Use*. Congressional Research Service: **2017**.
858. Longo, S., d'Antoni, B. M., Bongards, M., Chaparro, A., Cronrath, A., Fatone, F., Lema, J. M., Mauricio-Iglesias, M., Soares, A., Hospido, A., Monitoring and diagnosis of energy consumption in wastewater treatment plants. A state of the art and proposals for improvement. *Applied Energy* **2016**, *179*, 1251–1268.
859. Elimelech, M., Phillip, W. A., The future of seawater desalination: energy, technology, and the environment. *Science* **2011**, *333* (6043), 712–717.
860. Shemer, H., Semiat, R., Sustainable RO desalination – energy demand and environmental impact. *Desalination* **2017**, *424*, 10–16.

861. Kloppmann, W., Vengosh, A., Guerrot, C., Millot, R., Pankratov, I., Isotope and ion selectivity in reverse osmosis desalination: geochemical tracers for man-made freshwater. *Environmental Science & Technology* **2008**, *42* (13), 4723–4731.
862. Friedler, E., Lahav, O., Jizhaki, H., Lahav, T., Study of urban population attitudes towards various wastewater reuse options: Israel as a case study. *Journal of Environmental Management* **2006**, *81* (4), 360–370.
863. Borokhov Akerman, E., V. Simhon, M., Gitis, V., Advanced treatment options to remove boron from seawater. *Desalination and Water Treatment* **2012**, *46* (1–3), 285–294.
864. Segal, H., Birnhack, L., Nir, O., Lahav, O., Intensification and energy minimization of seawater reverse osmosis desalination through high-pH operation: temperature dependency and second pass implications. *Chemical Engineering and Processing – Process Intensification* **2018**, *131*, 84–91.
865. Nir, O., Lahav, O., Coupling mass transport and chemical equilibrium models for improving the prediction of SWRO permeate boron concentrations. *Desalination* **2013**, *310*, 87–92.
866. Birnhack, L., Voutchkov, N., Lahav, O., Fundamental chemistry and engineering aspects of post-treatment processes for desalinated water – a review. *Desalination* **2011**, *273* (1), 6–22.
867. Nir, O., Lahav, O., Single SWRO pass boron removal at high pH: prospects and challenges. In *Boron Separation Processes*, Kabay, N., Bryjak, M., Hilal, N., eds. Elsevier: **2015**, pp. 297–323.
868. Vinson, D. S., Schwartz, H. G., Dwyer, G. S., Vengosh, A., Evaluating salinity sources of groundwater and implications for sustainable reverse osmosis desalination in coastal North Carolina, USA. *Hydrogeology Journal* **2011**, *19* (5), 981–994.
869. Catling, L., Abubakar, I., Lake, I., Swift, L., Hunter, P., *Review of Evidence for Relationship between Incidence of Cardiovascular Disease and Water Hardness.* University of East Anglia and Drinking Water Inspectorate: **2005**.
870. World Health Organization, *Calcium and Magnesium in Drinking Water: Public Health Significance.* World Health Organization: **2009**.
871. Rosborg, I., Nihlgaˊ, B., Gerhardsson, L., Sverdrup, H., Concentrations of inorganic elements in 20 municipal waters in Sweden before and after treatment – links to human health. *Environmental Geochemistry and Health* **2006**, *28*, 215–229.
872. Sedlak, D. L., The unintended consequences of the reverse osmosis revolution. *Environmental Science & Technology* **2019**, *53*, 3999–4000.
873. Jiang, L. H., P., Chen, J., Liu, Y., Liu, D., Qin, G., Tan, N., Magnesium levels in drinking water and coronary heart disease mortality risk: a meta-analysis. *Nutrients* **2016**, *8*, 5.
874. Calderon, R., Hunter, P., Epidemiological studies and the association of cardiovascular disease risks with water hardness. In *Calcium and Magnesium in Drinking-Water. Public Health Significance,* Cortuvo, J., Bartram, J., eds. World Health Organization: **2009**, pp. 110–144.
875. Catling, L. A., Abubakar, I., Lake, I. R., Swift, L., Hunter, P. R., A systematic review of analytical observational studies investigating the association between cardiovascular disease and drinking water hardness. *Journal of Water and Health* **2008**, *6* (4), 433–442.
876. Kozisek, F., Regulations for calcium, magnesium or hardness in drinking water in the European Union member states. *Regulatory Toxicology and Pharmacology* **2020**, *112*, 104589.
877. Yang, C.-Y., Chiu, H.-F., Calcium and magnesium in drinking water and the risk of death from hypertension. *American Journal of Hypertension* **1999**, *12* (9), 894–899.
878. Ben Zaken, S., Simantov, O., Abenstein, A., Radomysky, Z., Koren, G., Water desalination, serum magnesium and dementia: a population-based study. *Journal of Water and Health* **2020**, *18* (5), 722–727.
879. Koren, G., Shlezinger, M., Katz, R., Shalev, V., Amitai, Y., Seawater desalination and serum magnesium concentrations in Israel. *Journal of Water and Health* **2016**, *15* (2), 296–299.
880. Shlezinger, M., Amitai, Y., Akriv, A., Gabay, H., Shechter, M., Leventer-Roberts, M., Association between exposure to desalinated sea water and ischemic heart disease, diabetes mellitus and colorectal cancer: a population-based study in Israel. *Environmental Research* **2018**, *166*, 620–627.
881. Shlezinger, M., Amitai, Y., Goldenberg, I., Shechter, M., Desalinated seawater supply and all-cause mortality in hospitalized acute myocardial infarction patients from the Acute Coronary Syndrome Israeli Survey 2002–2013. *International Journal of Cardiology* **2016**, *220*, 544–550.
882. Rosen, V. V., Garber, O. G., Chen, Y., Magnesium deficiency in tap water in Israel: the desalination era. *Desalination* **2018**, *426*, 88–96.
883. Avni, N., Eben-Chaime, M., Oron, G., Optimizing desalinated sea water blending with other sources to meet magnesium requirements for potable and irrigation waters. *Water Research* **2013**, *47* (7), 2164–2176.
884. Penn, R., Birnhack, L., Adin, A., Lahav, O., New desalinated drinking water regulations are met by an innovative post-treatment process for improved public health. *Water Supply* **2009**, *9* (3), 225–231.

885. Lesimple, A., Ahmed, F. E., Hilal, N., Remineralization of desalinated water: Methods and environmental impact. *Desalination* **2020**, *496*, 114692.
886. Huang, Y., Wang, J., Tan, Y., Wang, L., Lin, H., Lan, L., Xiong, Y., Huang, W., Shu, W., Low-mineral direct drinking water in school may retard height growth and increase dental caries in schoolchildren in China. *Environment International* **2018**, *115*, 104–109.
887. Stein, S., Yechieli, Y., Shalev, E., Kasher, R., Sivan, O., The effect of pumping saline groundwater for desalination on the fresh–saline water interface dynamics. *Water Research* **2019**, *156*, 46–57.
888. Russak, A., Sivan, O., Yechieli, Y., Trace elements (Li, B, Mn and Ba) as sensitive indicators for salinization and freshening events in coastal aquifers. *Chemical Geology* **2016**, *441*, 35–46.
889. Thurber, M. C., *Coal.* Polity Press: **2019**.
890. Yao, K., Meng, M., China expects to lay off 1.8 million workers in coal, steel sectors. www.reuters.com/article/us-china-economy-employment/china-expects-to-lay-off-1-8-million-workers-in-coal-steel-sectors-idUSKCN0W205X.
891. Wacket, M., Germany to phase out coal by 2038 in move away from fossil fuels. www.reuters.com/article/us-germany-energy-coal/germany-to-phase-out-coal-by-2038-in-move-away-from-fossil-fuels-idUSKCN1PK04L.
892. Varadhan, S., Coal-fired plants around New Delhi running despite missing emissions deadline. www.reuters.com/article/us-india-pollution-coal/coal-fired-plants-around-new-delhi-running-despite-missing-emissions-deadline-idUSKBN1Z01VD.
893. Kuriakose, S., Lewis, J., Tamanini, J., Yusuf, S. *Accelerating Innovation in China's Solar, Wind and Energy Storage Sectors.* World Bank Group: **2007**.
894. Marshall, J. S., Renewables beat coal for first time in 130 years. www.eenews.net/greenwire/2020/05/28/stories/1063257289.
895. Brandt, J. E., Lauer, N. E., Vengosh, A., Bernhardt, E. S., Di Giulio, R. T., Strontium isotope ratios in fish otoliths as biogenic tracers of coal combustion residual inputs to freshwater ecosystems. *Environmental Science & Technology Letters* **2018**, *5* (12), 718–723.
896. Groom, N., Special Report: Millions of Abandoned Oil Wells Are Leaking Methane, a Climate Menace. www.reuters.com/article/us-usa-drilling-abandoned-specialreport/special-report-millions-of-abandoned-oil-wells-are-leaking-methane-a-climate-menace-idUSKBN23N1NL.
897. International Energy Agency (IEA), *Renewable 2020: Analysis and Forecast to 2025.* International Energy Agency (IEA): **2020**.

Index

$\delta^{11}B$ · 174–175, *see* boron isotopes

4-methylcyclohexane methanol (MCHM) · 46
24 × 7 Power for All · 79, *see* India
2-n-Butoxyethanol · 187, *see* hydraulic fracturing

$^{87}Sr/^{86}Sr$ · 50, 172, *see* strontium

abandoned oil and gas wells · 187, 241, 258, 261
abandoned wells · 111, 113, 214, 241, 258
Abiotic gas · 141
Absheron Peninsula · 14, 81
acid mine drainage (AMD) · 11, 79–80, 229, 232, 256–257, 261
acid rain · 75, 247, *see* air pollution
acidity · 46, 49, 68, *see* pH
acid mine drainage · *see* coal mining
acids · 118, 163, 176, *see* hydraulic fracturing
acrylamide · 163
adsorption · 152, 178, 180, 198, 206
Africa · 216, 233, 243
agricultural · 5–7, 115, 130, 134, 156, 215, 217, 247, 249, *see* irrigation
agricultural sector · 6, 115, 130, 134, 215
agriculture · 4, 9, 27–28, 134, 215–216, 254
air pollution · 232, 247
air quality · 14–15, 32, 79, 248, 257, *see* air pollution
Alabama · 109, 199
Alaska · 9, 83, 121, 195
Alberta · 90, 116–118, 152, 192–193, *see* Canada
Algeria · 81, 136, 139, 148, *see* oil
alkaline · 35, 39, 47, 49, 68, 255
alkalinity · 118, 146, 171
alkanes · 125, 143, 176, *see* hydrocarbons
Allegheny River · 182, *see* Pennsylvania
alternative water source · 120, 134
aluminum · 49, 61, 162, *see* metals
Amazon · 110
American Petroleum Institute · 86, *see* crude oil

ammonium · 105–106, 110, 162–163, 176, 178, 182, *see* OPW
anaerobic · 57, 64, 191, 193, 237
annulus · 192, 197
anthracite · 16, 34, 51, 61, 236
Anthropocene · 2, 7
anthropogenic · 2, 13, 38, 67, 126, 146, 215–216, 219, 237–241, 260
API gravity · 86, *see* crude oil
Appalachia · 32, 38, *see* Appalachian Basin
Appalachian Basin · 31, 40, 47–48, 101, 105, 110–111, 169–170, 172–173, 176, 187, 190, 192, 233
aquatic system · 57, 66, 68, 80, 105
aquifers · 12, 44–45, 68–69, 111, 113, 119, 122–123, 135, 141, 146, 157, 169, 186–188, 190–194, 198–201, 205–206, 215, 255, 259
Aral Sea · 10
Argentina · 81, 90, 139, 148
argon · 191
aridity · 8, 216
Arkansas · 191
arsenic · 36, 63–66, 68–69, 124, 193, 201, 254, 258
arsenite · 254, *see* Arsenic
Asia · 83, 137, 215–216, 251, 256
Athabasca Oil Sands Region · 119, 126, *see* oil sands
Athabasca River · 117–118, *see* Canada
atmosphere · 2, 13, 33, 63, 68, 114, 119, 126, 142, 218, 233–236, 238, 241–243, 246, 248, 258
atmospheric circulation · 216
Australia · 15, 20, 31–32, 90, 139, 148, 156, 195, 199, 202, 214, 216, 232
Azerbaijan · 14, 81, 248, *see* oil

Bakken · 87, 90, 94, 98, 101, 105–107, 110–111, 114, 122, 130, 132, 162, 243–244, *see* tight oil
ban · 210–211
barium · 105, 110, 169, 174, 176, 180, 182, 247, *see* metals
Barnett · 146, 161, 169, 191, 193, *see* shale gas

Barnett Shale · 161, 191–193, *see* Texas
base–exchange reactions · 174, *see* adsorption
baseline study · 194
batch leaching · 64
Bemidji · 123–125, *see* Minnesota
beneficial use · 11, 97, 115, 204, 206, *see* reuse
benzene · 86, 104, 120, 176, *see* BTEX
Bernard Plot · 144, *see* carbon isotopes
beryllium · 36, 64
bicarbonate · 46, 146, 171, 193, 200, 206, 247, 255
bioaccumulation · 68, 135, 183
biocides · 106, 162
biodegradable · 162
biodegradation · 104, 124, 143
biodiversity · 57
biofuels · 4
Biogenic gas · 141
biomass burning · 237, 240
bis(2-ethylhexyl) phthalate · 186
bitumen · 86, 116, 118–119, 126, *see* crude oil,
bituminous · 16, 34, 51, 61, 236, *see* coals
booms and busts · 15
boric acid · 254, *see* boron
boron · 36, 38, 63–66, 68, 70, 102, 105, 113, 115, 118,
    162, 169, 174, 187, 201, 254–255, 258
boron isotope · 69, 174, 256
bottom ash · 60, 63–64, 70, *see* CCRs
BP global dataset · 30, 137–138, 226–227, 234, 244,
    259, *see* British Petroleum
Br/Cl ratio · 101, 169, 188, *see* tracer
brackish · 94, 118, 161, 204, 250, 252–253
Brazil · 2, 139, 148, 233
breakers · 162, *see* hydraulic fracturing
brines · 11, 100–102, 110, 163, 169–170, 176, 184,
    187–188, 190, 252–253, 258, *see* evaporation
Britain · 14, 136, *see* England
British Columbia · 192
British Petroleum (BP) · 16, 29
British thermal unit, BTU · 16
bromide · 76, 101, 105, 108, 135, 174, 176, 178, 180,
    182
bromine · 75, 108
BTEX · 104, 120, 124, 126, 176, 246, *see* contaminants
buffering capacity · 66
Bureau of Ocean Energy Management, Regulation and
    Enforcement · 131, *see* MMS
butane · 143, 153, *see* hydrocarbons

Ca-chloride · 101, *see* brines
cadmium · 68, 70, 247
calcite · 200
calcium · 61, 105, 162, 169, 174, 200, 206, 255
California · 7, 81, 87, 96, 101, 103, 105–107, 109, 115,
    134, 218
Canada · 3, 15, 20, 86, 90, 116–117, 126, 130,
    138–139, 148, 152, 187, 191–193, 195, 201,
    210–211, 233, 258

carbon · 2, 13, 16, 32–34, 45, 57, 61, 63, 75, 77–79, 86,
    88, 104, 118–119, 124, 126, 143–145, 171, 181,
    187, 190, 194, 199, 218, 234, 237, 239, 241–243,
    245, 257, 260
carbon capture and storage · 257
Carbon Capture and Storage (CCS) · 33
carbon dioxide · 2, 13, 78, 88, 124, 143, 145, 218, 234,
    237, 239, 241–243, 245
carbon emissions · 32–34, 77, *see* global warming
carbon isotope · 118, 144–145, 199, 239–240
carcinogenic · 70, 163
cardiovascular · 255
casing · 185, 192, 197, 210, 213
Cawelo Water District · 115, 134, *see* Kern County
CCR disposal sites · 77, *see* impoundments, landfills
cement · 61, 111, 126, 153, 163, 192–193
Central Valley · 106–107, 109, 115, *see* California
centralized waste treatment (CWT) · 181
Changning · 148, 157, 165, 168, *see* shale gas
chemical additives · 46, 161–163, 175–176, 213, *see*
    frac water
chemical hazards · 11
Chemical Oxidation Demand (COD) · 120, *see*
    wastewater
Chile · 233
China · 2–3, 5, 12, 15–16, 20–21, 29, 31, 35–36, 41,
    44, 46, 50, 52, 54, 56–62, 64, 67, 69, 73–74, 78–79,
    83, 90, 97, 110, 114, 130, 137–139, 147–148,
    154–157, 176, 179, 188, 191, 195, 198, 202–203,
    212, 214, 216–217, 221, 224, 226, 230, 233, 235,
    250, 255–257, 260
chloride · 101–103, 105, 108, 110, 113, 115, 118,
    161–162, 169–172, 176, 179–180, 182, 186, 201
chlorination · 76, 108, 180, *see* treatment
chlorofluorocarbons · 234
chromium · 67, 69, 86, 124, 247
Class II injection wells · 99, 113, 167, *see* disposal
clay minerals · 152
clay stabilizers · 163, *see* hydraulic fracturing
Clean Air Act · 132
Clean Air Act (CAA) · 248
Clean Air Act Amendments · 75
Clean Power Plan · 78
clean water · 7, 28, 135, 218–219, 233, 261
Clean Water Act (CWA) · 132
climate change · 2–3, 7–8, 13, 15, 34, 78, 83, 217,
    234–235, 237, 241, 256, 260–261, *see* global
    warming
climate crisis · 21, 33, 139, 218, 242, 257, 259
climate models · 218
CO_2 reduction · 143 , *see* methanogenesis
coal · 1–6, 9–10, 13–18, 21–22, 24, 27, 29–31,
    33–35, 38–42, 45–46, 48–52, 54, 56–61, 64–72,
    74–76, 78–81, 126, 137, 139, 141, 145, 158–159,
    187, 189, 194–195, 197, 199–201, 205, 217,
    220–221, 224–226, 229, 232–234, 236, 241,
    256–258, 261

coal ash · 10–11, 24, 34, 38, 60–62, 64–69, 71, 75–76, 80, 220, 224, 226, 231, 258, *see* coal combustion residuals (CCR)
coal ash impoundments · 11, 62, 67, 70–71, 230–231
coal burning · 14, *see* coal combustion
coal combustion · 2, 11, 32, 34, 36, 38–39, 45, 52–53, 57, 60, 63, 73–76, 79–80, 159, 236, 241, 261
coal cumbustion residues (CCRs) · 60, *see* coal ash
coal deposits · 32, 34, 38, 61, 64, 69, *see* coals
coal mine drainage (CMD) · 47, *see* acid mine drainage
coal mines · 9, 31, 40, 44–45, 51–52, 78, 194, 229, 232, 256–257
coal mining · 1–2, 5–6, 10–11, 15, 24, 31, 34, 38, 40–42, 45–46, 49–52, 72, 74, 77–80, 220–221, 227, 230–231, 241, 256
Coal mining · 42, 45, 74, 232, *see* coal
coal peak · 33
coal seams · 145, 197–199
coal slurry · 41, *see* coal
coalbed methane (CBM) · 12, 139, 145, 194, 198–199, 209, 214, 241, *see* unconventional gas
coal-fired power plants · *see* coal plants
coalification · 34, 145, *see* coal
coals · 11, 16, 24, 34–37, 39–41, 45–46, 50–51, 61, 63–65, 69, 75, 145, 198, 236
coastal aquifers · 45, 215, 255
cobalt · 49, 124, *see* metals
Cold Lake · 87, *see* oil shale
Colorado · 101, 113, 115, 179, 182, 185, 187, 190, 205, 248–249
Colorado Oil and Gas Conservation Commission (COGCC) · 249
combined cycle · 159–160, *see* electricity
combined-cycle gas turbines (CCGTs) · 57, *see* natural gas
compressor stations · 157, *see* natural gas
Congress · 10, 77, 83
connate water · 100–101, *see* fossil water
conservative elements · 101
contaminated groundwater · 68, 125, 191, 206
contaminated water · 18, 46, 52
contamination · 3, 6–7, 9–12, 24, 28, 38, 45–46, 49–50, 53, 66–70, 72, 83, 97, 99, 105–106, 108–111, 118–119, 121, 123–124, 126, 129, 133, 135, 146, 167, 179–180, 184–187, 189, 191–193, 199, 206, 211, 213–215, 219, 229, 231, 233, 238, 246–247, 256–257
conventional gas · 12, 136, 141, 145–146, 158, 164, 168, 175, 177, 183, 186, 209, 213
conventional oil · 11, 84, 87, 89, 91, 94–97, 102, 106, 110–111, 113–114, 116, 127–128, 130, 133–134, 163, 176, 180, 182–184, 187, 189, 192, 242, *see* crude oil
conventional wells · 130, 193
cooling · 1, 4–6, 11, 18, 22, 27–28, 34, 52–54, 56–58, 64, 73, 80, 120, 126–127, 153, 159, 200, 207, 213, 219–220, 224–225, 227, 233, 249, 258

cooling power generation · 53, *see* thermoelectric plants
cooling systems · 22, 54, 56–58, 80, 159, *see* coal plants
cooling water · 6, 11, 53, 74
Corporate Social Responsibility (CSR) · 132
corrosion inhibitors · 163, *see* hydraulic fracturing
COVID-19 pandemic · 82, 84, 199, 259
cracking · 106, 119–120
cross-linked gel · 161–162, *see* hydraulic fracturing
crosslinker · 106, *see* hydraulic fracturing
crosslinkers · *see* hydraulic fracturing
cubic meter per megawatt per hour (m^3/MWh) · 18, *see* electricity

dam · 67
Davernay Shale · 90, *see* Canada
Deep Horizon · *see* oil spill
Deepwater Horizon · 82–83, 121, *see* oil spill
deep-well injection · 97, 113, 130, 168, 176, 212, 231
degree of evaporation · 101
deicing · 114, 133, 184, 212, 214, 232
Denver-Julesburg Basin · 179, 190, *see* Colorado
Department of Environmental Quality (DEQ) · 51, *see* Pennsylvania
desalination · 1, 4, 13, 18, 116, 215, 218, 250, 252–253, 260–261
desalination plants · 218, 251, 254
desalinization · 181
desorption · 174, 195, *see* adsorption
diesel · 86, 123, 186, 236, 250, *see* crude oil
dilution factor · 52, 109, 125, 181, 185, 202, 204, 206, 250
disclosure · 210–211, 214
Disi Water Conveyance · 5
disinfection byproduct · 108, 180, *see* treatment
disinfection byproducts · 76, 105, 108–109, 135, 182
disposal of oil wastewater · 109
disputed waters · 67
dissimilatory bacterial sulfate reduction · 193
dissolution · 49, 101, 168, 206
dissolved inorganic carbon · 171, 200, *see* DIC
dissolved organic matter (DOC) · 104–105, 107, 114–115
dissolved oxygen (DO) · 57
distillation · 116, 119, 126, 250, 252
dolomite · 200, 255
domestic · 77, 82–83, 94, 108, 134, 138, 156, 180, 186, 189, 193, 216, 248–249, 251, 254, 261
domestic wastewater · 94, 251, 254
drilling · 3–5, 10–11, 15, 18, 22, 24, 78, 82–83, 85, 94, 116, 122, 127, 130, 132, 138–139, 146, 153, 168, 184–185, 189–190, 192–193, 207, 209–211, 220, 222, 242, 248
drought · 13, 133, 216, 218
droughts · *see* climate change
dry cooling · 56–57, 159, 225, *see* thermoelectric plants

dry natural gas · 143, 147
dust suppression · 114, 133, 135, 184, 212, 214, 232

Eagle Ford · 87, 90, 98, 161, 191, 244–245, 249, *see*
　　shale
Earthjustice · 67
earthquakes · *see* siesmicity
ecological effects · 57, 72
ecological standards · 110, 112
ecological threshold · 115, 183–184
Ecuador · 9, 83
effluents · 11, 38, 45, 48–49, 57, 67, 69–70, 72, 76,
　　109, 182, 232, 255, *see* wastewater
Egypt · 3
El-Niño · 216
electric vehicles · 257
electrical conductivity · 46, 50, 52, *see* salinity
electricity · 1–2, 4–6, 11–13, 15–18, 22, 24, 26–27, 29,
　　31, 41, 43, 45, 52–53, 56, 77–80, 125, 139,
　　153–154, 158, 160, 207, 217, 220, 224–225, 227,
　　235–236, 249, 251, 256, 260–261
electricity intensity · 18, 252–253
electrostatic precipitators (ESPs) · 60, *see* CCRs
Elk River · 46, *see* West Virginia
emission · 2–3, 11, 13, 32, 45, 57, 59–60, 68, 74–75,
　　79, 106, 119, 126, 142, 214, 219, 234–235, 238,
　　240–241, 243, 246, 257, 260–261
endocrine disrupting activities · 163
energy policy · 10
Energy Policy Act · 10, 210
energy transitions · 78, 257
England · 14, 81
enhanced oil recovery (EOR) · 88, 90, *see* conventional
　　oil
environmental calamities · 82, *see* water contamination
environmental externalities · *see* legacy
environmental impacts · 3, 9–10, 13, 49, 61, 76, 79,
　　82–83, 94, 118, 121, 150, 181, 189, 206, 210–211,
　　256
environmental risks · 11, 106, 168, 189, *see*
　　contamination
epidemiologic studies · 255
Estimated Ultimate Recovery (EUR) · 150–151
ethane · 143, 145, 153, 190, 238–240, *see*
　　hydrocarbons
ethylbenzene · 104, 120, 176, *see* BTEX
ethylene glycol · 162
Euphrates · 215
Eureka Resources · 181, *see* wastewater
Europe · 2–3, 20, 29, 60, 78, 137, 216–217, 231–232,
　　235, 256
eutrophication · 57
evaporated seawater · 101, 104, 134, 168–169, 171,
　　213
evaporation · 6, 53, 97, 100–101, 105, 159, *see* salts
exploration · 1–4, 8–12, 15, 18, 21, 24, 38, 81, 83, 85,
　　87–88, 91, 93–94, 98, 100, 110, 113–114, 121,
　　127, 131–134, 139, 142, 147, 155–156, 163,

167–168, 185, 189, 207, 210–211, 213–214, 219,
　　222–223, 229, 233, 236, 239, 242–243, 245–246,
　　256, 259
explosion · 189
Exxon Valdez · *see* oil spill

Fayetteville · 146, 169, 191, *see* shale gas
Federal Oil and Gas Leasing Reform Act · 132
fermentation · 143, 200, 237
ferric oxides · 193, *see* oxides
Five-Year plans · 212, *see* China
flammability · 189
flammable · 106, 189, *see* gas
flaring · 9, 13, 114, 210, 242–243, 245, 248
flowback · 11, 24, 85, 88, 94–97, 113–114, 134–135,
　　139, 146, 151, 157, 163, 165, 168, 170, 172, 186,
　　209, 213, 230, *see* hydraulic fracturing
flowback and produced water (FPW) · 11, 24, 88,
　　94–96, 114, 134, 139, 151, 157, 163, 165, 168,
　　170, 172, 209, 213
flue gas · 60, 63–64, 75, *see* coal plants
flue gas desulfurization (FGD) · 60, *see* coal ash
fluids migration · 186
fluoride · 63, 215, 236, 255
fly ash · 60–61, 63–64, 70–71, 75–76, 236, *see* CCRs
food chain · 59
food security · 215
formaldehyde · 163
formation water · 87–88, 95, 100–102, 106, 111, 118,
　　134, 146, 151–152, 163–164, 168, 170–172, 174,
　　176, 178, 187, 199, 201, 213, *see* produced water
Fort McMurray · 118, *see* Alberta
fossil fuels · 1–3, 5–6, 8, 10, 13–19, 22, 25, 27–28, 41,
　　78–79, 120, 123, 217, 219–221, 223, 225–227,
　　229–235, 237, 240, 256–258, 261
fossil water · 1–3, 5–6, 8, 10, 13–19, 22, 25, 27–29, 53,
　　77–79, 81, 100, 139, 205, 217, 219–221, 224,
　　226–227, 230, 232–234, 236–237, 239–240,
　　256–259, 261
frac chemicals · 106, 114, 146–147, 150, 161, 187,
　　212–213, *see* hydraulic fracturing
frac fluids · 133, 162, 186, *see* hydraulic fracturing
frac water · 12, 114, 161, 163, 172, 174, 186, 213, *see*
　　hydraulic fracturing
FracFocus · 212
fracking chemicals · 88, 94, *see* hydraulic fracturing
fractures · 88, 146, 150, 152–153, 161, 185, 189, 213,
　　see hydraulic fracturing
France · 139, 211
Franklin · 181, *see* Pennsylvania
free gas · *see* natural gas
free-gas · 192
freshening · 174, 255
friction reducers · 162, *see* hydraulic fracturing
Fruitland Formation · 199, *see* San Juan
fugitive atmospheric emissions · 218
fugitive emission · 236
fugitive gas · 114, 192

fugitive leaking · 194
fugitive methane · 193, *see* stray gas
Fukushima · 256, *see* Japan
Fuling · 148, *see* shale gas

Gansu Province · 110, *see* China
Garfield County · 113, 187, *see* Colorado
gas turbines · 57, 159, *see* electricity
gas wetness · 143, *see* natural gas
Gasland · 110, 189, *see* hydraulic fracturing
gasoline · 86, 119, 123, 186, 236
geochemical characteristics · 169, 203
geochemical evolution · 101
geochemical fingerprint · 145, 187, 213, *see* isotopes
geochemical tracer · 69, *see* water contamination
geochemical tracers · 175, 187
geochemistry · 10–11, 68, 101, 169, 171, 187, 191, 194
geogenic · 67, 216, *see* contamination
geological history · 101–102, 170
geothermal · 27, 78, 141, 258
Germany · 32–33, 78, 211, 218, 257
global · 1–5, 10, 12–13, 15–16, 19, 21–23, 29–30, 33, 38–40, 45, 52, 57, 61–62, 64, 75, 77–79, 84, 95–96, 116, 121, 126, 130, 132, 134, 136–138, 147, 156, 194–195, 214–218, 221–222, 224–225, 229, 231–236, 242–244, 249, 251, 255, 258–259, 261
global climate · *see* climate change
global electricity · 52, 217, 224, 250, 252
global fluxes · 38
Global Gas Flaring Initiative · 243, *see* World Bank
global warming · 7, 13, 32–33, 217, 219, 226, 233, 235–236, 256, 260, *see* climate change
global wastewater · 230, 251
Global Water Cycle · 13, 215
global water resources · 13, 130, 260
Gravity Recovery and Climate Experiment (GRACE) · 216
greenhouse gases (GHGs) · 3, 13, 33, 78, 216–219, 225, 233, 236–237, 243, 246, 248–249, 257
Greenpeace · 67
Ground Water Protection Council · 212
groundwater · 4–5, 7, 12, 27, 44–45, 47, 50, 63, 66–68, 72, 77, 94, 108, 113, 115, 117–118, 122–125, 135, 141, 146, 161, 169, 171, 180, 186–187, 189, 191–195, 197, 199–201, 205–206, 214–215, 218, 231, 247, 250, 254, 258
groundwater pollution · 12
groundwater quality · 77, 123–124, 186, 188, 193–194
groundwater table · 44, 67, 123, 198, 216
Gulf · 5, 9, 81–83, 104, 121, 131, 200, 253
Gulf Coast · 83, 177
Gulf of Mexico · 131
gypsum · 61, 168, 206

H$_2$S · 106, 120, 153–154, 158, 193, *see* hydrogen sulfide
halides · 104, 106, 108, 176, 180, *see* salts

halite · 101, 168, *see* NaCl
Halliburton loophole · 210, *see* regulations
halogens · 61, 75, *see* salts
Hart · 181, *see* Pennsylvania
Haynesville · 145–146, 153, 167, 169, 191, *see* shale gas
hazardous waste · 133, 232
heart disease · 255
heavy metals · 247
helium · 143, 191
heteroatoms · 86, *see* hydrocarbons
hexane · 143, *see* hydrocarbons
hexavalent chromium · 67, 69, *see* water contamination
horizontal drilling · 3, 147, *see* shale gas
horizontal well · 146–147, 155, *see* hydraulic fracturing
Horn River Basin · 152, *see* Canada
Horseshoe Canyon/Belly River Group · 201, *see* Canada
human development · 9
human health · 15, 46, 63, 66, 70, 75, 106, 116, 120, 133, 135, 161, 163, 213, 227, 255, 258
hydraulic fracturing · 1, 3–6, 9–11, 18, 21–22, 78, 85, 88, 91, 94–95, 97, 99, 102, 106, 114, 122, 127, 130, 132–135, 138–139, 146–147, 150–151, 153, 155–156, 160–165, 167, 169–171, 174–175, 181, 185–187, 189, 192, 198, 207, 209–212, 220, 222, 231
hydrocarbons · 2, 86–87, 95, 104, 116, 119–121, 124, 126, 141, 143–145, 153, 164, 176, 186, 190–191, 194, 234, 241, 243, 246
hydrogen · 106, 118, 120, 143, 194, 200, 255–256
hydrogen sulfide · 106, *see* H2S
hydrogeological · 190
hydrogeology · 10, 122, 194
hydrological balance · 205
hydrological system · 5, 25
hydrology · 10, 233
hydropower · 4, 27, 260

Illinois · 31, 171
imbibition · 152, 170–171, *see* hydraulic fracturing
impaired water · 11–13, 18, 24, 28, 45–46, 51–52, 60, 72–74, 108–110, 115, 121, 125, 129–130, 133, 135, 140, 180–181, 183–184, 202, 204, 206, 218, 229, 231–233, 247, 256, 261
impoundment · 68, 72, 77, 206
impoundments · 11, 61–62, 67, 70, 76–77, 80, 205–206, 232, 258
India · 2, 7, 16, 20, 31–32, 50, 53, 60–62, 64, 73, 79, 83–84, 139, 156, 203, 215–216, 233, 235, 257
Indian Ocean · 217
Indo-Gangetic Basin · 215, *see* India
Indonesia · 20
Indus · 215
industrial · 5, 7, 119, 153, 156, 217, 234, 237, 249
Industrial Revolution · 2, 14, 81
infrastructure · 136–137, 168, 212, 242, 244
ingrowth · 111, *see* radioactive decay

insoluble · 124, *see* contaminants
international · 9, 16, 33, 69, 77, 81, 131, 217, 222, 232, 244
International Convention for the Prevention of Pollution from Ships (MARPOL) · 131, *see* oil spills
International Energy Agency (IEA) · 4–5, 22–23, 57, 139, 210, 216, 220, 222, 227, 242, 250–252, 260
international waters · 9
iodide · 105, 108, 135, 176, 178, 180, 182
iodine · 108, 169, 182
Iran · 3, 20–21, 81–82, 136–137, 215, 244
Iraq · 20, 81–82, 244, *see* oil
iron · 46, 49, 61, 86, 105, 124, 163, 193, 247, *see* metals
iron oxides · 124
irrigation · 4, 27, 115, 134, 186, 199, 205, 215, 250–251, 254
isotope fractionation · 143, 239, 256
isotope ratio · 69, 145, 174, 239–240, *see* tracer
isotopes · 69, 102, 113, 118, 171, 175, 187, 214, 239, 254–255
Israel · 3, 5, 23, 53, 138, 254–255
Italy · 137

Japan · 2, 84, 256
jet fuel · 86, *see* crude oil
Joaquim Valley · 87, *see* California
Jordan · 3, 5–6, 23, 215
Josephine · 181–182, *see* Pennsylvania
joules, J · 17

Kansas · 136
Kentucky · 38, 49, 113
Kern County · 115, *see* California
kerogen · 87, 116, 143–144, 152, *see* oil
kilowatt-hour (kWh) · 16, *see* electricity
Kingston · 66, *see* TVA
Kuwait · 20, 251, 254

La Niña · 216
landfills · 60, 62, 67, 70, 72, 76–77, 80, 141, 237, 258, *see* CCRs
Latin America · 3
leaching · 45–46, 64, 66, 68, 73, 77, 118, 178, 206, 261
lead · 63, 68, 70, 105, 110, 126, 139, 201, 247, 256, 260, *see* metals
leaking · 11–12, 63, 66, 69, 72, 108, 111, 119, 123, 146, 167, 180, 184–187, 189, 192–194, 206, 233, 238–239
leaks · 106, 130, 180, 184–186, 213, 242, 261
Lebanon · 138
legacy · 8–9, 11, 25, 52, 74, 79–80, 106, 111, 113, 187, 199, 214, 232, 241, 243, 257
life cycles · 18, 28, 34, 44, 72, 127, 139, 159, 207, 238, *see* water use

lignite · 31, 34, 61, 236, *see* coal
limestone · 171
linear gel fluid · 161–162, *see* hydraulic fracturing
liquid to solid ratio (L/S) · 64
liquified petroleum gas (LPG) · 125
liter per gigajoules (L/GJ) · 18, *see* water intensity
lithium · 66, 102, 105, 118, 187
London · 14
Longmaxi Formation · 148, *see* Sichuan Basin
Longmaxi Shale · 171–172, 175, *see* Sichuan Basin
Long-Range Transboundary Air Pollution (LTRAP) · 232
Lycoming County · 192, 194, *see* Pennsylvania

magnesium · 105, 200, 206, 255
magnesium deficiency · 255
manganese · 105, 110, 124, 193, *see* metals
man-made · 12, 45–46, 114, 161, 213, 216, 233, 254–255, *see* Anthropogenic
man-made chemicals · 46, 161
Mannville Formation · 201, *see* Canada
Mannville Group · 192, *see* Canada
Marcellus · 105, 110, 114, 145–146, 153, 157, 161–162, 165, 167, 169, 171–172, 175, 179, 181, 183–185, 187, 190–191, 193–194, *see* shale
Marcellus Shale · 146, 153, 157, 161–162, 165, 167, 169, 171–172, 174–175, 179, 181, 185, 190–192, *see* hydraulic fracturing
marine ecology · 121
Martha oil field · 113, *see* Kentucky
Maximum Contaminant Limit (MCL) · 193
Medina Group · 169, *see* Appalachian Basin
Mediterranean · 137–138, 254
Mekong · 215
mercury · 32, 36–37, 61, 63–65, 70, 75, 126, 236
Mercury and Air Toxic Standards (MATS) rule · 75
mercury emission · 32, 61, 75–76, 126, *see* air pollution
metalloids · 36, 47, 51, 64, 66, 68, 104–105, 176, *see* metals
metals · 35, 45–46, 49, 51, 64, 66, 68, 76, 104–105, 108, 115, 119, 124, 126, 135, 176, 180, 182–183, 213, 232, 247, 254, *see* water contamination
meteoric water · 69, 100–101, 146, 168–169, 171, 191, 255
methane · 2–3, 10, 12–13, 15, 124, 139, 141, 143–144, 146, 152–153, 164, 171, 188–189, 191–195, 199, 202, 209, 213–214, 218, 234, 236, 238, 240–243, 246, 248–249, 258
methane emission · 239, 241
methane leaking · 238
methanogenesis · 143, 145, 171, 199, 237
Mexico · 9, 81, 83, 99, 104, 121, 131, 133, 139, 148, 182, 185, 199, 205, 214–215, 243
Middle East · 3–4, 7–8, 23, 53, 139, 215–216, 242, 250–251
mineral rights · 131, 210
mineralization · 45–46, 49–50, 255, *see* salinization

Minerals Management Service (MMS) · 131
mining · 4–5, 11, 18, 22, 24, 27, 31, 38, 40–42, 46, 51–52, 61, 73, 77–79, 92, 116–118, 220–221, 230–231, 256–257
Minnesota · 123
mixing · 59, 70, 102, 145, 171, 174, 187
mobilization · 38, 45, 47, 51, 64, 66, 100, 105, 123–124, 174, 194, 200, 247, 258
molybdenum · 36–37, 64–65
monitoring wells · 67, 77, 186, 191
Montana · 109, 199–200, 205, 214
Montney and Horn River · 192, 194, *see* shale
Montney Shale · *See* Canada
moratorium · 210–211
Morocco · 137
mountaintop mining (MTM) · 38, 50 *See* coal
Mud River · 50, *see* West Virginia
multi-effect distillation (MED) · 250, 252–253
multi-stage flash distillation (MSF) · 250, 252–253, *see*
municipal wastewater · 181, 251
municipal wastewater treatment plants (WWTP) · 181, *see* wastewater
municipal water · 120

NaCl · 101, *see* salts
naphthalene · 104, 163
National Oceanic and Atmospheric Administration (NOAA) · 237
National Pollutant Discharge Elimination System (NPDES) · 132, 181
National Research Council (NRC) · 121, 196, 201, 203, 205
National Shale Gas Development Plan · 212, *see* shale gas, China
natural gas · 1–3, 6, 9–10, 12–13, 15–18, 22–23, 27, 29, 32, 57, 77–78, 80, 99, 110, 114–115, 125, 136–139, 141–145, 147, 150–151, 153–154, 156–159, 161, 163–166, 168, 171, 176, 181–186, 189–192, 194, 199, 202–204, 207, 210, 212–214, 217, 220–222, 224–225, 227, 230–231, 234, 236, 239–240, 242, 244–245, 250, 256–258
Natural gas · 17, 27, 136, 139, 141, 154, 163, 232
Natural gas combined cycle plant (NGCC) · 159, *see* electricity
natural gas condensates · 153
Natural Resources Defense Council · 211
naturally occurring · 146
naturally occurring contaminants · 194
naturally occurring radioactive elements · *see* wastewater
Naturally Occurring Radioactive Material · 104–105, 176, *see* NORMs
neon · 191
Netherlands · 2, 136, 193
New Mexico · 122
New York · 114, 181, 184, 189, 211
nickel · 70, 86, 124, 126
Niger Delta · 83, 110, 233, 242, 247–248, *see* Nigeria

Nigeria · 9, 81, 83, 105, 244, 247–248
Nile · 215
nitrate · 124, 216, 247
nitrogen · 75, 86, 119, 143, 145, 243, 246
nitrous oxide · 234
NO_2 · 75
noble gas · 185, 190, 194, 214
non-hazardous · 210
non-renewable · 205
nonrenewable · *see* sustainability
NORM · 64, 104, 176, 180, 183, *see* Naturally Occurring Radioactive Material
North Africa · 8, 23, 148, 250–251
North America · 139, 187, 216, 235, 242, 251
North Carolina · 67, 69–70, 254–255, 258, *see* NC
North Dakota · 101, 106, 109–111, 122, 125, 185, 233, 243–244
Northern Sea · 86
nuclear · 4, 6, 27, 31, 52, 256, 258
nutrients · 104–105, 108, 134, 176, 180, 213

Obama Administration · 75, 77–78, 248
ocean · 96, 121, 254
offshore · 95, 222
Ogoniland · 248
Ohio · 168
oil · 1, 3–5, 9–10, 13–16, 18, 21–22, 24, 32, 75, 78–79, 81–84, 86–88, 90–92, 94–96, 99, 101–102, 104–105, 107–111, 114, 116–118, 120–123, 125–129, 131, 133–134, 136, 139, 141, 143–144, 146, 148, 153, 156–157, 161–163, 167–168, 170–172, 175, 179–184, 187, 189–190, 192–194, 199–200, 210–212, 217, 221–222, 224, 226, 229, 231, 233, 236, 238, 241–243, 246, 248, 250, 256, 258–259, 261
oil and gas industry · 81, 132, 161, 210, 238, 241, 245, 258
oil and gas regulation · 131–132
Oil and Gas Royalty Management Act · 131
oil and gas wells · 110–111, 128, 139, 142, 172, 175, 181, 187, 192, 194, 241, 258
oil basins · 90, 98, 101, 103, 106–107, 134
oil combustion · 126–127
oil exploration · 11, 85, 87–88, 91, 94, 110, 114, 128, 130, 134, 168, 242, 245
oil industry · 14, 81, 106, 130, 132, 136
Oil Pollution Act (OPA) · 131, *see* oil spills
oil prices · 15, 82, 84
oil produced water (OPW) · 96–97, 102, 104–111, 114, 128, 133, 176, 231
oil production · 12–14, 18, 20–21, 83, 85, 87–89, 93–94, 96, 100, 106, 109–111, 115–116, 118, 121–122, 125, 127, 129–131, 133–134, 136, 167, 220, 222–223, 231, 242–245, 247–248
oil refining · 12, 86, 119, 126, 134, 220, 226, *see* crude oil, *see* oil
oil sand · *see* crude oil
oil sands · 116–118, 126, 130, *see* crude oil

oil sands process-affected water (OSPW), 118, *see* oil sands

oil separation · 116, *see* crude oil

oil shale · 84, 87, *see* crude oil

oil spills · 12, 110, 121–123, 125, 132, *see* crude oil

oil tankers · 83, 131

oilfield waters · 101

Oklahoma · 86, 110, 133

Oman · 3

once-through · 53–54, 56, 80, 159, *see* thermoelectric plants

onshore · 95, 97, 99, 108, 121, 125, 127–128, 130, 222, 230–231, 242

open loop · 53–54, 56, 74, 154, *see* thermoelectric plants

organic acids · 104, 118

organic compounds · 104, 119, 125, 186, 246

organic matter · 34, 106, 108, 116, 120, 141, 143–145, 151, 163, 171, 176, 195

organic-rich sediments · 34

osmotic effects · 152

Outer Bank · 255

Outer Continental Shelf Lands Act · 132

outfall · 70, 121, 183

overexploitation · 45, *see* sustainability

oxidation · 46, 63, 66, 68, 75, 145, 191, 193, 200, 237, 239, 257

oxidation-reduction potential (ORP) · 66

oxides · 61, 75, 193, 246

oxidizing · 66, 68

oxygen · 34, 49, 57, 86, 104, 118, 124, 143, 162, 171, 233, 254–255

Pacific · 137, 251

Pakistan · 3, 79, 156, 215–216

Paris Agreement · 77, 79, 261, *see* climate change

particulate matter (PM) · 32, *see* air pollution

Pavillion Field · 185, *see* Wyoming

peak oil · 15, 144

peat · 34, 36

Pennsylvania · 14, 46, 51, 76, 81, 105–106, 109–110, 113–114, 133, 176, 180–190, 192–194, 212, 214, 233, 238, 241

Pennsylvania Department of Environmental Protection · 181

pentane · 143, 153, *see* hydrocarbons

perception · 10, 94

permeability · 87–88, 95, 103, 122, 144, 150, 161, 163, 198, 206

permeates · 254–255, *see* desalination

Permian · 87, 90, 93–94, 98–99, 101, 106–107, 109, 114, 122, 128, 134, 147, 153, 156, 161, 165, 171, 174, 243–244, 249, *see* tight oil

Permian Basin · 90, 93, 99, 101, 107, 109, 114, 122, 128, 134, 153, 156, 165, 243, 246, 249, *see* tight oil

petroleum · 14, 16, 18, 20, 27, 29, 81–82, 84, 86, 119, 121, 125–126, 129, 137, 141, 153, 234, *see* crude oil

petroleum coke · 126

pH · 46, 48–49, 64, 66, 68, 106, 162, 247, 254–255

phenols · 104, 120

Piceance · 198, 201–202, 205, *see* Coalbed methane

pipeline burst · 123, *see* spills

pipelines · 82, 136, 157, 242

pits · 119, 180

plant toxicity · 115

plume · 124, 191, 193, 247

Poland · 29, 139, 191, 256

polycyclic aromatic hydrocarbon (PAH) · 104, 119, 246, *see* crude oil

Polycyclic Aromatic Hydrocarbons · 246, *see* PAHs

population growth · 7, 215, 217, 219

potable water · 134, 156, 213, 253, 255

potassium · 162, 186

Powder River · 195, 199–203, 205–206

power generation · 16, 28, 34, 52, 56, 73, 78–80, 217, 220, 224–226, 229, 235, 261, *see* electricity

power plants · 5, 16, 29, 41, 53, 57, 60, 67, 75–76, 78, 93, 126, 159, 220, 227, 257, *see* electricity

precipitation · 7, 36, 51, 101, 115, 163, 169, 180–181, 200, 216, 218, 237

prevention of pollution · 131

primary energy · 4–6, 22, 29, 220, 249

processing · 5–6, 11, 18, 22, 24, 28, 34, 40–41, 44–46, 57, 69, 73, 79, 92, 126–127, 139, 153–154, 158, 207, 217, 220–221, 230, 232–233, 238, 249, 260

produced water · 9, 85, 87, 90, 92, 94–96, 98, 101–102, 104, 108, 113, 115, 128–129, 133–135, 154, 157, 163, 165–166, 168, 170–172, 175–178, 180–181, 183–185, 188, 197–200, 202–204, 206, 209, 211, 213, 230–231

propane · 143, 145, 153, 190, *see* hydrocarbons

proppant · 162, *see* hydraulic fracturing

proprietary · 161, *see* hydraulic fracturing

proxy · 64, 193

Prudhoe Bay · 83, *see* Alaska

public debate · 76, 156

public perception · 94, 134, 213

public scrutiny · 210

publicly owned treatment works (POTWs) · 181, *see* wastewater

pyrite · 46, *see* sulfide minerals

Qatar · 21, 81, 136, 138, 254

quad · 16, 25–26

radioactive · 11, 36, 64, 97, 105, 108, 135, 169, 172, 178, 183, 213, 258

Radioactive elements · 64, *see* NORMs

radioactivity · 12, 36, 83, 106, 180, *see* Radium

radium · 63–64, 102, 105–110, 114–115, 133, 163, 169, 175–178, 180, 182–183, *see* radioactivity, NORMs

Raton · 195, 201–202, 205, *see* Coalbed methane

recharge · 51, 191, 199–200, 205

recirculation · 53, *see* closed-loop
recovery and extraction · 220
recovery ratio · 253, *see* desalination
recovery stage · 11, 153, 157, 220, 224, 226
recycling · 114, 117, 159, 168, 231, 250, *see* reuse
redox state · 64, 66, 194
reducing · 32–33, 52, 66, 68, 75, 78, 106, 124, 141,
    152, 157, 162, 172, 193, 200, 248, 261
reduction · 2, 7, 20, 33, 40, 66, 68, 74, 79, 114–115,
    121–122, 130, 143, 146–147, 161, 193, 199–200,
    217–219, 240, 247, 256–257, 260
reductive dissolution · 193
regulations · 11–12, 74–76, 80, 82, 105, 131–133, 187,
    204, 209–211, 213, 232, 248
regulatory framework · 210
remediation · 121, 124, 220, 233, 243, 260
renewable energy · 13, 21, 28, 52, 80, 217, 226–227,
    257–259, *see* solar, wind
renewable water · *see* sustainability
replenishment · 51, 117, 216
Resource Conservation and Recovery Act · 76, 133,
    232, *see* US EPA
resource curse · 9
reuse · 11, 61–62, 90, 94, 96, 100, 114, 133–134, 159,
    167, 176, 199, 204, 212, 230, 249, 251, 254, 261
reverse osmosis · 5, 18, 251, *see* desalination
riverbed sediments · 108, 180, 182
RO membrane · 254–255, *see* desalination
Romania · 233
Ruhr region · 78, *see* Germany
Russia · 20, 32, 84, 90, 116, 136–137, 148, 195, 214,
    243–244

Safe Drinking Water Act (SDWA) · 132, 210
Sahel · 217, *see* Africa
Salina evaporite formation · 169, *see* Appalachian Basin
saline–fresh water interface · 255, *see* seawater
    intrusion
salinity · 49–50, 94, 100–102, 105–106, 109, 114–115,
    152, 157, 161, 164, 169–171, 174, 176, 178,
    187–188, 191, 199–200, 202, 204, 206, 247, 253,
    255, *see* salts, *see* water contamination
salinization · 108, 113, 169, 180, 187, 205–206, 215,
    255, *see* contamination
salt caverns · 154, 157, *see* natural gas
salt dissolution · 171
salts · 12, 46, 50, 70, 101, 104, 108, 110, 114, 120,
    134, 163, 168, 176, 180–182, 184, 206, 213, 253,
    see water contamination
San Juan · 195, 199, 201–202, 204–205, *see* Coalbed
    methane
sandstone · 87, 116, 141, 171, 179
Santa Barbara · 9, 82
Saskatchewan · 90, *see* Canada
satellite · 216, 240, 242
saturated zone · 122, 124
Saudi Arabia · 5, 20, 81, 84, 130, 251, 254
scale inhibitors · 163, *see* hydraulic fracturing

scrubber · 63, 75, 126
seawater evaporation · 101
seawater intrusion · 45, 215, *see* water contamination
seawater reverse osmosis · 251, *see* SWRO
secular equilibrium · 36
sedimentary rocks · 35–37, 49, 64–65, 141, 143–144
seismicity · 12, 99, 109, 114, 130, 133, 168
selenium · 8, 36–37, 47, 50, 63–66, 68–69, 72, 193,
    198, 247
Seven Sisters · 81, *see* oil
shale · 2–3, 6, 10, 12, 15, 20–21, 78–81, 87–88, 91, 94,
    98, 102, 105, 114, 133, 138–139, 141, 144,
    146–148, 151–154, 156, 158, 160–165, 168,
    170–172, 176, 178, 181, 184–187, 189, 191–193,
    195, 199–200, 207, 210–214, 220, 223, 225, 230,
    236–238, 242
shale formations · 88, 103, 145–147, 169, 175, 194
shale gas · 2–3, 6, 10, 12, 15, 21, 78–80, 88, 94, 102,
    114, 139, 141, 146–148, 151, 153–154, 156–158,
    160–165, 168, 171–172, 174, 176, 181, 184–187,
    189–195, 199, 207, 209–214, 222–223, 225, 230,
    236–238, *see* unconventional
shale plays · 147, 149, 157, 162, 165, 191, *see* shale
    gas
shale revolution · 15, 133, 138, 165–166, 236, *see*
    hydraulic fracturing, *see* shale gas
shallow aquifers · 187, 190–191, 199, 205, *see*
    hydrogeology
Siberia · 110, 136
Sichuan Basin · 3, 114, 139, 148, 152, 154, 156–157,
    161, 165, 171, 174, 188
slick water · 161–162, *see* hydraulic fracturing
sludge · 250, *see* wastewater treatment
SO_2 · 32, 75
socio-economic analysis · 10
soil · 40–41, 64, 77, 82, 97, 106, 108, 110–111, 115,
    135, 180, 184, 204, 206, 237, 247
soil amendments · 206
solar · 21, 27, 78, 80, 226, 257–258, 260
soluble · 64, 66, 68–69, 123–124, 164, 172, 193, 200,
    see contaminants
soluble elements · 64, 66
South Africa · 29, 38, 148, 190, 211, 233
South Korea · 2, 256
Soviet Union · 136–137, 256, *see* Russia
Spain · 137
speciation · 66
spherical ash particles · 63, *see* CCRs
spills · 9, 11, 66, 82, 85, 94, 106, 108–110, 121–123,
    125, 129–132, 135, 163, 180, 182, 184–185, 189,
    207, 213, 229, 232–233, 258, 261, *see*
    contamination
spreading · 15, 109, 114, 180, 184, 191, 205, 232, *see*
    wastewater
Steam Electric Power Generating Effluent Guidelines ·
    76, *see* US EPA
steam engines · 29
steam turbines · 159, *see* electricity

stray gas · 9, 12, 186, 190–193, 214, 238, *see* contamination

stray gas contamination · 186, 190–192, 194, *see* hydraulic fracturing, *see* shale gas

stream ecology · 206

strontium · 50, 66, 68, 102, 105, 118, 169, 171–172, 176, 183, 187, 193

structural integrity · 77, 193, *see* oil and gas wells

structure fill · 61, *see* CCRs

subsurface mining · 34, 38, 40–41, *see* coal

sulfate · 46, 49–50, 61, 106, 162, 187, 193, 200, 206, 247, *see* salts

sulfate reduction · 193, 200

sulfur · 34, 40, 45, 47, 61, 63, 69, 75, 86, 119, 126, 143, 232–233, 243, 246, 257

sulfur minerals · 45, 257, *see* Acid Mine Drainage

surface mining · 9, 12, 24, 27, 31, 34, 38, 40–42, 45, 47, 49, 51, 62–63, 66–68, 70, 76–77, 79–80, 88, 97–99, 105, 108–110, 115–116, 118, 120, 129–130, 133, 135, 146–147, 152, 161–164, 171, 180–182, 184, 186–187, 189, 192–193, 197, 205–206, 213, 215–216, 218, 222, 226, 231, 234–235, 247, 250

surface water · 9, 12, 27, 49–51, 62, 67–68, 76, 97–99, 105, 108–109, 115, 118, 129–130, 133, 135, 161, 171, 180–182, 184, 187, 205–206, 215, 218, 231, 247, 250

surfactants · 106, 163

sustainable development · *see* sustainability

Sutton Lake · *see* NC

SWRO · 250–251, 253, *see* seawater reverse osmosis

Syria · 138

Taiwan · 256

tankers · 82, 136, *see* oil

Tennessee · 40, 49, 66, 69, 76

tertiary wastewater treatment · 251

Texas · 86, 90–91, 99, 103, 109, 128, 133–134, 156, 191–192, 194, 214, 243, 245, 248

Texas Railroad Commission (TRRC) · 248

thermal pollution · 6, 11, 57, 74, 80, 213, 233

thermoelectric plants · 4–6, 18, 22, 27–28, 52, 57, 80, 219, 224–225, 233, *see* coal plants

Thermogenic gas · 141

thorium · 36, 63, 105, 172–173, 179

tight gas · 141, 146, 186, *see* shale gas

tight oil · *see* crude oil

tight sand gas · 139, 164, *see* unconventional gas

tight sand oil · 15, *see* tight oil

Titusville · 14, 81, *see* Pennsylvania

toluene · 104, 120, 176, *see* BTEX

total dissolved salts (TDS) · 49, 100, 103, 107, 110, 115, 118, 164, 169, 177, 184, 193, 199, 201–202, 204, 206, *see* salinity

toxic elements · 36, 61, 70, 80, 105–106

toxic metals · 116

toxicity effects · 66, 104

trade secret · 212

transboundary rivers · 215

transport · 1, 5, 13, 18, 22, 40, 46, 66, 73, 75, 82, 97, 114, 124, 127, 136, 139, 143, 153, 158, 162, 167, 181, 184, 186, 207, 218, 232, 237, 242, 247, 249, 254

transportation sector · 82, 234, 259

troposphere · 237, 247

tube wells · 215

Tunisia · 137

Turkey · 3

Turkmenistan · 136–137

TVA (Tennessee Valley Authority) · 66, 69, 76

USA · 2–3, 6, 10, 12, 15–17, 20, 25–26, 28–29, 31–32, 35–38, 40–42, 49–52, 54, 56, 58, 60–61, 64, 66–68, 70, 72–76, 78–82, 84, 86, 88–89, 91, 94–97, 101–103, 105–108, 110–112, 114, 119–121, 123–128, 131, 133–137, 139, 142, 146–148, 151, 153–159, 164–165, 169, 176–177, 179–181, 183–186, 189–190, 194, 199, 201–203, 205, 207, 209–210, 212–214, 216–217, 220, 222, 225, 230–231, 233, 236–238, 240–242, 244–245, 248–249, 256–257, 260

US Department of Energy (DOE) · 16, 25, 245

US Energy Information Administration (EIA) · 2, 16, 50, 139, 180, 238, 246

US Environmental Protection Agency (EPA) · 40, 52, 60, 66, 70, 72, 74–76, 107, 109, 112, 119–120, 177, 186

US Geological Survey · 25–26, 33, 51, 68, 102–103, 123–124, 177

Uinta · 198, 201, 205, *see* Coalbed methane

unconventional · *see* shale gas

unconventional energy resources · 13, 15, *see* shale gas, tight oil

unconventional gas · 136, 139, 141, 158, 178, 193, 209–210, 214, *see* shale gas

unconventional oil · 10–11, 15, 18, 23, 84–85, 88–89, 91, 93–94, 98, 100, 102, 106, 110, 113–114, 121, 125, 128–130, 133–135, 167, 175, 193, 212, 231, 238–239, 241, 244–245, *see* tight oil

unconventional oil exploration · 90, 94, 128, *see* tight oil

Underground Injection Control · 132, 210

UN Environment Programme · 248

United Arab Emirates (UAE) · 3, 20, 251

United Nations · 7

unlined ponds · 186

unsaturated zone · 123, 206, *see* aquifer

uranium · 7, 36, 63–64, 105, 170–171, 173, 179, 216

Utica · 146, 161, 179, *see* shale gas

Valdez oil spill · 9, 82, 121

vanadium · 38, 86, 126

Venezuela · 81, 116, 243, *see* oil

venting · 210, 248

Virginia · 40, 49

volatile elements · 63–64, 68–69, 236

Warrior Coal Basin · 199, *see* Alabama
Washington DC · 10
waste disposal · 11, 24, 73, 183, 220
Waste Framework Directive · 232
waste management · 8
wastewater · 1, 4, 8, 12–13, 18, 38, 45–46, 52, 57, 66, 68, 70–72, 74–76, 85, 94–95, 98, 106, 108–109, 114, 118, 120–121, 127, 129–130, 133–134, 153, 157, 163, 167, 175–176, 178, 180–185, 187, 194, 205, 207, 210, 212, 218, 226, 229, 231, 243, 250, 252, 254, 259, 261, *see* oil and gas wastewater
wastewater disposal · 85, 99, 127, 130, 183, 185, 207, 249, *see* injection
wastewater intensity · 18, 70, 96
wastewater treatment plants · 181, 250
water combustion · 236
water consumption · 5–6, 11, 22, 24, 28, 41, 44, 53–54, 56, 73–74, 80, 91, 118–119, 127, 158, 160, 207–208, 219–220, 222, 224–225, 227, 256, 258
water flooding · 88, 90, 127, *see* EOR
water footprint · 12–13, 22, 42, 56, 79, 85, 88, 91, 94–95, 108, 111, 114, 118, 128–129, 133, 139, 159, 168, 180, 208, 225, 227, 231, 236, 256–257, 261
water impaired · 11, 23, 72–73, 127, 129, 207, 213, 232, *see* water contamination
water intensity · 6, 11, 13, 18, 28, 41, 45, 51–53, 56, 69–70, 72, 79–80, 90–92, 94, 96, 99, 109–110, 115–117, 120, 125–127, 130, 133, 135, 153–154, 158–160, 164–165, 181–185, 202–204, 206–207, 218, 220, 222–223, 226–227, 230, 232, 247, 256, 258
water management · 1, 218, 261
water quality · 1, 3, 6–7, 10–11, 16, 23–24, 28, 45–47, 49–50, 60, 67, 73, 77, 80, 82–83, 94, 100, 106, 112, 123, 131–133, 135, 138, 161, 169, 178, 182, 184, 188, 190–191, 204, 206, 211, 213, 215, 218, 229, 232–233, 247, 254, 256–257, 260, *see* water contamination
water–rock interactions · 101, 168, 171, 200
water scarcity · 7, 13, 31, 44–45, 53, 156, 199, 213, 216, 243, 251, 254
water use · 4, 6, 10–11, 13, 16, 22, 25, 28, 38, 41, 44, 52, 54, 56, 58, 63, 69, 73, 80, 88, 90, 93–94, 98, 118, 120, 128–129, 134, 139, 147, 153, 156, 158–159, 164, 171, 186, 206, 208, 210, 213, 215, 217, 219–220, 222–223, 226–227, 229, 233, 251, 257, 261
water vapor · 13, 235
water withdrawal · 5–6, 9, 11, 13, 22–23, 28, 34, 41, 43, 52, 54, 56–57, 72, 80, 91–92, 126–127, 154, 158–159, 207, 210, 213, 216, 218, 220, 222–223, 225–226, 229, 233, 256, 258
Weiyuan · 148, *see* shale gas
West Texas Intermediate (WTI) · 86, *see* crude oil
West Virginia · 31, 38, 46, 49–50, 113–114, 182, 184–185, 187, 192, 194
wet natural gas · 143
wet tower · 53, *see* cooling thermoelectric
Williston Basin · 90, *see* tight oil
wind · 21, 27, 52, 78, 80, 226, 247, 257–258, 260
Woodford · 146, 169, *see* shale gas
World Bank · 242–243, 245
World Coal Association · 40
World Health Organization (WHO) · 247
Wyoming · 186, 199–200, 205–206, 241

xylenes · 104, 120, *see* BTEX

zero discharge policy · 121, *see* wastewater
Zhaotong · 148, *see* shale gas
zinc · 105, 247, *see* metals

Printed in the United States
by Baker & Taylor Publisher Services